Global Initiatives for Waste Reduction and Cutting Food Loss

Aparna B. Gunjal
Asian Agri Food Consultancy Services Ltd, India

Meghmala S. Waghmode
Annasaheb Magar Mahavidyalaya, India

Neha N. Patil
Annasaheb Magar Mahavidyalaya, India

Pankaj Bhatt
Dolphin (P.G) College of Biomedical and Natural Sciences Dehradun, India

A volume in the Advances in Environmental Engineering and Green Technologies (AEEGT) Book Series

IGI Global
DISSEMINATOR OF KNOWLEDGE

Published in the United States of America by
IGI Global
Engineering Science Reference (an imprint of IGI Global)
701 E. Chocolate Avenue
Hershey PA, USA 17033
Tel: 717-533-8845
Fax: 717-533-8661
E-mail: cust@igi-global.com
Web site: http://www.igi-global.com

Library of Congress Cataloging-in-Publication Data

Names: Gunjal, Aparna B., 1981- editor.
Title: Global initiatives for waste reduction and cutting food loss / Aparna B. Gunjal, Meghmala S. Waghmode, Neha N. Patil, and Pankaj Bhatt, editors.
Description: Hershey, PA : Engineering Science Reference, [2019] | Includes bibliographical references.
Identifiers: LCCN 2018036332| ISBN 9781522577065 (h/c) | ISBN 9781522577072 (eISBN)
Subjects: LCSH: Food industry and trade--Waste minimization. | Food industry and trade--Waste disposal. | Agricultural wastes--Management. | Salvage (Waste, etc.).
Classification: LCC TD793.9 .G56 2019 | DDC 363.72/8--dc23 LC record available at https://lccn.loc.gov/2018036332

This book is published in the IGI Global book series Advances in Environmental Engineering and Green Technologies (AEEGT) (ISSN: 2326-9162; eISSN: 2326-9170)

British Cataloguing in Publication Data
A Cataloguing in Publication record for this book is available from the British Library.

Advances in Environmental Engineering and Green Technologies (AEEGT) Book Series

ISSN:2326-9162
EISSN:2326-9170

MISSION

Growing awareness and an increased focus on environmental issues such as climate change, energy use, and loss of non-renewable resources have brought about a greater need for research that provides potential solutions to these problems. Research in environmental science and engineering continues to play a vital role in uncovering new opportunities for a "green" future.

The **Advances in Environmental Engineering and Green Technologies (AEEGT)** book series is a mouthpiece for research in all aspects of environmental science, earth science, and green initiatives. This series supports the ongoing research in this field through publishing books that discuss topics within environmental engineering or that deal with the interdisciplinary field of green technologies.

COVERAGE

- Cleantech
- Pollution Management
- Biofilters and Biofiltration
- Contaminated Site Remediation
- Air Quality
- Renewable Energy
- Electric Vehicles
- Green Technology
- Alternative Power Sources
- Radioactive Waste Treatment

IGI Global is currently accepting manuscripts for publication within this series. To submit a proposal for a volume in this series, please contact our Acquisition Editors at Acquisitions@igi-global.com or visit: http://www.igi-global.com/publish/.

The Advances in Environmental Engineering and Green Technologies (AEEGT) Book Series (ISSN 2326-9162) is published by IGI Global, 701 E. Chocolate Avenue, Hershey, PA 17033-1240, USA, www.igi-global.com. This series is composed of titles available for purchase individually; each title is edited to be contextually exclusive from any other title within the series. For pricing and ordering information please visit http://www.igi-global.com/book-series/advances-environmental-engineering-green-technologies/73679. Postmaster: Send all address changes to above address.

Titles in this Series

For a list of additional titles in this series, please visit:
https://www.igi-global.com/book-series/advances-environmental-engineering-green-technologies/73679

Building Sustainability Through Environmental Education
Lynn A. Wilson (SeaTrust Institute, USA) and Carolyn N. Stevenson (Purdue University Global, USA)
Engineering Science Reference • ©2019 • 335pp • H/C (ISBN: 9781522577270) • US $195.00

Unique Sequence Signatures in Plant Lipolytic Enzymes Emerging Research and Opportunities
Nihed Ben Halima (University of Sfax, Tunisia)
Engineering Science Reference • ©2019 • 197pp • H/C (ISBN: 9781522574828) • US $145.00

Handbook of Research on the Adverse Effects of Pesticide Pollution in Aquatic Ecosystems
Khursheed Ahmad Wani (Government Degree College Bijbehara, India) and Mamta (Jiwaji University, India)
Engineering Science Reference • ©2019 • 538pp • H/C (ISBN: 9781522561118) • US $295.00

Harnessing Marine Macroalgae for Industrial Purposes in an Australian Context Emerging ...
Göran Roos (Economic Development Board of South Australia, Australia) Anthony Cheshire (Balance Carbon Pty Ltd, Australia) Sasi Nayar (South Australian Aquatic Sciences Centre, Australia) Steven M. Clarke (South Australian Research and Development Institute, Australia) and Wei Zhang (Flinders University, Australia)
Engineering Science Reference • ©2019 • 291pp • H/C (ISBN: 9781522555773) • US $195.00

Advanced Methodologies and Technologies in Engineering and Environmental Science
Mehdi Khosrow-Pour, D.B.A. (Information Resources Management Association, USA)
Engineering Science Reference • ©2019 • 506pp • H/C (ISBN: 9781522573593) • US $295.00

Transitioning Island Nations Into Sustainable Energy Hubs Emerging Research and ...
Catalina Spataru (UCL Energy Institute, UK)
Engineering Science Reference • ©2019 • 172pp • H/C (ISBN: 9781522560029) • US $135.00

For an entire list of titles in this series, please visit:
https://www.igi-global.com/book-series/advances-environmental-engineering-green-technologies/73679

IGI Global
DISSEMINATOR OF KNOWLEDGE

701 East Chocolate Avenue, Hershey, PA 17033, USA
Tel: 717-533-8845 x100 • Fax: 717-533-8661
E-Mail: cust@igi-global.com • www.igi-global.com

Editorial Advisory Board

Table of Contents

Detailed Table of Contents

process for the production of different industrially valued enzymes like amylases, cellulases, pectinases, proteases, phytases, and a few others using a wide range of microorganisms from different types of food waste like kitchen waste and food processing waste.

Food waste is the most challenging issue humankind is facing worldwide. Food waste, which consists of carbohydrates, proteins, lipids, and inorganic compounds, is a biodegradable waste discharged from food processing industries, households, and hospitality sectors. The management of food waste is very important. The food waste generated is usually incinerated or dumped in open areas which may cause severe health and environmental issues. The management of food waste can be done by conversion to different value-added products, for example, phytochemicals, bioactive compounds, food supplements, livestock feed, dietary fibers, biopigments and colorants, emulsifiers, edible and essential oils, biopreservatives, biofertilizers, biofuels, and single cell proteins. The value-added products from food waste will be very eco-friendly. The chapter will focus on different value-added products from food waste.

To link food demand and reduction in food waste, proactive approaches should be taken. Perishable food is mainly fruits and vegetables, waste from different processing industries like pulses, meat products, oil products, dairy products, and fishery byproducts. Conventional food waste management solution is land filling which is not sustainable as it generates global warming gases like methane and carbon dioxide. To reduce food waste, the process known as "food valorization" has become another solution to landfilling, the concept which is given by European Commission in 2012, meaning food processing waste conversion to value-added products. In this chapter the study focuses on production of industrially important enzymes from food waste which could be one of the reactive solutions. Different enzymes like pectinase, peroxidase, lipase, glucoamylase, and protease can be produced from food waste.

three main aspects (i.e., recycle, reuse, and reduce [R3]). Building on this familiar concept of "R3" will impact environmental protection and more fully recognize the impacts of the food and agriculture wasted. Thus, in the chapter, the authors highlight the sustainable utilization of waste generated from coffee and cocoa processing for the development of value-added products.

The food industry generates a huge amount of waste annually around the globe from a variety of sources. Approximately one third of all food produced today goes to landfill as waste. The food waste is not only a humanitarian problem, but also a serious economic and environmental pollution problem. The global volume of food wastage has been reported to around 1.3bn tones worth to about $165 bn. In India, about 40% of the food produced is wasted, which is estimated to about Rs. 50,000 crores worth every year. The important types of food wastes generated are agricultural residue, processed food, fruit and vegetable processing, marine food, dairy processing, meat and poultry, hotel and restaurant, etc. The food industrial waste can be converted into byproducts mainly based on the processing of fruits and vegetables and allied food manufacturing, supply and distribution, livestock feed, using it as source of bioactive compounds, useful bioenergy production, artificial fertilizer and decomposed manure, a variety of chemicals, antioxidant, nutraceuticals, etc.

Food wastage is a huge crisis arising in today's world. An extensive amount of waste generation has become a serious concern of our society in the past years that affects developing and developed countries equally, and according to the Food and Agriculture Organization (FAO), as much as one-third of the food intentionally grown for human consumption is never consumed and is therefore wasted, with significant environmental, social, and economic ramifications. By wasting food, we also waste the time and energy that we have used to produce the food and as well our natural resources and the limited available agricultural land will be used up which could be handled in a much better and sustainable way. Additionally, waste has a strong financial impact and affects the environment including the overall greenhouse gas emission. In an increasingly resource-constrained world, it is imperative to reduce the high environmental, social, and economic impacts associated with this type of waste.

Due to existing application gap, the diverse microorganisms and their processes in food waste have not been maximally explored or harnessed. This chapter addresses the possible application of "-omics" technologies to understand food waste microbial composition and metabolic processes, to stimulate future research in mining food waste for important microorganisms and bioactive compounds. The chapter highlights potential "-omics" procedures for food waste assessments. These innovative, culture-independent, high-throughput technologies have already revolutionized diverse fields of research and human endeavors. This chapter also introduces the concept of food wastomics to detect, identify, and measure the different molecules and microorganisms that are present and expressed in food waste such as DNA, RNA, proteins, lipids, and metabolites. This knowledge will create a greater understanding of how discarded food is degraded and what useful products or organisms may be harvested from it. Finally, the chapter recommends the integration of food wastomics into foodomics.

In September 2015, the United Nations General Assembly adopted a set of 17 Sustainable Development Goals (SDGs), which include Target 12.3's call for halving food waste and reducing food losses worldwide by 2030. One-third of all food produced each year in the world is never eaten, while at the same time nearly 800 million people go hungry. This situation urges us to take immediate resolutions and steps towards reduction of food waste and food loss. This cannot be done by one person or overnight. This requires systematic analysis in various layers and collective and appropriate effort. This target can be achieved altogether by various sectors including government organizations; non-government organizations; and private companies in collaboration with schools; colleges; universities; research institutes; religion-based organizations such as temples, churches, etc.; and charity-based organizations. This book chapter will discuss the various steps that can possibly be adopted and implemented to address the serious issue of reduction of food waste and food loss.

Games are ordered activities, generally undertaken for recreation. The design elements of these games are being used by people all around the planet to make this world a better place. The opportunities for gamification are being discussed in this chapter along with the use of a decision-making method as both have been applied to the study using a local running mobile application as platform for encouraging students enrolled in various educational institutions to promote avoidance of mess food wastage and in gaining confidence to integrate to use this approach to fight the cause of this global malady in their everyday life. The overarching issue of student mentality about food wastage is being discussed along with how to merge gamification with digital technology in this aspect and its participatory design. This provides the background for addressing points of using a gameful system to foster empowerment and connection among the students of NIT Jalandhar where this case was studied and the proposed approach was implemented.

Foreword

Food is lost or wasted throughout the supply chain, from initial production down to final household consumption. Part of the world is still suffering and frequently dying without having enough nutritional food. Nevertheless, the actual challenge is ahead. The world food crises in 2050 is not that much of far away.

The crucial problem is that the peoples are having lack of sufficient motivation, ability and opportunity to reduce food waste, with problems specifically in the lack of knowledge, lack of planning, and financial background.

Food waste is a rich source of various vital components such as protein, carbohydrate (hemicellulose, cellulose, starch, and sugar like sucrose, fructose, and glucose), oil, mineral, and fat that can be used in a wide range of enzymatic and microbial processes. As well, organic matter with high biodegradability rates can be used to produce promising feedstock for biogas production. Food waste has also been used as a sole microbial feedstock to produce various bio products such as ethanol, methane (CH_4), hydrogen (H_2), organic acids, enzymes and biopolymers. Food waste also can be used to produce organic manures, and animal feed (e.g.- swine feed, and fish feed) as well. Food waste also reduces by developing value-added products such as cut fruits, oil, juices and salads etc.

At present there is a high need to address the food waste. In fact, the food waste management should be initiated, from the kitchen, canteen, food processing and restaurant waste, and as well as the municipal solid waste (MSW). According to FAO, roughly one-third of the edible parts of food produced for human consumption, gets lost or wasted per year, and that would be about 1.3 billion tons of food in the form of fruits, bakery, bread, vegetables, dairy products, and meat is lost every year through the food supply chain in world- wide.

Anyhow, many of strategies apply in globally to achieve the ‘zero waste concept’. Process designing on the zero-waste concept has given a new turn to industries to meet the need of the consumers with no longer waste generation.

The book entitled “Global Initiatives for Waste Reduction and Cutting Food Loss” is a timely needed one and present a comprehensive account of this concept.

I consider it is a privilege to write the forward of this timely publication on “Global Initiatives for Waste Reduction and Cutting Food Loss”. I wish to take this opportunity to congratulate the editors for their enthusiasm and commitment towards analysing and reporting some valuable aspects towards such an important global issue. I have no doubt; this book will be an important milestone in this direction.

Wish you all the very best!

Chandrakantha Mahendranathan
Eastern University, Sri Lanka

Chandrakantha Mahendranathan *is the Vice Chancellor at the Eastern University, Sri Lanka. She has been a Senior Academic at this Institution for last 23 years and has been engaged in teaching undergraduate and post-graduate students. She is also involved in several research projects in the field of Postharvest Pathology and Technology, Plant Pathology and Environmental Microbiology and published her findings in many journals and also at several national and international conferences and symposiums, held in USA, Australia, Turkey, Dubai, India, Singapore, and Sri Lanka.*

Foreword

Waste generation and management in the world is a very serious issue from the point of environmental protection and human, animal and plant health. The imbalance between food wastes generated over the globe versus food scarcity must be addressed to avoid economic losses that incur for food waste management and unavailability of food for socio-economically deprived sector. Each year approximately 1.3 billion tons of food goes to waste at a global level. The food waste accounts roughly US$ 680 and 310 billion in developed and developing countries respectively. Food waste is also a major source of greenhouse gas emissions.

The book *Global Initiatives for Waste Reduction and Cutting Food Loss* gives detailed information and knowledge about the management and minimization of food waste. The book also highlights conversion of food waste to different value-added products viz., compost; biogas; biofuels; industrially important enzymes. It also gives details on fermented foods and their production. Molecular approaches for value-added products from lignocellulosic food waste are also mentioned. Anaerobic digestion as a sustainable approach towards food waste management is also discussed.

The policy initiatives, regulations and education campaigns related to food waste management are also focused. The omics approaches for food waste management are discussed to provide new insights to understand food waste decomposition for enhanced recovery of value-added products.

The book gives an idea about necessary and important approaches to be taken for decreasing food wastage. This book is useful to college students; researchers and other scientists and is excellent guide that discusses challenges of food waste management and describes sustainable and effective solutions.

Neelu N. Nawani
Dr. D. Y. Patil Biotechnology and Bioinformatics Institute, India

Neelu N. Nawani *is Professor and In-charge of Microbial Diversity Research Centre at Dr. D.Y. Patil Biotechnology and Bioinformatics, Dr. D. Y. Patil Vidyapeeth, Pune, India. She completed her Ph.D in Microbiology in 2002 from Savitribai Phule Pune University (formerly University of Pune). Her research areas of interest are Environmental and Health Biotechnology. She has mobilized nearly 15 crores from national and international funding agencies for her research and has 80 publications to her credit which includes research articles, book chapters and review articles. She has six patents with two technologies commercialized for benefit of society. She has presented more than 100 research papers in national and international Conferences. She is also member of many societies viz., Biotech Research Society of India, Association of Microbiologists of India and Indian Women Scientist Association. She has received Young Scientist Fellowship, Department of Science and Technology, New Delhi, India; Linnaeus-Palme exchange teacher to University of Skovde, Sweden; Gandhian Young Technology Innovation Award, SRISTI, BIRAC-GYTI, Department of Biotechnology and National Innovation Fund, India for the innovation "A multipurpose low-cost biological air purifier" under socially relevant innovation and many other accolades.*

Preface

The food waste is a very serious issue and which needs to be given attention. The management of food waste has to be done in many countries. Globally, on an average 1.3 billion tones of food is thrown worldwide. The problem of food waste and loss needs solutions to be solved. The book *Global Initiatives for Waste Reduction and Cutting Food Loss* deals with various approaches to the food waste management. The authors have contributed on various aspects related to food waste management.

Chapter 1 describes industrially important enzymes from food waste and their applications in industries. It also describes food waste disposal techniques and solid state fermentation for production of enzymes using microorganisms with food waste as substrate.

Chapter 2 describes in detail the management of food waste by conversion to different value-added products viz., ethanol; biogas; glycoalkaloids; oligomers; single cell protein; fertilizer; organic acids; colors and dyes and polyhydroxybutyrate.

Chapter 3 describes management of food waste by conversion to enzymes viz., glycoamylase; protease and lipase which can be alternative solution to landfilling.

Chapter 4 describes the molecular mechanism along with transcriptome, proteome and recombinant DNA technology for value-added products from lignocellulosic food waste.

Chapter 5 describes in detail various fermented foods viz., milk-based; cereal and legume-based; cereal/pulses and butter milk-based, vegetable, bamboo shoot and unripe fruits-based fermented foods. The chapter also describes the health benefits of fermented foods.

Chapter 6 describes the nutritional value and health effects of fermented foods. It also describes the fermentation technology and their potential to minimize food wastage by conversion to value-added products and also their benefits.

Chapter 7 describes the types and sources of food waste and impact of food waste on the environment. It further describes the role of microbes in composting - eco-friendly way to food waste management by value-addition.

Chapter 8 mainly describes in detail sustainable utilization of waste generated from coffee and cocoa processing for the development of different value-added products.

Chapter 9 focuses on classification of food waste; global food loss and various food processing industries. It also describes the measures to reduce the food loss mainly by recycling.

Chapter 10 mentions about food waste management by 3R principle; i.e. reduce, recycle and reuse. It also describes various policies and regulations related to food waste. The chapter also describes minimization of food waste by conversion to biofuels and bioenergy.

Chapter 11 addresses application of "omics" technology to understand food waste microbial composition and metabolic processes. It also introduces the concept food wastomics to identify and measure different molecules and microorganisms present in food waste.

Chapter 12 mentions the various steps that can be implemented for reduction of food waste and food loss.

Chapter 13 describes the opportunities for gamification, design along with use of decision-making method for encouraging students in educational institutions to promote avoidance of mess food wastage.

The target audience for the book will be students from schools, colleges and universities and researcher's worldwide working on food waste management. The book will provide a guide for the reduction of food loss and management and also help researchers develop new ideas for study on various aspects related to food loss and management.

The editors are thankful to all the authors for their timely contribution in the form of book chapters.

Aparna Baban Gunjal
Asian Agri Food Consultancy Services Ltd, India

Meghmala Sheshrao Waghmode
Annasaheb Magar Mahavidyalaya, India

Neha N. Patil
Annasaheb Magar Mahavidyalaya, India

Pankaj Bhatt
Dolphin (P.G) Institute of Biomedical and Natural Sciences Dehradun, India

Chapter 1
Enzymes Production From Food Waste and Their Application

Vismaya N. Kumar
Cashew Export Promotion Council of India (CEPCI), India

Binod Parameswaran
National Institute for Interdisciplinary Science and Technology, India

Sharrel Rebello
St. Joseph's College, India

Ashok Pandey
Indian Institute of Toxicology Research, India

Sindhu Raveendran
National Institute for Interdisciplinary Science and Technology, India

Embalil Mathachan Aneesh
St. Joseph's College, India

Prabhakumari C.
Cashew Export Promotion Council of India (CEPCI), India

ABSTRACT

The chapter reviews the fermentation-based production of industrially important enzymes from food waste (FW). Nearly one-third of the food produced globally is wasted and poses serious problems regarding its disposal. A number of dumping systems have been developed in the nations worldwide which has later become a threat to the environment. This problem is both of an environmental and economic concern. Recent developments in the area have revealed the application of bioremediation as the best way to dispose food waste. Composting and anaerobic digestion of the organic waste are gaining importance for the better use of household-level waste rather than just dumping it in landfill sites. This chapter focuses mainly on the processing waste.

DOI: 10.4018/978-1-5225-7706-5.ch001

different types of FW, its disposal techniques, optimization of the fermentation process for the production of different industrially valued enzymes like amylases, cellulases, pectinases, proteases, phytases, and a few others using a wide range of microorganisms from different types of food waste like kitchen waste and food

INTRODUCTION

According to a study by the Food and Agricultural Organisation (FAO), nearly 1.3 billion tons of foods produced globally are lost along the food supply chain. Food waste (FW) can be defined as a biodegradable organic waste discharged from different sources including food processing industries, households and hospitality sectors (FAO, 2012). The rise in population of the world is the main reason behind the increasing amount of FW which poses serious threat to the environment in the form of pollution, health risks and scarcity of dumping land. Among the well-developed nations, the United State, the United Kingdom and Japan are placed on top of the food waste list globally (Melikoglu et al. 2013a). About 96 billion pounds of food was wasted in the US and it amounts to 27% of the total food available for human consumption.

Food waste has been categorized depending on various factors based on the type of food as cereals, fruits, meat, fish, drinks, its nutrient composition as carbohydrates and fat based FW (Russ and Meyer-Pittroff 2004), chemical composition depending on its C, H, N, O, S and Cl content (Bernstad and La Cour Jansen 2012), storage temperature into ambient, chilled or frozen (Bernstad and La Cour Jansen 2012), household level waste as cooked and uncooked, unpackaged and packaged food waste. Food waste was classified into organic comprising fruits and vegetables, catering waste, animal by-products, packaging, mixed food waste and domestic waste by Lin et al. 2013.

In a study that reviewed the characteristics of treated food waste, loss of carbon, nutrients and other compounds during treatment, energy recovery through incineration, hazardous emission from composting, storage and use of bio-fertilizers and chemical fertilizers were found to be a direct cause that leads to global warming and other major environmental and health hazards. One of the main characteristics of food waste, that many other wastes lack is that food waste can be subjected to biological processes like before its disposal. But these processes can result in toxic emission that has negative impact on the environment and will affect its property for the recovery of nutrients and energy recovery by different alternative treatment techniques (Bernstad and La Cour Jansen 2012).

Inappropriate disposal techniques may lead to problems to both human and animal health and also incur serious economic, environmental and biological losses. Figure 1 depicts the propects and consequences of various food waste disposal methods. Composting and vermi-composting are better methods than incineration. The disadvantage of composting is that it is a slow process and requires large space and produces foul odors (Sharholy et al. 2008). Disposing at landfill sites is another more commonly employed system of waste disposal. But land filling poses the most serious environmental impacts compared to the other treatment options such as anaerobic digestion or composting and therefore should only be thought of as a last aid (Sharholy et al. 2008). Land filling can be employed for wastes that are not suited for recycling or biological processing which are non-biodegradable or inert. Under such circumstances where the safe and hygienic removal of food waste has become a serious issue, a best option is to utilize FW for its significant potential in the production of value-added products like enzymes, fuel, and various other products. Figure 3 gives an outline of the production of value-added products from food wastes. This paper reviews about the production of industrially important enzymes using FW as substrates. This method is not only important in an eco-friendly point of view but also economically valued.

Figure 1. Different FW disposal techniques and its advantages and disadvantages

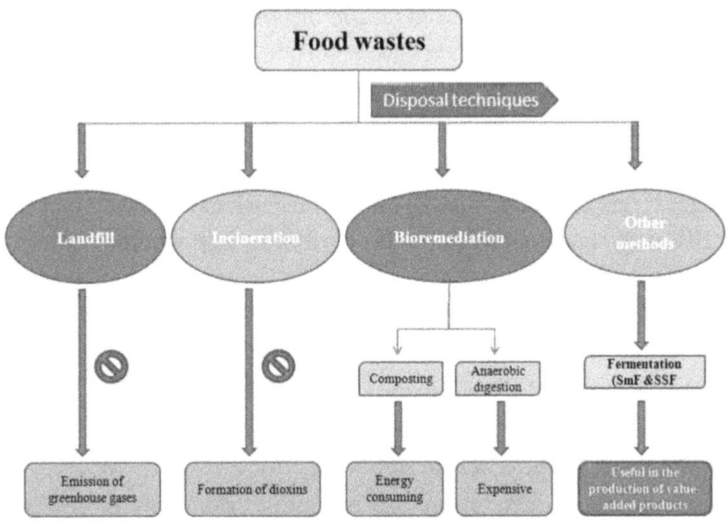

Types of Food Waste

Food wastes mainly include cereals, vegetables, meat, egg, fruits, bakery and dairy products which are rich sources of carbohydrates, proteins, lipids and other inorganic compounds (Paritosh et al. 2017). The different FW can be categorized into two main headings – pre-consumer and post-consumer. Pre-consumer type includes all the food processing and industry wastes whereas post-consumer type comprises the household or kitchen waste (Lin et al. 2013). Organic crop residues (fruits and vegetables), catering wastes (non-factory used cooking oil), slaughter waste, mixed food waste and domestic waste are the different types of FW that is being dumped into the environment. Agro -wastes from industries are a rich source of nutrients and various bioactive compounds. Such wastes can be used as raw material instead of dumping them as mere wastes for the production of industrially important products. It provides suitable growth conditions for the microorganisms. The microorganisms can use the waste as substrate for their growth by fermentation processes (Sadh et al 2018). Domestic food waste is another major class of FW. Domestic food storage, domestic food preparation behaviors, and cooking large amounts of food for special events were found to be the causes of accumulation of household food waste (Romani et al. 2018).

Food Wastes Disposal Techniques

Dumping of FW has become a serious issue of both environmental and economic concern. Different disposal methods have been adopted like composting, anaerobic digestion, co-digestion with sludge, landfill and dryer-incineration which elicited the potential use of food wastes in the production of energy in the form of biogas and heat energy(Kim et al. 2013). Figure 1 depicts the different methods of FW disposal and the pros and cons of each method. Among the various methods, incineration and landfill of FW were considered to be the least desirable whereas composting and anaerobic digestion were considered most useful (Girotto et al 2015). Disposal of FW to landfill sites result in its anaerobic decomposition and production of greenhouse gases like carbon dioxide. It also results in loss of energy content at the landfill sites which has also gained importance (Melikoglu et al. 2013a). Direct landfill of FW was banned in South Korea in 2005 in order to reduce the contamination of soil and groundwater by the leachate of dumped food waste (Kim et al. 2013). In a study, it was found that incineration of a mixture of domestic food waste leads to formation of dioxins which is considered as an environmental pollutant (Katami and Prefectural 2004).

Bioremediation of FW

Bioremediation can be defined as a biological process in which microorganisms either transform or convert environmental contaminants to innocuous end products. Bioremediation is the most useful technique that solves the problems related to dumping of food wastes and also other major environmental concerns like pesticides, herbicides and other chemicals which are converted into non-toxic substances by use of microorganisms (Thassitou and Arvanitoyannis 2002). Bioremediation of food wastes has taken the scenario to a different view point rendering food wastes less harmful than industrial waste. Besides being harmless, FW has become a very useful media for the production of biofertilizers and biogas. In recent years, food wastes are considered a valuable resource for the production of value-added bio-products such as methane, ethanol, hydrogen, enzymes, fuels (Uc and Trzcinski 2014). In a study by Tsai, Liu, and Yang in 2007, FW were used for the production of bio-fertilizers using microorganisms. Ngoc et al. 2014 in a research evaluated the bioenergy potential of FW of 21 countries which accounted to about 833,555KWh/year and is likely to solve the energy need in future. According to Markley, anaerobic digestion and composting are the efficient methods of bioremediation of FW (Markley 1998) which gets converted to non-toxic or usable products. Biological degradation of the organic matter of FW under aerobic and anaerobic conditions are the basis of both bioremediation processes (Cerda et al. 2017).

Composting

Composting reduces the problems caused by food wastes to the environment and is more advantageous than landfill of food wastes. It is considered as a sustainable and eco-friendly method of food waste management which can effectively replace landfill of waste thereby reducing environmental pollution (Schaub 1996). Food waste is the best source of raw compost agent but because of its high moisture content and low physical structure, bulking agents like sawdust or yard wastes should be mixed with raw FW prior to composting (Risse and Faucette 2017). The major factors that affect the process of composting are carbon to nitrogen ratio which should be 30:1, moisture content of 60%, optimum level of aeration, smaller particle size, pH ranging from 6.0-7.8, temperature from 50-158 degrees F (depending on the microorganism acting on the FW) above which may char the compost or create conditions best suited for spontaneous combustion. An increase or decrease in these optimum values may cause slowing down of the decomposition process and create foul odors.

Anaerobic Digestion

Anaerobic digestion (AD) is the microbial breakdown of organic matter present in materials like food wastes, animal manure, fats, oils, and greases, industrial organic residuals, and sewage sludge under strictly maintained anaerobic conditions (US EPA). The process of AD involves four main steps: hydrolysis, acidogenesis, acetogenesis and methanogenesis. The organic content of FW is acted upon by the fermentative bacteria and gets converted to smaller molecules which are later subjected to hydrolysis by hydrolytic enzymes forming organic acids like acetate, propionate and butyrate leaving NH_3, CO_2 and H_2S as by-products. In the next step, the above mentioned products are further converted to acetate, H_2 and CO_2 which in turn are used by methanogens for the production of methane (Zhang et al. 2014). Anaerobic digestion of FW collected in the City of San Francisco, California, resulted in the production of 435mL/g of methane gas after 28 days incubation at 50±2°C which accounted for about 73% of biogas produced (Zhang et al. 2007).

Submerged Fermentation (SmF) vs. Solid State Fermentation (SSF)

Enzyme production is an emerging field of biotechnology which has been made economical with the introduction of food wastes as substrate in the production process. A wide range of food waste can be utilized for the production of enzymes by fermentation, some of which are enlisted in Table 1. Submerged fermentation (SmF) was the widely used technique for the production of enzymes which has now been succeeded by a more efficient and advantageous method, Solid State Fermentation (SSF) (Rodr and Sanrom 2005). Recent studies have shown that Solid State Fermentation (SSF) has become a very promising tool for the production of enzymes from organic waste thereby reducing the environmental and economic issues associated with its disposal (Rodr and Sanrom 2005). A comparative study by Viniegra-gonzález et al. 2003, it was shown that enzyme production is higher in SSF than in SmF. They compared the productivity of three enzymes namely invertases, pectinases and tannases and the results clearly proved that SSF is more efficient than SmF in the production of enzymes. Selection of microorganisms is the most crucial step in SSF because in a report by Pandey 1992 it was shown that a single microorganism can produce about 18 different enzymes. *Aspergillus niger* is found to be a potent strain that can produce a wide range of useful enzymes from varieties of food industry wastes. A list of the different enzymes produced by *Aspergillus niger* is shown in Figure 2. Similarly, a single enzyme can be produced by different microbial cultures.

Table 1. Different enzymes produced by microorganisms and their substrate (FW)

Enzyme produced	Food waste utilized	Microorganism	Reference
Amylase	Potato peel	*Bacillus subtilis* K-18	(Mushtaq et al. 2016)
	Potato peel	*Aspergillus niger*	(Shahid et al. 2016)
	Coffee waste	*Neurospora crassa* CFR 308	(Murthy, Naidu, and Srinivas 2009)
Glucoamylase	Waste bread, savory, waste cake, cafeteria waste, fruits, vegetables and potatoes	*Aspergillus awamori*	Kiran, Trzcinski, and Liu 2014
	Waste bread	*Aspergillus awamori*	(Melikoglu et al. 2013b)
	Tea waste	*Aspergillus niger* NCIM 1248	(Selvakumar and Pandey 1998)
Phytase	Potato waste	*Aspergillus ficuum*	(Tian and Yuan 2016)
	Soybean meal	*Aspergillus niger*	(Saithi and Tongta 2016)
Cellulase & xylanase	FW from Municipal Solid Waste (MSW)	*Aspergillus niger*	(Tian et al. 2017)
Cellulase	Vegetable waste	*Trichoderma atroviride*	(Bairagi, Sciences, and Haryana 2016)
	Vinegar industry waste	*Trichoderma koningii* AS3.4262	(Liu and Yang 2007)
	Agricultural & kitchen waste	*Aspergillus niger* NS2	(Bansal et al. 2012)
Protease	Waste bread	*Aspergillus awamori*	(Melikoglu et al. 2013b)

Figure 2. Overview of the production of enzymes from food wastes

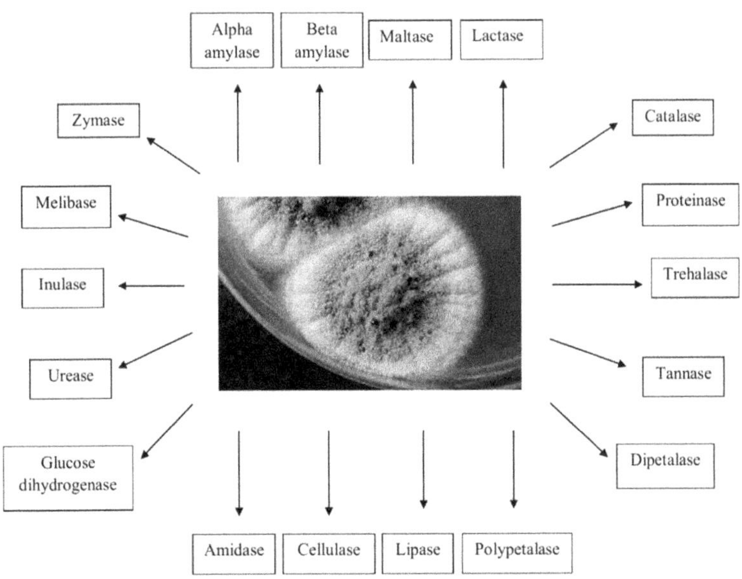

Fermentation Based Optimization of Enzymes From Food Waste

About 30% of the total cost of production of enzymes is attributed to the cost of the raw materials and thus, large-scale production of enzymes is very expensive (Ravindran and Jaiswal 2016). Various studies have shown the production of different enzymes like amylase, glucoamylase, cellulose, pectinase from different samples of food waste like fruit waste (Silva et al. 2002), bread waste (Asad et al. 2001), coffee waste (Murthy, Naidu, and Srinivas 2009), potato peel (Mushtaq et al. 2016), agriculture and kitchen waste (Bansal et al. 2012), tea waste (Selvakumar and Pandey 1998), and many other substrates. Figure 3 shows an illustration of the process of production of enzymes from food wastes. Amylases are a group of industrial enzymes that hydrolyse starch and constitutes about 25% of the enzymes that has potential applications the different industries like sugar, textile, paper, brewing, distilling industries and pharmaceuticals (Sudharhsan, Senthilkumar, and Ranjith 2007). A wide range of microbes are involved in the production of amylase which includes bacteria and molds like *Bacillus* sp. *Aspergillus niger* (Fogarty and Kelly 1990). Fungi are the most commonly used source for the production of amylase enzyme using food waste as substrate especially *Aspergillus* species among which *A. niger* was widely used (Mathew et al. 2016; Kwatia et al 2017).

In a statistical media optimization studies, it was shown that the amount of enzyme produced in a bioreactor and shake flasks (161 & 133 U/ml) were same but the time taken by shake flask was longer. Excessive foaming in a bioreactor was prevented by the presence of potassium in the medium and a low inoculum size but agitation was not much important. pH was found to be a crucial parameter

Figure 3. Process to synthesize value added products from food wastes

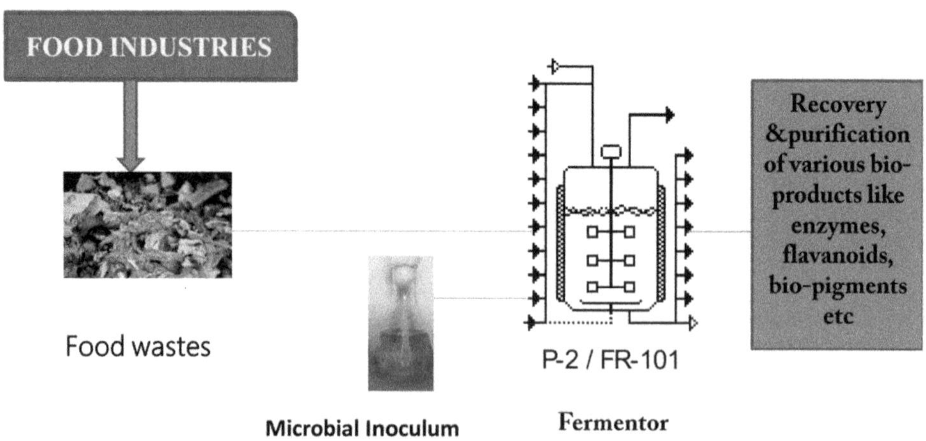

and the concentration of the enzyme increased above pH 7.5 (Gigras, Sahai, and Gupta 2002). SSF of papaya waste (peel & seed stone) at pH 6.0 using 0.850 mm particle size when incubated at 37 °C for 48 hrs produced amylase activity at the rate of 41.22 IU/min (Sharanappa et al. 2011). In another study on the production of α-amylase by solid state fermentation by *A. niger* using potato peel as substrate, the maximum amylase activity recorded under optimum conditions was 1262.27 ± 2.11 U/g. They optimized the different cultural parameters and found that the maximum enzyme productivity was obtained after 72 hrs incubation at 30 °C, pH 5.0 and 5% moisture and inoculum size. 1.5% peptone, 0.75% NH_4NO_3, 1.25% soluble starch were also utilized as nitrogen and carbon source respectively (Shahid et al. 2016). Table 2 enlists the optimum values of various cultural parameters, the enzyme activity observed and the different modes of fermentation applied.

Cellulase is another industrially important enzyme with very many applications whose production has been made low-cost by solid state or submerged fermentation, utilizing food waste as substrate. It hydrolyses the cellulose-rich organic materials in the fermentation media. Different microorganisms have the potential to produce cellulase of which fungi are the most efficient producers including *Aspergillus, Penicillium and Trichoderma*. The optimum values of the cultural conditions for maximum enzyme yield of cellulase by SSF of vegetable waste by *Trichoderma atroviride* as studied by Bairagi were at 30°C with a pH of 6, incubation time of 5days, sucrose as carbon source, yeast extract as nitrogen source and tween-80 as detergent source (Bairagi 2016). A wide range of kitchen and agriculture waste can be utilized as substrates for the production of cellulase by SSF which includes corn cobs, carrot peelings, composite, grass, leaves, orange peelings, pineapple peelings, potato peelings, rice husk, sugarcane bagasse, saw dust, wheat bran, and wheat straw (Bansal et al. 2012).

Table 2. Enzymes produced by Aspergillus niger

1 Alpha amylase 2. Beta amylase 3. Maltase 4. Lactase 5. Catalase 6. Proteinase 7. Trehalase 8. Tannase 9. Dipetalase 10. Polypetalase	11. Lipase 12. Cellulase 13. Amidase 15. Glucose dihydrogenase 16. Urease 17. Insulase 18. Melibase 19. Zymase (trace)

(Adapted from Pandey, 1992)

Pectinase is widely used in the fruit juice industry with various aspects, especially to reduce blur in fruit juice (Aysun et al 2016). They hydrolyse pectin present in the organic residues of food waste (Kalaichelvan 2014). The substrates used for pectinase production include cassava waste, citrus peel, orange waste, agriculture waste (Ahmed et al. 2015; Hachemi, Nouani, and Benchabane 2015; Jahan et al. 2017). A report on the production and optimization of pectinase from cassava waste using *Bacillus sp*., the maximum enzyme activity was obtained at 35 °C, pH of 6.5 and incubation time of 72h (Kalaichelvan 2014). An 87.5565 U/ml enzyme activity was shown by submerged fermentation using *Enterobacter sp*. PSTB-1 as inoculums and mango waste as substrate. The incubation conditions provided were a media with pH 6, incubation temperature 37 °C for a period of 48h (Purna and Saritha 2016). Various other reports on the standardized culture condition like incubation temperature, time and pH of the media required for maximum pectinase production are given in Table 2.

Proteases (EC 3.4.21–24) are a class of enzymes which hydrolyzes proteins and catabolizes it by hydrolysis of the peptide bonds that link amino acids in a polypeptide chain forming proteins. Proteases are also known as peptidyl-peptide hydrolase and constitutes about 60–65% of the enzyme market globally (Souza et al. 2015). Proteases are commercially important enzymes that have a wide range of applications in various industrial, biotechnological, medicinal and basic research fields. *Bacillus* sp. is the major source which secretes a variety of soluble extracellular enzymes. In a study on the SSF by *Bacillus* sp. on coffee waste, statistics-based contour and 3-D plots were generated to evaluate the variations in the response surface and understand the relationship between the culture conditions and enzyme yield. The maximum enzyme yield of protease of 920 U/mL was obtained after 60 h of incubation with 3.0 g/L of coffee waste at pH 8 and temperature 37°C (Kandasamy, Muthusamy, and Balakrishnan 2016). Various other reports on the optimization of culture conditions are provided in Table 2.

Applications

The enzymes produced from food waste have wide applications in various industries such as detergents, food processing units, textile industry, environmental remediation etc. However, commercial production of enzymes from food wastes is meager in quantity due to alternate microbial sources and cost effective biotechnological processes. However, enzymes from plant food processing units such as Horse radish have been used in the production of horse radish peroxidase; an enzyme used in free or immobolilized form in an array of applications in biochemistry, non-radioactive chemiluminiscence, diagnostic applications and in the remediation of xenobiotics (Bansal 2013,Pandey et al 2017). In addition, peroxidases from potato waste, tomato waste and onion waste find much utility in the treatment of effluents containing dyes, phenolics etc (Makris 2015). The use of Polyphenol oxidases particularly from potato and mushroom waste also find application in bioremediation (Makris 2015).

The use of amylase in detergents enhances the ability to remove tough stains and making it eco-friendly. Amylases are a type of enzymes that has application in detergent industry. It is a main ingredient of detergents and it is found that almost 90% of liquid detergents contain these enzymes (Hmidet et al 2009; Mitidieri et al 2006). Amylases are widely used in food-processing industries such as baking, brewing, production of cakes, fruit juices and starch syrups (Couto et al 2006). The α- amylases have been widely used in the baking industry as well as in textile industry for the process called desizing in which sizing agents like starch can be removed from the woven fabric in a wet-process. Desizing involves the removal of starch from the fabric which prevents breaking of the warp thread during the weaving process. The α-amylases remove selectively the size and do not attack the fibers (Ahlawat et al 2009).

The application of cellulases in the pulp and paper industry has increased in importance. Alteration of fiber properties to improve drainage in paper mills is done using a mixture of cellulases and hemicellulases. Cellulases or a mixture of cellulases and xylanases are useful for removal of ink from different types of paper wastes. Ink can be removed from fiber surface by the partial hydrolysis of carbohydrate by using cellulases and hemicellulases (Kuhad et al 2011). Textile wet processing is another application of cellulases (Hebeish and Ibrahim 2007; Karmakar and Ray 2011).

Pectinases are one of the industrially important enzymes having applications in fruit and textile industries. Alkaline pectinases are also important in textile industry and in the pretreatment of pectic wastewater discarded from fruit juice industries. Pectinases play an foremost role in the clarification of beer, fruit juices like grapes, apple, mango and also used in coffee and tea fermentation and removal of the mucilage coat of the coffee beans (Kashyap et al. 2001).

Proteases have a wide range of applications in food, pharmaceutical, detergent and leather industries. Proteases play important role in food biotechnology, food processing and detergent industry, leather industry, pharmaceutical industry and bioremediation processes. Meat tenderization with proteases like papain is an example of the large scale application of enzymatic hydrolysis. Proteases are used for fish processing aids like deskinning and descaling. The main use of proteases in dairy industry is in the coagulation of milk to manufacture cheese. The enzymes have various applications such as cheese making, baking, preparation of soya hydrolysates and meat tenderization. Spontaneous development of oxidized flavors is a common problem in dairy products which is prevented by the action of proteases like trypsin. The application of proteases as alternatives to toxic chemicals such as sodium sulphide has been proved in improving leather quality and reducing environmental pollution. Proteases have widespread application in pharmaceutical sectors uses like in the treatment of burns and wounds, various forms of lymphocytic leukemia, as digestive aids to cure certain lytic enzyme deficiency syndromes (Souza et al. 2015).

CONCLUSION

This paper is a review on the enzymes produced from food wastes. The different types of food waste being accumulated in the environment can be useful to some industries in the production of industrially important enzymes. The problems caused by the increasing amount of food waste and its disposal techniques prove the relevance of this review. Excessive accumulation of food waste is a result of the ever-increasing population in the world. Fermentation of different food waste using microorganisms and the optimization of the cultural conditions for fermentation have been discussed. Applications of these enzymes in various industries also add to the importance of this review paper.

ACKNOWLEDGMENT

Vismaya N Kumar acknowledges BCIL for funding (File No: BCIL/HRD/BITP/STIP). Sharrel Rebello acknowledges SERB for National Post Doctoral Fellowship (File No: PDF/2015/000472). Raveendran Sindhu acknowledges Department of Science and Technology for sanctioning a project under WOS-B scheme. Parameswaran Binod and Raveendran Sindhu acknowledge EPFL, Lausanne, Switzerland for visiting fellowship. PB acknowledges the financial support by the Department of Science and Technology (DST), New Delhi under INNO-INDIGO/INDO-NORDEN project (Sanction No. DST/IMRCD/INNO-INDIGO/INDO-NORDEN/2017(G).

REFERENCES

Ahlawat, S., Dhiman, S. S., Battan, B., Mandhan, R. P., & Sharma, J. (2009). Pectinase production by *Bacillus subtilis* and its potential application in biopreparation of cotton and micropoly fabric. *Process Biochemistry*, *44*(5), 521–526. doi:10.1016/j.procbio.2009.01.003

Ahmed, I., Muhammad, A., Muhammad, A., Zain, A., & Muhammad, T. (2015). Bioprocessing of citrus waste peel for induced pectinase production by *Aspergillus niger* ; Its purification and characterization. *Journal of Radiation Research and Applied Sciences*, 1–7.

Anaerobic Digestion. (n.d.). Retrieved from https://www.epa.gov/anaerobic-digestion

Asad, M. J., Asghar, M., Sheikh, M. A., & Sultan, J. I. (2001). Production of α-amylase by *Aspergillus niger* and its partial purification. *Pakistan Journal of Agricultural Sciences, 38*, 3–4.

Aysun, H., Mercimek T., & Filiz U. T. (2016). Extracellular Pectinase Production and Purification from a Newly Isolated *Bacillus subtilis* Strain Extracellular Pectinase Production and Purification from a Newly Isolated *Bacillus subtilis* Strain. *International Journal of Food Properties, 19*(11), 2443–50.

Bairagi, S., & Rohtak, H. (2016). Optimization of Cellulase Enzyme from Vegetable Waste by Using *Trichoderma atroviride* in Solid State Fermentation. *Journal of Environmental Science, Toxicology and Food Technology*, *10*(5), 68–73.

Bansal, N., & Kanwar, S. S. (2013). Peroxidase (s) in environment protection. *The Scientific World Journal, 2013*, 714639. doi:10.1155/2013/714639 PMID:24453894

Bansal, N., Rupinder, T., Raman, S., & Sanjeev, K. (2012). Production of Cellulases from Aspergillus niger NS-2 in Solid State Fermentation on Agricultural and Kitchen Waste Residues. Waste Management, 32(7), 1341–46.

Bernstad, A., & Jansen, J. (2012). Review of Comparative LCAs of Food Waste Management Systems - Current Status and Potential Improvements. *Waste Management (New York, N.Y.)*, *32*(12), 2439–2455. doi:10.1016/j.wasman.2012.07.023 PMID:22922048

Cerda, A., Adriana, A., Xavier, F., Raquel, B., Teresa, G., Antoni, S., & Alejandra, C. (2017). Composting of Food Wastes : Status and Challenges. *Bioresource Technology*, *248*(Part A), 57-67.

Couto, S. R., & Sanromán, M. A. (2006). Application of solid-state fermentation to food industry—a review. *Journal of Food Engineering*, *76*(3), 291–302. doi:10.1016/j.jfoodeng.2005.05.022

Fogarty, W. M., & Kelly, C. T. (1990). Recent Advances in Microbial Amylases. In W. M. Fogarty & C. T. Kelly (Eds.), *Microbial Enzymes and Biotechnology*. Dordrecht: Springer. doi:10.1007/978-94-009-0765-2_3

Gigras, P., Vikram, S., & Rani, G. (2002). Statistical Media Optimization and Production of ITS-Amylase from *Aspergillus oryzae* in a Bioreactor. *Current Microbiology*, *45*(3), 203–208. doi:10.100700284-001-0107-4 PMID:12177743

Girotto, F., Luca, A., & Raffaello, C. (2015). Food Waste Generation and Industrial Uses : A Review. *Waste Management (New York, N.Y.)*, *45*, 32–41. doi:10.1016/j.wasman.2015.06.008 PMID:26130171

Hachemi, N., Nouani, A., & Benchabane, A. (2015). Bioconversion of Oranges Wastes for Pectinase Production Using *Aspergillus niger* under Solid State Fermentation. *International Journal of Bioengineering and Life Sciences*, *9*(9), 1004–1009.

Hebeish, & Ibrahim, N.A. (2007). The impact of frontier sciences on textile industry. *Colourage*, *54*, 41–55.

Hmidet, N., Ali, N. E. H., Haddar, A., Kanoun, S., Alya, S. K., & Nasri, M. (2009). Alkaline proteases and thermostable α-amylase co-produced by *Bacillus licheniformis* NH1: Characterization and potential application as detergent additive. *Biochemical Engineering Journal*, *47*(1-3), 71–79. doi:10.1016/j.bej.2009.07.005

Jahan, N., Faiza, S., Afsheen, A., Talat, Y. M., Shah, A., & Qader, U. I. (2017). Utilization of Agro Waste Pectin for the Production of Industrially Important Polygalacturonase. *Heliyon (London)*, *3*(6), e00330. doi:10.1016/j.heliyon.2017.e00330 PMID:28664192

Jasani, H., Nimita, U., Darshan, D., & Jagdish, P. (2016). Isolation, Optimization and Production of Cellulase by *Aspergillus niger* from Agricultural Waste Isolation. *Journal of Pure & Applied Microbiology*, *10*(2), 1159–1166.

Johnvesly, B., Manjunath, B. R., & Naik, G. R. (2002). Pigeon Pea Waste as a Novel Inexpensive Substrate for Production of a Thermostable Alkaline Protease from Thermoalkalophilic *Bacillus* Sp. JB-99. *Bioresource Technology*, *82*(1), 61–64. doi:10.1016/S0960-8524(01)00147-X PMID:11848379

Kalaichelvan, P. T. (2014). Production and Optimization of Pectinase from *Bacillus* Sp. MFW7 Using Cassava Waste. *Asian Journal of Plant Science and Research*, *2*(3), 369–375.

Kandasamy, S., Govarthanan, M., & Senthilkumar, B. (2016). Optimization of Protease Production from Surface-Modified Coffee Pulp Waste and Corncobs Using *Bacillus* Sp . by SSF. *3 Biotech*, *6*(2), 1–11.

Karmakar, M., & Ray, R. R. (2011). Current trends in research and application of microbial cellulases. *Res J Microbiol*, *6*(1), 41–53. doi:10.3923/jm.2011.41.53

Kashyap, D. R., Vohra, P. K., Chopra, S., & Tewari, R. (2001). Applications of Pectinases in the Commercial Sector : A Review. *Bioresource Technology*, *77*(3), 215–227. doi:10.1016/S0960-8524(00)00118-8 PMID:11272008

Katami, T., & Gifu, P. (2004). Formation of Dioxins from Incineration of Foods Found in Domestic Garbage. *Environmental Science & Technology*, *38*(4), 1062–1065. doi:10.1021/es030606y PMID:14998019

Kim, M. H., Han, B. S., Yuleum, S., In, T. J., & Jung, W. K. (2013). Evaluation of Food Waste Disposal Options in Terms of Global Warming and Energy Recovery : Korea. *International Journal of Energy and Environmental Engineering*, *4*(1), 1. doi:10.1186/2251-6832-4-1

Kiran, E. U., Antoine, P. T., & Yu, L. (2014). Glucoamylase Production from Food Waste by Solid State Fermentation and Its Evaluation in the Hydrolysis of Domestic Food Waste. *Biofuel Research Journal*, *3*, 98–105. doi:10.18331/BRJ2015.1.3.7

Kuhad, R. C., Gupta, R., & Singh, A. (2011). Microbial cellulases and their industrial applications. *Enzyme Research*, *2011*, 1–10. doi:10.4061/2011/280696 PMID:21912738

Kwatia, S., & Isaac, W. O. (2017). Optimization of Amylase Production by *Aspergillus niger* Cultivated on Yam Peels in Solid State Fermentation Using Response Surface Methodology. *African Journal of Biochemistry Research*, *11*(7), 34–42. doi:10.5897/AJBR2017.0941

Lin, C. S. K., Pfaltzgraff, L. A., Herrero-Davila, L., Mubofu, E. B., Abderrahim, S., Clark, J. H., ... Thankappan, S. (2013). Food waste as a valuable resource for the production of chemicals, materials and fuels, Current situation and global perspective. *Energy & Environmental Science*, *6*(2), 426–464. doi:10.1039/c2ee23440h

Liu, J., & Jichu, Y. (2007). Cellulase Production by *Trichoderma koningii* AS3. 4262 in Solid-State Fermentation Using Lignocellulosic Waste from the Vinegar Industry. *Food Technology and Biotechnology*, *45*(4), 420–425.

Makris, D. P. (2015). Recovery and applications of enzymes from food wastes. Food Waste Recovery, 361-379. doi:10.1016/B978-0-12-800351-0.00016-X

Mathew, J. J., Prem, J. V., & Sajeshkumar, N. K. (2016). Amylase Production by *Aspergillus niger* through Submerged Fermentation Using Starchy Food Byproducts as Substrate Amylase Production by *Aspergillus niger* through Submerged Fermentation Using Starchy Food Byproducts as Substrate. *International Journal of Herbal Medicine*, *4*(6), 34–40.

Melikoglu, M., Carol, S., Ki, L., & Colin, W. (2013). Analysing Global Food Waste Problem : Pinpointing the Facts and Estimating the Energy Content. *Central European Journal of Engineering*, *3*(2), 157–164.

Melikoglu, M., Lin, C., & Webb, C. (2013). Food and Bioproducts Processing Stepwise Optimisation of Enzyme Production in Solid State Fermentation of Waste Bread Pieces. *Food and Bioproducts Processing*, *91*(4), 638–646. doi:10.1016/j.fbp.2013.04.008

Mitidieri, S., Martinelli, A. H. S., Schrank, A., & Vainstein, M. H. (2006). Enzymatic detergent formulation containing amylase from *Aspergillus niger*: A comparative study with commercial detergent formulations. *Bioresource Technology*, *97*(10), 1217–1224. doi:10.1016/j.biortech.2005.05.022 PMID:16112858

Murthy, P. S., Naidu, M. M., & Srinivas, P. (2009). Production of α -Amylase under Solid-State Fermentation Utilizing Coffee Waste. *Journal of Chemical Technology and Biotechnology (Oxford, Oxfordshire)*, *84*(8), 1246–1249. doi:10.1002/jctb.2142

Mushtaq, Q., Muhammad, I., Fouzia, T., & Javed, I. (2016). Potato Peels : A Potential Food Waste for Amylase Production. *Journal of Food Process Engineering*.

Ngoc, T., Bao, D., Biswarup, S., & Chin-chao, C. (2014). Food Waste to Bioenergy via Anaerobic Processes. *Energy Procedia*, *61*, 307–312.

Pandey, A. (1992). Recent process developments in solid-state fermentation. *Process Biochemistry*, *27*(2), 109–117. doi:10.1016/0032-9592(92)80017-W

Pandey, V. P., Awasthi, M., Singh, S., Tiwari, S., & Dwivedi, U. N. (2017). A Comprehensive Review on Function and Application of Plant Peroxidases. *Biochemistry and Analytical Biochemistry*, *6*(01), 308. doi:10.4172/2161-1009.1000308

Paranthaman, R., Alagusundaram, K., & Indhumathi, J. (2009). Production of Protease from Rice Mill Wastes by *Aspergillus niger* in Solid State Fermentation. *World Journal of Agricultural Sciences*, *5*(3), 308–312.

Paritosh, K., Sandeep, K. K., Monika, Y., Nidhi, P., Aakash, C., & Vivekanand, V. (2017). Food Waste to Energy : An Overview of Sustainable Approaches for Food Waste Management and Nutrient Recycling. *BioMed Research International*, *2017*, 1–19. doi:10.1155/2017/2370927 PMID:28293629

Purna, M., & Reddy, K.V. (2016). Effects of the Culture Media Optimization on Pectinase Production by *Enterobacter* Sp . PSTB-1. *3 Biotech*, *6*(2), 1–11.

Ravindran, R., & Amit, J. (2016). Microbial Enzyme Production Using Lignocellulosic Food Industry Wastes as Feedstock: A Review. *Bioengineering*, *3*(4), 30. doi:10.3390/bioengineering3040030 PMID:28952592

Rodr, S., & Sanrom, M. A. (2005). Application of Solid-State Fermentation to Ligninolytic Enzyme Production. *Biochemical Engineering Journal*, *22*(3), 211–219. doi:10.1016/j.bej.2004.09.013

Romani, S., Silvia, G., Richard, P. B., & Ada, M. B. (2018). Domestic Food Practices: A Study of Food Management Behaviors and the Role of Food Preparation Planning in Reducing Waste. *Appetite*, *121*, 215–227. doi:10.1016/j.appet.2017.11.093 PMID:29155173

Russ, W., & Meyer-Pittroff, R. (2004). Utilizing waste products from the food production and processing industries. *Critical Reviews in Food Science and Nutrition*, *44*(1), 57–62. doi:10.1080/10408690490263783 PMID:15077881

Sadh, P. K., Surekha, D., & Joginder, S. D. (2018). Agro - Industrial Wastes and Their Utilization Using Solid State Fermentation : A Review. *Bioresources and Bioprocessing*, *5*(1), 1–15. doi:10.118640643-017-0187-z

Saithi, S., & Anan, T. (2016). Phytase Production of *Aspergillus niger* on Soybean Meal by Solid - State Fermentation Using a Rotating Drum Bioreactor. *Italian Oral Surgery*, *11*, 25–30.

Saxena, R., & Singh, R. (2011). Amylase production by solid-state fermentation of agro-industrial wastes using *Bacillus* sp. *Brazilian Journal of Microbiology*, *42*(4), 1334–1342. doi:10.1590/S1517-83822011000400014 PMID:24031761

Selvakumar, P., & Pandey, A. (1998). Biosynthesis Of Glucoamylase From *Aspergillus niger* By Solid-State Fermentation Using Tea Waste As The Basis Of A Solid Substrate. *Bioresource Technology*, *65*(1-2), 83–85. doi:10.1016/S0960-8524(98)00012-1

Shahid, M. G., Muhammad, N., Muhammad, I., & Quratulain, S. (2016). Partial Characterization of α-amylase Produced from *Aspergillus niger* using Potato Peel as Substrate, Punjab Univ. *Journal of Zoology (London, England)*, *33*(1), 22–27.

Sharanappa, A., Wani, K. S., & Pallavi, P. (2011). Bioprocessing of food industrial waste for α-amylase production by solid state fermentation. *Int. J. Adv. Biotechnol. Res*, *2*(4), 473–480.

Sharholy, M., Kafeel, A., Gauhar, M., & Trivedi, R. C. (2008). Municipal Solid Waste Management in Indian Cities - A Review. *Waste Management (New York, N.Y.)*, *28*(2), 459–467. doi:10.1016/j.wasman.2007.02.008 PMID:17433664

Sharma, K. H., Burnwal, P. K., Dubey, L., & Rao, R. J. (2013). Optimization and Production of Cellulase from Agricultural Waste. *Research Journal of Agriculture and Forestry Sciences*, *1*(1), 18–20.

Silva, D., Silva, M., Roberto, S., & Eleni, G. (2002). Pectinase Production By *Penicillium viridicatum* Rfc3 By Solid State Fermentation Using Agricultural Wastes And Agro-Industrial By-Products. *Brazilian Journal of Microbiology*, *33*(4), 318–324. doi:10.1590/S1517-83822002000400008

Souza, P. M. D., Bittencourt, M. L. D. A., Caprara, C. C., Freitas, M. D., Almeida, R. P. C. D., Silveira, D., ... Magalhães, P. O. (2015). A biotechnology perspective of fungal proteases. *Brazilian Journal of Microbiology*, *46*(2), 337–346. doi:10.1590/S1517-838246220140359 PMID:26273247

Sudharhsan, S., Sivaprakasam, S., & Karunasena, R. (2007). Physical and Nutritional Factors Affecting the Production of Amylase from Species of *Bacillus* Isolated from Spoiled Food Waste. *African Journal of Biotechnology*, *6*(4), 430–435.

Thassitou, P. K., & Arvanitoyannis, I. S. (2002). Bioremediation : A Novel Approach to Food Waste Management. *Trends in Food Science & Technology*, *12*(5-6), 185–196. doi:10.1016/S0924-2244(01)00081-4

Tian, M., Alvan, W., Tuhin, K.G., Georg, H., & Qiuyan, Y. (2017). Production of Cellulase and Xylanase Using Food Waste by Solid-State Fermentation. *Waste and Biomass Valorization*, 1-8.

Tian, M., & Qiuyan, Y. (2016). Optimization of Phytase Production from Potato Waste Using *Aspergillus ficuum*. *3 Biotech, 6*(2), 1–8.

Tsai, S., Ching-piao, L., & Shang-shyng, Y. (2007). Microbial Conversion of Food Wastes for Biofertilizer Production with Thermophilic Lipolytic Microbes. *Renewable Energy*, *32*(6), 904–915. doi:10.1016/j.renene.2006.04.019

Uc, E., & Antoine, P. T. (2014). Enzyme Production from Food Wastes Using a Biorefinery Concept. *Waste and Biomass Valorization*, *5*(6), 903–917. doi:10.100712649-014-9311-x

Viniegra-gonzález, G., Ernesto, F., & Cristobal, N. (2003). Advantages of Fungal Enzyme Production in Solid State over Liquid Fermentation Systems. *Biochemical Engineering Journal*, *13*(2-3), 157–167. doi:10.1016/S1369-703X(02)00128-6

Zhang, C., Haijia, S., Jan, B., & Tianwei, T. (2014). Reviewing the Anaerobic Digestion of Food Waste for Biogas Production. *Renewable & Sustainable Energy Reviews*, *38*, 383–392. doi:10.1016/j.rser.2014.05.038

Zhang, R., Hamed, M. E., Karl, H., Fengyu, W., Guangqing, L., Chris, C., & Paul, G. (2007). Characterization of Food Waste as Feedstock for Anaerobic Digestion. *Bioresource Technology*, *98*(4), 929–935. doi:10.1016/j.biortech.2006.02.039 PMID:16635571

Chapter 2
Value-Added Products From Food Waste

Baban Baburao Gunjal
Sunrise Biotech Organisation, India

ABSTRACT

Food waste is the most challenging issue humankind is facing worldwide. Food waste, which consists of carbohydrates, proteins, lipids, and inorganic compounds, is a biodegradable waste discharged from food processing industries, households, and hospitality sectors. The management of food waste is very important. The food waste generated is usually incinerated or dumped in open areas which may cause severe health and environmental issues. The management of food waste can be done by conversion to different value-added products, for example, phytochemicals, bioactive compounds, food supplements, livestock feed, dietary fibers, biopigments and colorants, emulsifiers, edible and essential oils, biopreservatives, biofertilizers, biofuels, and single cell proteins. The value-added products from food waste will be very eco-friendly. The chapter will focus on different value-added products from food waste.

INTRODUCTION

The problem of food waste is increasing, involving all sectors of waste management from collection to disposal. Global food waste is approximately 1.3 billion tons per year (Kojima & Ishikawa, 2013). It is estimated that more food is wasted in the industrialized countries compared to the developing nations on per-capita basis (Gustavsson et al., 2011). The wastes generated from food processing industries are shown in Table 1.

DOI: 10.4018/978-1-5225-7706-5.ch002

Recently, there is great emphasis on the recovery, recycling and reconditioning of food waste. The efforts are made to convert food waste into value-added products (Laufenberg et al., 2003). The food waste can be converted into useful value-added products viz., phytochemicals, bioactive compounds, food supplements, livestock feed, dietary fibres, biopigments and colourants, emulsifiers, edible and essential oils, biopreservatives, biofertilizers, biofuels and single cell protein. India's share in some agricultural and horticultural produce is shown in Table 2.

The potential of vegetable wastes for production of value added products and for the generation of biofuels is an efficient mode of food waste management. Strategies for efficient waste management must be adopted. The best approach for the waste management is reduction of the waste at its source. Socio-economic aspect of waste generation and handling also has to be considered for adopting an efficient strategy of integrated waste management. Food waste is generated as a part of human society at small domestic level and at large industrial level. In developed countries, the waste management practices followed are viz., sanitary landfills, composting, incineration etc. Wastes are collected and mostly dumped in open or burnt in open (Sandra, 2006). This has serious impact on both environment and human health. When dumped in open or in landfills, food wastes get decomposed by the action of various microorganisms. This produces different gases like methane and carbon dioxide both of which contribute to the greenhouse effect leading to global warming (Brown & David, 1994).

The problem of food waste must be solved by converting the waste into various value-added products which will be very eco-friendly and also effective. The various value-added products from carrot, onion, pea, tomato and sugar beet are mushroom, biomethane, biohydrogen, single cell protein, biogas, bioethanol, mushroom, vinegar, α-L-arabinofuranosidase, organic acids, oligomers, fertilizers, glycoalkaloids, animal feeds, etc.

Table 1. Wastes generated from food processing industries

Food processing industry	Waste materials
Animal products	Skins, hides, blood, fats, horns, hairs, bones, liver, intestines
Poultry processing	Skin, blood, fats, hairs, feathers, bones, liver, intestines, wings, trimmed organs
Marine products processing	Shells, roes, trimmed parts, pincers
Cereals and pulse processing	Husk, hull, chaff, stalks
Fruits and vegetable processing	Skin, peels, stones, fibre, pith
Nuts	Shells, coir, pith
Spices and condiments	Hulls, stalks.

(Rao, 2010)

Table 2. India's share in some agricultural and horticultural produce

Fruit / Vegetable	Global production (%)
Mangoes	54
Cauliflower	30
Bananas	23
Green peas	36
Onions	10

Various Value-Added Products From Food Waste

The management of food waste can be done by converting into value-added products. There are many uses of exotic fruits, olives and tomatoes for the production of antioxidants, fibers, phenols, polyphenols and carotenoids. Dairy by-products and slaughter house waste can be potential source for lactic acid and protein extraction (Mirabella et al., 2014).

Fruit processing produce large amount of waste products viz., seeds, kernels, flesh and peels; which contain valuable compounds in higher quantities (Mirabella et al., 2014). Recovery of value-added products from passion fruit processing waste (Mirabella et al., 2014); and pineapple stem (Canteri et al., 2010); amino acids and phenolic compounds from mango seeds (Upadhyay et al., 2012); polyphenols, carotenoids, vitamins, enzymes and dietary fibres from mango peels (Abdalla et al., 2007); coconut protein powder from coconut processing industry waste (Ajila et al., 2010) etc. have been reported.

About 50% of the cheese-whey, the by-product of cheese manufacturing, is converted into value added products viz., whey powder, whey protein, whey permeate, bioethanol, biopolymers, hydrogen, methane, single cell protein (Siso, 1996; Yadav et al., 2015). Cheese-whey is known to consist of mostly lactose (4.5-5% w/v), soluble proteins (0.6-0.8% w/v), lipids (0.4-0.5% w/v) and mineral salts (8-10% of dried extract); less quantities of lactic acid, citric acid, non-protein nitrogen compounds and B vitamins are also reported (Prazeres et al., 2012).

Ethanol Production From the Food Waste

There is a report where food waste with a composition of 23.3% w/w total reducing sugars, 34.8% w/w starch, and 1.6% w/w fibres has been converted for the ethanol production (Zhang & Richard, 2011). Similarly, Moon et al. (2009) also studied ethanol production from food waste with high starch (30.1% w/w) and fibre (14.9% w/w) contents, with a total of 17.6% w/w reducing sugars. Matsakas et al. (2014)

reported ethanol yield of 108 g/kg dry material from household food waste. The bioethanol yield from various food residues and wastes is shown in Table 3.

Food Waste to Biogas

Food wastes are good source of biogas that can be used in plants. Due to high moisture content of food wastes, bioconversion technologies such as anaerobic digestion are more suitable compared to thermo-chemical conversion technologies. Many researchers have studied the potential of food waste for biogas production. Recently, Chinese researchers studied the anaerobic digestion of food waste resulting in final total solids and volatile solids. Due to this, the quantity of wasted food decreases and also produces clean biogas (Yang et al., 2015; Zhang et al., 2014).

Glycoalkaloids From Potato Waste

In countries like Ireland, the potato production in 2010 was 450,000 tonnes. Industrial processing of potatoes generates large quantities of peel which creates disposal, sanitation, and environmental problems. The potato peels are rich in glycoalkaloids such as α-solanine and α-chaconine; carbohydrates; starch; and proteins.

Marine Food Waste for Production of Oligomers

The fish waste is a potential source of high-value biochemicals, such as biopolymers (chitin, chitosan), pigments (carotenoid, astaxanthin), minerals, and proteins. The different value-added products from the marine food wastes are viz., chitin and chitosan oligomers.

Table 3. Bioethanol yield from various food residues and wastes

Food residue/wastes	Country	Bioethanol yield (%)	Reference
Switch grass	USA	72	(Asli et al., 2008)
Corn steep liquor	USA and Brazil	No yield	(Ruanglek et al., 2006)
Brewer's yeast autolysate	Thailand	88	(Ruanglek et al., 2006)
Waste potato	Finland	87	(Liimatainen et al., 2004)
Oil palm empty fruit	Malaysia	Not specified	(Ibrahim et al., 2012; Razak et al., 2013)

Fruit Wastes for Single Cell Protein Production

Fruit wastes have been used as substrates for the production of Single Cell Protein (SCP) by many researchers. Sweet orange residues have been used for SCP production (Nwabueze & Oguntimein, 1987). Rahmat et al. (1995) used apple pomace for the production of single cell protein from *Kloechera apiculata* and *Candida utilis* so as to improve stock feed. Pineapple cannery effluent has been utilized for SCP production by Nigam (1998). Essien et al. (2005) utilized banana peel as a substrate for mould growth and biomass production.

Food Wastes as Substrates for the Production of Organic Fertilizers

After food wastes are degraded by aerobic microorganism, the secondary pollution is avoided, and has good environmental benefits. The degradation products can be divided into organic fertilizer, bio-organic fertilizer and soil conditioner, which contain plant growth promoting substances. This will reduce the use of chemical fertilizers and therefore pollution also will be reduced.

Food Processing Wastes as Substrates for the Production of Animal Feeds

The food processing wastes can be used in the production of animal feed. Soya bean cake obtained after extraction of the soya bean milk and groundnut cake obtained from the groundnut processing industry have been used for the production of animal feeds.

Vegetable Oil and Its By-Products

The vegetable oil industry generates number of waste products. India produces about 70 lakh tons of vegetable oil. The phytosterols can be obtained from vegetable oils during refining process, which has nutraceutical value. The value-added products from vegetable oil industry is shown in Table 4.

High Value By-Products From Fruits and Vegetables

Peel is the major waste generated from fruits. The peel after fruit processing undergoes rapid changes in quality for the generation of secondary value-added products. They can be used for the production of secondary products viz., pectin, mucilage, gum, anthocyanin, carotenoids, antioxidants, antimicrobials and fermented products. Fruit peels are the best source for compounds such as polyphenols, flavonoids,

Table 4. Value-added products from vegetable oil industry

Oil seeds	Primary product	Secondary product	Value-added products
Soybean	Soyabean oil	Gum sludge, deodorizer distillate, soap stock, deoiled cake	Soy protein, isoflavones, lecithin, tocopherol, fatty acids
Cotton seed	Cotton seed oil	Gum sludge, deodourizer distillate, soap stock, deoiled cake	Protein, gossypol, fatty acids, lecithin
Sunflower	Sunflower oil	Gum sludge, deodourizer distillate, soap stock, deoiled cake	Sunflower seed protein, lecithin, tocopherol concentrate, wax, fatty acids
Rape-mustard seed	Mustard oil	Gum sludge, deodourizer distillate, soap stock, deoiled cake	Mustard seed protein, glucosinolate concentrate, lecithin, tocopherols
Palm fruit	Palm oil, palm kernel oil	Gum sludge, deodourizer distillate, soap stock, deoiled cake	Beta-carotene, tocotrienol, palm stearin, palmitic and oleic acids, lecithin, carotenoids
Rice	Rice bran oil and chemically refined oils	Gum sludge, wax sludge, deodorizer distillate, fatty acids, soap stock, deoiled cake	Rice bran protein, coenzyme Q10, rice bran fibre for food purposes, lecithin, oryzanol, tocotrienols, squalene, phytosterols, wax, fatty acids

tannins, catechins and vitamins. The potential anthocyanin sources are blackcurrant, chokeberry, eggplant, orange, blackberry, vaccinium, raspberry, cherry, redcurrent, red grape, etc.

There are also many by-products which can be obtained from grape processing from seeds (grape seed oil) and skin (resveratrol, polyphenols). The use of grape skin colour powder includes beverages, sauces, baking and red wine. The extract after being processed with beverages, milk, chocolates, candy, etc. produces food products which are good for health. The grape waste can also be processed to get the value-added product polyphenolics which can be used as phytoceuticals.

Organic Acids From Food Waste

Lactic acid, acetic acid, oxalic acids are important because of their various applications existing from food industry to pharmaceuticals. They can be produced by fermentation process by the activity of different microorganisms. Various organic acids are produced by the microbial fermentation. Some of the examples are citric acid, lactic acid and acetic acid by *Aspergillus niger, Lactobacillus delbrueckii* and *Acetobacter aceti,* respectively (Sethi and Maini, 1999). Various substrates have been used for the citric acid production by fermentation process, but recently agricultural wastes have been used on high demand due to their abundant availability (Soccol et al., 2006). Among them, citric acid is an important chemical having worldwide demand due to its high usage and low toxicity. Beet molasses have been used for citric acid production. Zhang and Jin (2009) studied lactic acid production using potato starch waste where the lactic acid production obtained was 103.8 g/l in 48 h fermentation. Sugar beet molasses have been used as low-cost substrate for oxalic acid production using different reactors. High production was obtained in the reactor having nitrogen oxide (Guru et al., 2001). Acetic acid, another organic acid can be produced by carrots and white radish leafage. Carrots have been used as the substrate in the hydrothermal two stage production of acetic acid which resulted in high yield (Jin et al., 2005).

Natural Food Colours and Dyes

Natural colour has advantage over synthetic colours. These natural food colours are obtained from vegetable, animal or mineral. The natural colours come from sources viz., seeds, fruits, vegetables, leaves, algae and insects. The examples of commonly used natural colours are Annatto (seed), turmeric, beet juice (root), red cabbage (vegetable), spinach (leaf), anthocyanins, β-carotene, carmine, curcumin, canthaxanthin, etc.

Polyhydroxybutyrate Production From the Food Waste

Polyhydroxybutyrate (PHB) is a biopolymer used as a biodegradable thermoplastic material for waste management strategies and biocompatibility in medical devices (Gouda et al., 2001). It has wide applications viz., packaging, pharmaceuticals, chemical and cosmetic industries. Vegetable and food wastes have now been used as substrates for production of PHB in a cost effective way (Carucci et al., 2001; Koller et al., 2008). Rusendi and Sheppard (1995) reported the use of potato processing waste from the potato-chip plant for the production of PHB. Hafuka et al. (2011) reported 87% PHB yield from food wastes after 259 h of incubation.

CONCLUSION

Different value-added products viz., single cell protein; biogas; polymers; feed; organic fertilizers; organic acids; colours and dyes; PHB; etc. will be obtained from food waste. The value-added products from food waste will be very eco-friendly.

This will help the management of food waste which is very important. The different value-added products from food waste have many applications.

Future Possibilities

Food waste recycling will help mitigate greenhouse gas emissions. Food waste management also play important role to control environmental pollutants.

REFERENCES

Abdalla, A. M., Darwish, S. M., Ayad, E. E., & El-Hamahmy, R. M. (2007). Egyptian mango by-product 1. Compositional quality of mango seed kernel. *Food Chemistry*, *103*(4), 1134–1140. doi:10.1016/j.foodchem.2006.10.017

Ajila, C. M., Aalami, M., Leelavathi, K., & Rao, U. P. (2010). Mango peel powder: A potential source of antioxidant and dietary fibre in macaroni preparations. *Innovative Food Science & Emerging Technologies*, *11*(1), 219–224. doi:10.1016/j.ifset.2009.10.004

Asli, I., Jennifer, N. H., & Anthony, L. (2008). Aqueous ammonia soaking of switchgrass followed by simultaneous saccharification and fermentation. *Applied Biochemistry and Biotechnology*, *144*(1), 69–77. doi:10.100712010-007-8008-z PMID:18415988

Brown, K. A., & David, M. H. (1994). Using landfill gas: A UK perspective. *Renewable Energy*, *5*(5-8), 774–781. doi:10.1016/0960-1481(94)90086-8

Canteri, M. G., Scheer, A., Wosiacki, G., Ginies, C., Reich, M., & Renard, C. G. (2010). A comparative study of pectin extracted from passion fruit rind flours. *Journal of Polymers and the Environment*, *18*(4), 593–599. doi:10.100710924-010-0206-z

Carucci, A., Dionisi, D., Majone, M., Rolle, E., & Smurra, P. (2001). Aerobic storage by activated sludge on real wastewater. *Water Research*, *35*(16), 3833–3844. doi:10.1016/S0043-1354(01)00108-7 PMID:12230166

Essien, J. P., Akpan, E. J., & Essien, E. P. (2005). Studies on mould growth and biomass production using waste banana peel. *Bioresource Technology*, *96*(13), 1451–1456. doi:10.1016/j.biortech.2004.12.004 PMID:15939272

Gouda, M. K., Swellam, A. E., & Omar, S. H. (2001). Production of PHB by a *Bacillus megaterium* strain using sugarcane molasses and corn steep liquor as sole carbon and nitrogen source. *Microbiological Research Journal, 156*(3), 201–204. doi:10.1078/0944-5013-00104 PMID:11716209

Guru, M., Bilgesu, A. Y., & Pamuk, V. (2001). Production of oxalic acid from sugar beet molasses by formed nitrogen oxides. *Bioresource Technology*, *77*(1), 81–86. doi:10.1016/S0960-8524(00)00122-X PMID:11211079

Gustavsson, J., Cederberg, C., Sonesson, U., van Otterdijk, R., & Meybeck, A. (2011). Global food losses and food waste. Extent, causes and prevention. Rome: Academic Press.

Hafuka, A., Sakaida, K., Satoh, H., Takahashi, M., Watanabe, Y., & Okabe, S. (2011). Effect of feeding regimens on polyhydroxybutyrate production from food wastes by *Cupriavidus necator. Bioresource Technology*, *102*(3), 3551–3553. doi:10.1016/j.biortech.2010.09.018 PMID:20870404

Ibrahim, M. F., Abd-Azizi, S., Razak, M. A., Phang, L. Y., & Hassan, M. A. (2012). Oil palm empty fruit bunch as alternative substrate for acetone-butanol-ethanol production by *Clostridium butyricum* EB6. *Applied Biochemistry and Biotechnology*, *166*(7), 1615–1625. doi:10.100712010-012-9538-6 PMID:22391689

Jin, F., Zhou, Z., Moriya, T., Kishida, H., Higashijima, H., & Enomoto, H. (2005). Controlling hydrothermal reaction pathways to improve acetic acid production from carbohydrate biomass. *Environmental Science & Technology*, *39*(6), 1893–1902. doi:10.1021/es048867a PMID:15819253

Kojima, R., & Ishikawa, M. (2013). Prevention and recycling of food wastes in Japan: Policies and achievements. Kobe University.

Koller, R., Chiellini, B. E., Fernandes, E. G., Horvat, P., Kutschera, C., Hesse, P., & Braunegg, G. (2008). Polyhydroxy alkanoate production from whey by *Pseudomonas hydrogenovora. Bioresource Technology*, *99*(11), 4854–4863. doi:10.1016/j.biortech.2007.09.049 PMID:18053709

Laufenberg, G., Kunz, B., & Nystroem, M. (2003). Transformation of vegetable waste into value-added products: (A) the upgrading concept; (B) practical implementations. *Bioresource Technology*, *87*(2), 167–198. doi:10.1016/S0960-8524(02)00167-0 PMID:12765356

Liimatainen, H., Kuokkanen, T. K., & Ariainen, J. (2004). Development of bioethanol production from waste potatoes. In *Proceedings of the Waste Minimization and Resources Use Optimization Conference*. University of Oulu.

Matsakas, L., Kekos, D., Loizidou, M., & Christakopoulos, P. (2014). Utilization of household food waste for the production of ethanol at high dry material content. *Biotechnology for Biofuels*, *7*(1), 1–9. doi:10.1186/1754-6834-7-4 PMID:24401142

Mirabella, N., Castellani, V., & Sala, S. (2014). Current options for the valorization of food manufacturing waste: A review. *Journal of Cleaner Production*, *65*, 28–41. doi:10.1016/j.jclepro.2013.10.051

Moon, H. C., Song, I. S., Kim, J. C., Shirai, Y., Lee, D. H., Kim, J. K., ... Cho, Y. S. (2009). Enzymatic hydrolysis of food waste and ethanol fermentation. *International Journal of Energy Research*, *33*(2), 164–172. doi:10.1002/er.1432

Nigam, J. N. (1998). Single cell protein from pineapple cannery effluent. *World Journal of Microbiology & Biotechnology*, *14*(5), 693–696. doi:10.1023/A:1008853303596

Nwabueze, T. U., & Ogumtimein, G. B. (1987). Sweet orange (*Citrus sinensis*) residue as a substrate for single cell protein production. *Biological Wastes*, *20*(1), 71–75. doi:10.1016/0269-7483(87)90085-1

Prazeres, A. R., Carvalho, F., & Rivas, J. (2012). Cheese whey management: A review. *Journal of Environmental Management*, *110*, 48–68. doi:10.1016/j.jenvman.2012.05.018 PMID:22721610

Rahmat, H., Hodge, R., Manderson, G., & Yu, P. (1995). Solid substrate fermentation of *Kloechera apiculata* and *Candida utilis* on apple pomace to produce an improved stock-feed. *World Journal of Microbiology & Biotechnology*, *11*(2), 168–170. doi:10.1007/BF00704641 PMID:24414495

Rao, D. G. (2010). *Fundamentals of food engineering*. New Delhi: PHI Learning Private Ltd.

Razak, M. A., Ibrahim, M. F., Yee, P. L., Hassan, M. A., & Abd-Azizi, S. (2013). Statistical optimization of butanol Production from oil palm decantater cake hydrolysate by *Clostridium acetobutylicum* ATCC 824. *BioResources*, *8*, 1758–1770.

Ruanglek, V., Maneewatthana, D., & Tripetchkul, S. (2006). Evaluation of thai agroindustrial wastes for bio-ethanol production by *Zymomonas mobilis*. *Process Biochemistry*, *41*(6), 1432–1437. doi:10.1016/j.procbio.2006.01.010

Rusendi, D., & Sheppard, J. D. (1995). Hydrolysis of potato processing waste for the production of poly-ß hydroxybutyrate. *Bioresource Technology*, *54*(2), 191–196. doi:10.1016/0960-8524(95)00124-7

Sandra, C. (2006). Occupational and environmental health issues of solid waste management. The World Bank Group.

Sethi, V., & Maini, S. B. (1999). Production of organic acids. In V. K. Joshi & A. Pandey (Eds.), *Biotechnology: Food Fermentation, Microbiology, Biochemistry and Technology* (pp. 1259–1290). New Delhi: Educational Publishers and Distributors.

Siso, M. G. (1996). The biotechnological utilization of cheese whey: A review. *Bioresource Technology*, *57*(1), 1–11. doi:10.1016/0960-8524(96)00036-3

Soccol, C. R., Vandenberghe, L. S., Rodrigues, C., & Pandey, A. (2006). New perspectives for citric acid production and application. *Food Technology and Biotechnology*, *44*, 141–149.

Upadhyay, A., Chompoo, J., Araki, N., & Tawata, S. (2012). Antioxidant, antimicrobial, 15-LOX, and AGEs inhibitions by pineapple stem waste. *Journal of Food Science*, *77*(1), H9–H15. doi:10.1111/j.1750-3841.2011.02437.x PMID:22260109

Yang, L., Huang, Y., Zhao, M., Huang, Z., Miao, H., Xu, Z., & Ruan, W. (2015). Enhancing biogas generation performance from food wastes by high-solids thermophilic anaerobic digestion: Effect of pH adjustment. *International Biodeterioration & Biodegradation*, *105*, 153–159. doi:10.1016/j.ibiod.2015.09.005

Zhang, C., Su, H., Baeyens, J., & Tan, T. (2014). Reviewing the anaerobic digestion of food waste for biogas production. *Renewable & Sustainable Energy Reviews*, *38*, 383–392. doi:10.1016/j.rser.2014.05.038

Zhang, X., & Richard, T. (2011). Dual enzymatic saccharification of food waste for ethanol fermentation. *Proceedings of international conference on electrical and control engineering*. 10.1109/ICECENG.2011.6058308

Chapter 3
Industrially Important Enzymes Production From Food Waste:
An Alternative Approach to Land Filling

Madhuri Santosh Bhandwalkar
S. B. B. Alias Appasaheb Jedhe College, India

ABSTRACT

To link food demand and reduction in food waste, proactive approaches should be taken. Perishable food is mainly fruits and vegetables, waste from different processing industries like pulses, meat products, oil products, dairy products, and fishery byproducts. Conventional food waste management solution is land filling which is not sustainable as it generates global warming gases like methane and carbon dioxide. To reduce food waste, the process known as "food valorization" has become another solution to landfilling, the concept which is given by European Commission in 2012, meaning food processing waste conversion to value-added products. In this chapter the study focuses on production of industrially important enzymes from food waste which could be one of the reactive solutions. Different enzymes like pectinase, peroxidase, lipase, glucoamylase, and protease can be produced from food waste.

INTRODUCTION

Food Demand is rising globally in the proportion to rapid population growth. This leads to increase in Food production and ultimately in food waste or loss in food supply chain from initiation to final consumption. For sustainable food waste management, the waste hierarchy concept (1975) given by European Waste Policy

DOI: 10.4018/978-1-5225-7706-5.ch003

can be useful to categorize the food waste and treat them accordingly. Some of the indicators used in food waste classification are edibility (edible/nonedible), State (Eatable/uneatable), Origin (Animal/plant) and complexity (single product/complex product). Water activity is the main factor assisting to predict the presence of microorganisms spoiling the food. Among them bacteria need water activity 0.85 and molds (0.7-0.8) for growth. It is important to treat solid food waste and liquid food waste eco-friendly.

Liquid waste generally contains proteins, sugars, starches, and fats. The researchers focused on the study of different problems faced by developing and developed countries regarding food waste generation and management.

Out of total industrially important enzymes production food industry useful enzymes are 45%, detergent industry 35%, textile 10%, and leather 3%. For the production, fungi and bacteria are mainly used and others involve higher plants, higher animals, yeasts and *Streptomyces*. Chapter covers the information from food waste generation, classification, different strategies used to manage food waste, microbial sources used in different enzyme production, different food waste types with examples, future aspects and conclusion.

FOOD WASTE GENERATION

According to many researchers hospitality industries and households which are the end of the food supply chain are contributing food waste generation. Developed and developing countries are defined by the Gross National Income (GNI) index. This study reveals that though the developing countries have less food demand as compared to developed countries, the food waste generation contributed by both of them is equal in quantity. Many researchers have shown that most food is wasted at the end of the food supply chain. Before considering Food Waste (FW) it is important to distinguish between food loss and food waste. (As classified in Table 1)

Food loss is the one which occurs before completing food supply chain and transformed into a final product. Before that point only food spills, lost or reduce in nutritional value and volume also. Food waste is the one which occurs after completing food supply chain. It may occur before consumption or before spoiling it is left to spoil. Table 1 explains the food loss and food waste.

FOOD WASTE CLASSIFICATION

To carry optimal food waste management, classification of food waste using their types viz., fruits, cereals, drinks, fish, meat is necessary. Activities to be treated on

Table 1. Food loss and waste along the value chain

Production	Handling and storage	Processing and packaging	Distribution and Market	Consumption
During or immediately after harvesting on the farm	After produce leaves the farm for handling, storage and transport	During industrial or domestic processing and/ or packaging	During distribution to markets, including losses and wholesale and retail markets	Losses in the home or business of the consumer, including restaurants/ caterers
Fruits bruised during picking or threshing	Edible food eaten by pests	Milk spoiled during pasteurization and processing	Edible produce sorted out due to quality	Edible produce sorted out due to quality
Crops sorted out post- harvest for not meeting quality standards	Edible produce degraded by fungus or disease	Edible fruits or grains sorted out as not suitable for processing	Edible products expired before being purchased	Food purchased but not eaten
Crops left behind in fields due to poor mechanical harvesting or sharp drop in prices	Livestock death during transport to slaughter or not accepted for slaughter	Livestock trimming during slaughtering and industrial processing	Edible products spoiled or damaged in market	Food cooked but not eaten

Source: Bagherzadeh et.al, 2014 OECD, France

FW and manage them food classification is essential prerequisite. Also it decides a methodical procedure. Garcia-Garcia et al (2017) has explained these nine indicators. They are as Edibility, State, Origin, Complexity, Animal product presence, Treatment, Packaging. Food waste categorization also helps to monitor purposes like business or else so that assessing progress in the management and sustainability can easily be done. (Guillermo Garcia-Garcia et al., 2017). Figure1 describes Food Waste Management Decision Tree (FWMDT).

Best waste management option available after categorizing food waste are: Anaerobic digestion, composting or thermal treatment with energy recovery.

VARIOUS STRATEGIES FOR FOOD WASTE MANAGEMENT

According to many researchers hospitality industries and households which are the end of the food supply chain are contributing food waste generation. The best solution for handling food waste is to avoid the onset of food waste generation. As it is impossible to do so, there are various strategies which can be applied are described below.

1. **Animal Feeding:** FW generated from animal products like meat and bone meal, edible tallow, restaurant grease, feather meal, fish meal, dairy whey, dry milk are used as animal feed. FW has high coefficient variations which causes increase in digestibility coefficient and if it contains primary nutrients around $3/4^{th}$ part of total FW, can be recognized as feed additive. Also there are chances of causing diseases to animals, so less preferable solution for considering FW treatment.
2. **Anaerobic Digestion:** Anaerobic digestion is considered as one of the effective method to treat FW and getting biogas like methane. For enhancing anaerobic digestion of FW, pretreatment like physical methods with measures of mechanical grinding, ultrasound, microwave, thermal, pressure –depressure or chemical method with the use of acid and biological method using biological solubilization is done followed by co-digestion. Co-digestion is done to overcome inhibition of anaerobic digestion due to high lipid content. It is done by addition of organic substances like cattle manure (CM), green waste, and sewage sludge.

Balancing the nutrient imbalance and reducing energy expenses required for anaerobic digestion are the challenges for this treatment.

3. **Composting:** C:N ratio is significant (25:1 to 35:1 suitable for microbial activity)in the successful composting process as less ratio means excess nitrogen leads to ammonia production causing odor problems and high ratio means high carbon content leading to complete utilization of Nitrogen before carbon containing material decomposition. (Rynk. R., ed. 11992).Proper composting can give us sufficient temperature rise destructing pathogen, weeding seeds along with 40% volume reduction.(S.M. Schaub and J.J. Leonard)
4. **Incineration:** Previous studies suggest that incineration is used to produce thermal energy and can be used to reduce the volume of waste.(AnqiGao et al.).It is energy consuming as moisture content is high and incineration causes air pollution.

MICROBIAL SOURCES USED IN DIFFERENT FOOD WASTE FOR THE ENZYME PRODUCTION

As Microbial Growth is short, the requirements of industries for enzyme production using microbial source is fulfilled. The food waste is nutrient and organic rich source for production of valuable products like organic acids, methane, chemical, ethanol and most importantly enzymes using various technical processes.

There is tremendous potential in glucoamylase and protease enzymes for applying in pharmaceutical; food industrial processes. It has been demonstrated that glucoamylase and protease can be effectively produced from waste bread using solid state fermentation.

Scientists have produced the laccase production by fungus *Trametes hirsute* of family Polyporaceae using potato, orange, and apple peelings (Jasminailerdži). Laccase is lignin modifying enzyme. To remove phenolic compounds in the recovery process of enzymes laccase and peroxidase enzymes have been effective. It has been shown that by solid state fermentation three fungal enzymes (invertase, pectinase, and tannase) produced by *Aspergillus niger* produced higher yield compared to submerged fermentation. Physiological studies explain the reason behind this result. Such comparison studies are also helpful to get maximum enzyme production using fungus. Glucose oxidase enzyme production using the waste mycelium of *A.niger* is the other example. If *A. niger* waste mycelium is present in food waste containing sodium gluconate, potassium gluconate can be used for Glucose oxidase enzyme production. The study revealed that maintaining metal ions accelerates Glucose oxidase activity.

Scientists have been proposed an idea to valorize the molecules like amino acids and sugars recovered from bakery waste to feedstocks in bioconversion process. In this study they have given an innovative approach by using Bakery waste for valorization, biocolorant and enzyme production. Nutrient rich hydrolysate containing free amino nitrogen, sugars and phosphates was generated using bakery waste by *Aspergillus awamori* and *Aspergillus oryzae*. This hydrolysate was used for bio-colorant production and solid state fermentation for glucoamylase and protease enzyme production using filamentous fungus *Monascuspurpureus*.

Another interesting approach of linking discovery of renewable fuel and focus to use food waste as raw material for biofuels production has been done. Leonidas Matsakas and Paul Christakopoulos have been successful in ethanol production from enzymatically treated dried food waste using enzymes produced on-site. The ethanol production from source House hold food waste had been evaluated and also produced hydrolytic enzymes for cellulose hydrolysis in- house using thermophilic fungus *Myceliophthora thermophila.* After studying enzyme production it was optimized for an enzymatic activity up to 0.28 FPU/ml in the cell free broth. The enzymes produced in this way with 30% (w/w) were used to hydrolyze cellulose in obtaining sugars for ethanol production. A considerable amount of ethanol production was found to be produced when enzymes were added as compare to the control where no saccharification by enzymes was done.

TYPES OF VARIOUS FOOD WASTE USED FOR DIFFERENT ENZYMES

Food Waste Containing Bread, Savory, Waste Cakes, Cafeteria Waste, Fruits, Vegetables and Potatoes

Number of Enzyme recovery steps needed is proportional to degree of purity and safety needed for that enzyme. Saccharification of FW is the important step in bioconversion of polymers to their monomers. Glucoamylase is the industrially important enzyme. In situ enzyme production without downstream processing is a less expensive, productive method.(Merino and Cherry,2007,Wangetet..al.2010). This strategy has been used by many researchers.

EsraUckunKiran et.al, studied glucoamylase production using *Aspergillus awamori*by solid state fermentation. FW was bread, savory, waste cakes, cafeteria waste, fruits, vegetables, and potatoes. This study proved that cake waste was best substrate for glucoamylase production (Source: Biofuel Research Journal 3 (2014) 98-105).

For Glucoamylase production, food waste is sterilized and fermented after inoculating A.awarmori . Once fermentation is completed, enzyme is extracted with the addition of enzyme carrier and enzyme is stored.

In second step food waste in hydrolytic reactor,enzyme in proportion to food waste volume is added to give glucose rich bioliuid.

Thus initial pH of 7.9, initial moisture content of 69.6% was found to be optimum for glucoamylase production using *Aspergillus awamori*. The enzyme solution produced was used for Saccharifiation of starch present in FW.

FOOD PROCESSING WASTES

Protease Production

Industrial Protease production is contributing 65% of the global market. They are useful in food industry, detergent industry and pharmaceutical industry. The production of enzymes like proteases, lipases, chitinolytic enzymes, ligninolytic enzymes from Fish wastes (heads, viscera, chitinous material, wastewater, etc.) was studied in details by Faouzi Ben Rebah . *Bacillus cereus* Strain grown in fish processing waste based media was mostly used in protease production.

Fish raw materials used in growth media were heads and viscera of *Sardinella*, viscera from rainbow trout, swordfish, Acid hydrolyzed tuna waste, defatted tuna waste. Among these heads and viscera of *Sardinella* waste when used to grow

Pseudomonas aeruginosa showed protease activity maximum that is 7,800 U/ml (Table 2).

Lipases Enzymes

Lipases the next industrially important enzyme class can be produced in significant quantity using *Staphylococcus xylosus* growth in fish processing waste based media as mentioned in Table 2.

Table 2. Protease production by various microbial strains grown in fish processing waste based media

Fish raw materials	Preparation of the growth media	Microbial strains	Activity (U/ml)	References
Heads and viscera of *Sardinella*	Raw materials cooked, pressed, minced and dried (80 °C, 24–48 h)	*Pseudomonas aeruginosa MN7*	7,800	Triki-Ellouz et al. (2003)
Heads and viscera of *Sardinella*	Raw materials cooked, pressed, minced and dried (80 °C, 24–48 h)	*Bacillus subtilis*	720	Ellouz et al. (2001)
Viscera from rainbow trout, swordfish, squid and yellowfin tuna	For peptone preparation, raw materials were ground with water and supernatant recovered after centrifugation was processed	*Vibrio anguillarum*	35–68	Vazquez et al. (2006)
Viscera from rainbow trout, swordfish, squid and yellowfin tuna	For peptone preparation, raw materials were ground with water and supernatant recovered after centrifugation was processed	*Vibrio splendidus*	9–30	Vazquez et al. (2006)
Raw tuna waste	Raw materials cooked, bones removed, pressed to remove water and fat, pressed, minced and dried (80 °C, 24–48 h)	*Bacillus cereus*	74.77	Esakkiraj et al. (2009)
Defatted tuna waste	Extraction with chloroform/methanol	*Bacillus cereus*	134.57	Esakkiraj et al. (2009)
Acid-hydrolyzed tuna waste	Method described by Gao et al. (2006)	*Bacillus cereus*	60.37	Esakkiraj et al. (2009)
Alkali-hydrolyzed tuna waste	Method described by Batista (1999)	Bacillus cereus	65.96	Esakkiraj et al. (2009)

Source: Biotech (2013) 3:255–265

Table 3 gives the details about production of lipases.

Staphylococcus xylosus showed 19-28 U/ml lipase activity using Shrimp by-products, which is in considerable amount as compared to other raw materials. (Table 3).

Chitinolytic enzymes, Ligninolytic enzymes are equally important. It has been known that chitinases are useful in Single cell Protein production and isolating protoplast from yeast and fungi. (Dahiya et al. 2006). Shrimp waste from sea food can be a substrate for fermentative production of chitinase enzyme by solid state fermentation after mild treatment. (Y.L.Ramchandra et.al 2008) .They isolated *Oerskovia* Sp., *Sporolactobacillus* Sp.2, and *Sporolactobacillus* Sp.2 giving maximum of 56.8, 34.0, and 17.2 U/g, IDS chitinase production.

Ligninolytic enzymes are effective in xenobiotic substance removal and hence more study is needed to focus on this as xenobiotic is toxic to human and animal health.

Table 3. Production of lipase by different microbial species grown in fish processing by-products

Fish raw materials	Preparation of the growth media	Microbial strains	Lipase activity (U/ml)	References
Defatted tuna by-products	Extraction with chloroform/ methanol[a]	*Staphylococcus epidermidis*CMST Pi 2	12.63	Esakkiraj et al. (2010a)
Defatted tuna by-products	Extractionwith chloroform/ methanol[a]	*Staphylococcus epidermidis*CMST Pi 2	14.20	Esakkiraj et al. (2010a)
Tuna by-products	Raw materials were cooked, bones were removed, pressed to remove water and fat, pressed, minced and dried(80 °C, 24–48 h)[a]	*Staphylococcus epidermidis*CMST Pi 2	8.17	Esakkiraj et al. (2010a)
Shrimp by-products	Raw materials were boiled (100 °C for 20 min) in water and supernatants were recuperated by centrifugation[b]	*Staphylococcus xylosus*	19–28	Ben Rebah et al. (2008)
Cuttlefish by-products	Raw materials were boiled (100 °C for 20 min) in water and supernatants were recuperated by centrifugation[b]	*Staphylococcus xylosus*	5–9.50	Ben Rebah et al. (2008)
Tuna by-products	Raw materials were boiled (100 °C for 20 min) in water and supernatants were recuperated by centrifugation[b]	*Staphylococcus xylosus*	0–4	Ben Rebah et al. (2008)

a To the basal medium, fish powder obtained after processing was added at different proportions

b The supernatant was used as a nutrient source for lipase production

Source: 3 Biotech (2013) 3:255–265

FUTURE ASPECTS

As compare to chemical and physical degradation of food waste biological decomposing is effective and better for economic and ecological reason. To reduce food loss there is need to implement different policies like enabling export, reducing transport regulations, supporting food processing.

Enzymes, the high value added compounds have different challenges bound to its recovery from food waste. Considering all the parameters discussed in the chapter it interprets that more innovative techniques are needed to apply for recovery of enzymes. Nanotechnology is a promising technique to use in enzyme production using Food waste. Hydroxy nanoparticles can act as a chaperon; this property can be applied in the future.

CONCLUSION

All the enzymes aforementioned have been produced and they have high market potential. General techniques that are used in the recovery of enzymes can be applied to enzyme production using food waste. Isolation, Fractionation, precipitation, centrifugation, ultrafiltration, chromatography, electrophoresis and liquid - liquid extraction are the available methods for enzyme purification. It is not sufficient to use single method for isolating and purifying enzymes. Precipitation (salt precipitation, Solvent precipitation and isoelectric precipitation) is the initial step used for isolating enzyme. Later on chromatography techniques (Gel filtration chromatography, Ion exchange chromatography, Adsorption chromatography and Affinity chromatography) are used. Recombinant DNA technology has wide application in production of industrially important enzymes with desired characteristics, using food waste, resulting in cost effective manner.

REFERENCES

Ahamed, A., Yin, K., Ng, B. J. H., Ren, F., Chang, V. W.-C., & Wang, J.-Y. (2016). Life cycle assessment of the present and proposed food waste management technologies from environmental and economic impact perspectives. *Journal of Cleaner Production*, *131*, 607–614. doi:10.1016/j.jclepro.2016.04.127

Anto, H., Trivedi, U. B., & Patel, K. C. (2006). Glucoamylase production by solid-state fermentation using rice flake manufacturing waste products as substrate. *Bioresource Technology*, *97*(10), 1161–1166. doi:10.1016/j.biortech.2005.05.007 PMID:16006122

Ben Rebah, F., Frikha, F., Kammoun, W., Belbahri, L., Gargouri, Y., & Miled, N. (2008). Culture of Staphylococcus xylosus in fish processing by-product-based media for lipase production. Lett. *Applied Microbiology*, *47*(6), 549–554. doi:10.1111/j.1472-765X.2008.02465.x PMID:19120924

Ben Rebah & Miled. (2013). Fish processing wastes for microbial enzyme production: A review. *Biotech*, *3*, 255–265.

Ćilerdžić, Stajić, Vukojević, Duletić-Laušević, & Knežević. (2011). Potential of *Tramete Hirsuta* to Produce Ligninolytic Enzymes during Degradation Of Agricultural Residues. *BioResources*, *6*(3), 2885–2895.

Dahiya, N., Tewari, R., & Hoondal, G. S. (2006). Biotechnological aspects of chitinolytic enzymes: A review. *Applied Microbiology and Biotechnology*, *71*(6), 773–782. doi:10.100700253-005-0183-7 PMID:16249876

DeWitt, C. A. M., & Morrissey, M. T. (2002). Pilot plant recovery of catheptic proteases from surimi wash water. *Bioresource Technology*, *82*(3), 295–301. doi:10.1016/S0960-8524(01)00178-X PMID:11991080

Esakkiraj, P., Austin Jeba Dhas, G., Palavesam, A., & Immanuel, G. (2010a). Media preparation using tuna-processing wastes for improved lipase production by shrimp gut isolate *Staphylococcus epidermidis* CMST Pi2. *Applied Biochemistry and Biotechnology*, *160*(4), 1254–1265. doi:10.100712010-009-8632-x PMID:19430738

Esakkiraj, P., Rajkumarbharathi, M., Palavesam, A., & Immanuel, G. (2010b). Lipase production by *Staphylococcus epidermidis* CMST-Pi 1 isolated from the gut of shrimp *Penaeusindicus*. *Annals of Microbiology*, *60*(1), 37–42. doi:10.100713213-009-0003-x

Gaoa, A. (2017). Comparison between the technologies for food waste treatment. *Energy Procedia*, *105*, 3915–3921. doi:10.1016/j.egypro.2017.03.811

Garcia-Garcia, Woolley, & Rahimifard, Colwill, White, & Needham. (2017). A Methodology for Sustainable Management of Food Waste. *Waste and Biomass Valorization*, *8*, 2209–2227.

Kiran, Trzcinski, & Liu. (2014). Glucoamylase production from food waste by solid state fermentation and its evaluation in the hydrolysis of domestic food waste. *Biofuel Research Journal*, *3*, 98–105.

Moldes, D., Gallego, P. P., Rodriguez Couto, S., & Sanroman, M. A. (2003). Grape seeds: The best lignocellulosic waste to produce laccase by solid state cultures of Trameteshirsuta. *Biotechnology Letters*, *25*(6), 491–495. doi:10.1023/A:1022660230653 PMID:12882277

Ramchandra, Y. L., Padmalatha Rai, S., Sujan Ganapathy, P. S., Sudeep, H. V., & Krushnamurthy, N. B. (2008). Chitinase Production by Solid State Fermentation using Shrimp waste. *Asian Journal Of Microbiol.Biotech.Env.Sc.*, *10*(3), 615–620.

Rasit & Kuan. (2018). Investigation on the Influence of Bio-catalytic Enzyme Produced from Fruit and Vegetable Waste on Palm Oil Mill Effluent. *IOP Conf. Series: Earth and Environmental Science, 140*.

Rosales, E., Rodriguez Couto, S., & Sanhromán, M. A. (2007). Increased laccase production by *Tramete shirsuta* grown on ground orange peelings. *Enzyme and Microbial Technology*, *40*(5), 1286–1290. doi:10.1016/j.enzmictec.2006.09.015

Rosales, S., Rodríguez Couto, S., & Sanromán, A. (2002). New uses of food waste: Application to laccase production by *Tramete shirsuta*. *Biotechnology Letters*, *24*(9), 701–704. doi:10.1023/A:1015234100459

Schaub, S. M., & Leonard, J. J. (1996). Composting: An alternative waste management option for food processing industries. *Trends in Food Science & Technology*, *7*(8), 263–268. doi:10.1016/0924-2244(96)10029-7

Sharma, R., Chisti, Y., & Banerjee, U. C. (2001). Production, purification, characterization and applications of lipases. *Biotechnology Advances*, *19*(8), 627–662. doi:10.1016/S0734-9750(01)00086-6 PMID:14550014

Uçkun, E. (2014). Glucoamylase production from food waste by solid state fermentation and its evaluation in the hydrolysis of domestic food waste. *Biofuel Research Journal*, *3*, 98–105.

Unakal. (2012). Production of α-amylase using banana waste by *Bacillus subtilis* under solid state fermentation. *European Journal of Experimental Biology*, *2*(4), 1044–1052.

Wang, Q. H., Liu, Y. Y., & Ma, H. Z. (2010). On-site production of crude glucoamylase for kitchen waste hydrolysis. *Waste Management & Research*, *28*(6), 539–544. doi:10.1177/0734242X09354353 PMID:20015936

Wohlgemuth, Sigma-Aldrich, Buchs, & Switzerland. (2011). Product Recovery. Comprehensive Biotechnology, 2, 591-601.

Chapter 4
Recent Molecular Approaches for Development of Value-Added Products From Lignocellulosic Food Waste

Javed Abdulsalam Mulla
Institute of Bioinformatics and Biotechnology, India

ABSTRACT

The escalating global population has led to an ever-increasing demand for food processing industries, and as a result, the generation of huge amounts of food waste. The severity of this problem is augmented due to dawdling development of effective waste treatment and disposal strategies. In a quest of potential alternative bioenergy resources, lignocellulose is proven to be a good, abundantly available raw material on the land as a leftover of agricultural and industrial byproduct made up cellulose, hemicelluloses, and lignin. It is mostly utilized for biofuels, bio-ethanol production, and other value-added products. The development of the conversion of lignocellulosic biomass to fine chemicals still remains a big challenge. The deciphering molecular mechanism and effective cellulase and hemicellulases producing microorganisms might successfully be accomplished with transcriptome, proteome, and recombinant DNA technology; these are discussed in this chapter.

DOI: 10.4018/978-1-5225-7706-5.ch004

INTRODUCTION

Lignocellulose is a common term used to explain biomass of plant. Most of the food and food processing industry waste are lignocellulosic in nature with a global estimate of up to 1.3billion tons/year (Ravindran and Jaiswal, 2015). The many value added products are routinely generated fromreducing sugars obtained hydrolysis of Lignocellulose viz bioethanol, biogas, organic acids, enzymes and biosorbents. It is highly inexpensive renewable carbon resource having 75% of polysaccharide (Sun & Cheng, 2002).

In general it is composed of cellulose (40 to 50%) and hemicelluloses (25 to 30%) and lignin (10 to 20%) (Wyman et al., 1999). Lignin is a non carbohydrate polyphenolic compound. Cellulose hydrolysates comprise glucose and various levels of cellobiose and other glucose oligomers. On the other hand, hemicellulose hydrolysates are more complex mixtures as they include several hexoses (glucose, galactose, and mannose) and pentoses (xylose and arabinose) (Wiselogel et al., 1996). The food processing industry in the all over world is progressing at a very fast speed. Such an increasing industralisation can give rise to more waste that is ultimately left untreated due to lack of treatment options. The land filling could be the cheapest option for waste management by many industries. Incineration could be the one approach but it requires a lot of expenditure of energy recourses. However second one is composting of food waste are time consuming and sluggish. However improper disposal treatment of these waste leads to their putrefaction giving rise to toxic gases such as methane and leaching of other toxic liquids proving hazardous to the environment. Being the plenty and easy availability of food waste, exploitation of value added products from them is meagerly studies. Most of the waste generated from the food industry is lignocellulosic in nature, and thus can be used as potential substrates for the production of high value products.

One such problem that needs to be addressed immediately is the carbohydrate source used for enzyme production. Theoretically, it is possible to recycle cheap carbohydrate sources from industries and use it as a sugar source for enzyme production. However, the heterogeneous nature of biomass carbohydrate sources hinders them to be efficient nutrients leading to incompetent growth of the enzyme producing microorganisms. This is due to the fact that 5-C and 6-C sugars are absorbed by the microbe at different rates during fermentation (Abdel- Rahman et al., 2015). Furthermore these carbohydrate sources comprises of other substances that may act as inhibitors for microbial growth, and leads to poor fermentation yields and subsequently raising the production costs for the desired products.

Structure of Lignocellulose

As discussed earlier lignocellulose is composed of lignin and carbohydrates like cellulose, hemicellulose, pectin, ash, salts and minerals. Out of which cellulose is a polysaccharide that is made up of D-glucose having β (1→4) linkage forming linear chains. It is the main component of plant cell wall. Cellulose can exist in different polymorphs that are crystalline in nature. The naturally occurring cellulose is found in a polymorph structure called polymorph I. Three other polymorphs of cellulose have been attained via different pre-treatments viz. polymorphs II, III and IV. But the amorphous form of cellulose is highly soluble and can be easily digested by enzymes (Kulasinski, Keten, Churakov, Derome, &Carmeliet, 2014).

The second component is Hemicellulose, which is a mixed polymer made up of short chains of polysaccharide molecules. They made up of five different sugar monomers viz. D-xylose, L-arabinose (pentoses), D-galactose, D-mannose and D-glucose (hexoses) and constitute of 15-30% of plant cell. They are occurring in two forms O-acetyl-4-O-methylglucuronoxylans and O-acetyl galactoglucomannans depending upon the source of biomass. It also consists of Uronic acids, like α-D-glucuronic, α-D-4-O-methylgalacturonic and α-D-galacturonic acids. Hemicelluloses are linked to cellulose by hydrogen bonds and to lignin by covalent bonds (Sun et al., 2014).

Finally Lignin is the most abundant constituent in plant biomass. It consists of an aromatic and amorphous nature. The structure of lignin is totally dependent upon plant species it is derived from. The monomers are basically phenylpropane units that differ only in the substitution of methoxyl groups on the aromatic rings. The three main mono lignols (lignin monomcrs) that form thc lignin polymer are p-hydroxyphenyl alcohol (H), coniferyl alcohol (G), and synapyl alcohol (S). Softwoods usually contain more of the H subunit as compared to G and S where as in grass varieties all the mono lignols are found in equal proportions (Duval & Lawoko, 2014).

Pre-Treatment of Lignocelluloses

The enzymatic hydrolysis of cellulose is a complex process. Many factors influence this process and they can be broadly categorized in to structural features of lignocellulose and the mechanisms and interactions related to enzyme kinetics. The features that influence the difficulty in the degradation of biomass residues include hemicellulose, lignin content, available amorphous cellulose as compared to its crystalline counter-part, the degree of polymerization, acetyl groups, plant protein-enzyme interaction, the association of cellulose with hemicellulose and lignin, the accessible surface area to enzymatic degradation and porosity and the residual surface area of biomass (Singh et al., 2015).

Pre-treatments are necessary to carry out to tackle these issues by introducing in structural and compositional changes in lignocellulose. This is achieved by physical, chemical and even biological too. These pre-treatments help in removing lignin and hemicellulose to a certain extent and increase the porosity and concentration of amorphous cellulose. Novel pre-treatments for lignocellulose have been devised by several researchers over the past three decades; some of those are discussed in Table 1.

However hydrolysis by biological aspects may hamper due to lignin content. Research has shown that the removal of lignin facilitates cellulose hydrolysis. The mechanism by which lignin obstructs hydrolysis of cellulose is still not completely understood. This is due to the structure of lignin and the covalent bonding between lignin and cellulose prevents the carbohydrate to be exposed for enzymatic hydrolysis (Yuan et al., 2013).An addition to above problems lignin itself absorbs enzymes irreversibly. A study on the adsorption kinetics of enzyme to lignin and the effect of temperature revealed that higher temperatures accelerated the adsorption process attaining equilibrium. This may be a problem when conducting enzymatic hydrolysis at higher temperatures. The addition of surfactants was seen to reduce the adsorption of cellulase to lignin (Tu et al., 2009).

Table 1. List of various pre-treatment currently used for hydrolysis of biomass

Sr.No.	Name of pre-treatment	Reference
	Physical Pre-Treatment	
1	Grinding and milling	Silva et al. (2012)
2	Ultrasonic pre-treatment	Nakayama and Imai (2013)
3	Centrifugal grinding	Silva and Xavier (2011)
4	Extrusion pre-treatment	Yoo et al., (2011)
	Chemical Pre-Treatment	
1	Dilute-acid pre-treatment	Martin et al; (2015).
2	Acid-acetone pre-treatment	Qin et al; (2014)
3	Alkaline potassium permanganate pre-treatment	Ma et al; (2015)
	Biological Pre-Treatment	
1	Microbial consortium	Zhang et al., (2011)
2	By using fungal species	Vicuña et al; (2000)
3	Enzymatic pre-treatment	Brown & Chang; (2014)

Thus the efficient microbial strategies are to be constructed to overcome these difficulties and having maximum yield during downstream processes. The following part discussed the several genetic approaches used by researcher for effectively increase in utilization rate of biomass substrate.

Biotechnological Features of Lignocellulose Bioconversion by Using Co-Cultivation

Cellulose present in renewable lignocellulosic material is thought to be the most abundant organic substrate on earth as chemical feed stock however bioconversion of cellulosic substrates into glucose, is a complex process. (Kumar et al, 2008). The strategies behind regulation of cellulases and hemicellulases gene expression may be very helpful for increasing the production of these enzymes in their native producers. Fungi are broadly studied which synthesized cellulase and hemicellulases and mainly in *Aspergillus* and *Trichoderma*. The fungi produce these extracellular enzymes only when plant polymers are utilized as a carbon source. Thus the expression of these enzymes repressed in the presence of other easily degradable carbon source. Cellulases and hemicellulases convert lignocellulosic biomass into fermentable sugars for fuel production by the hydrolysis of polysaccharides is considered to be more vital event (Antonella et al, 2013). The filamentous ascomycete fungus *Neurospora crassa* is a model filamentous fungus, expresses and secretes enzymes required for plant cell wall deconstruction. *N. crassa* codes 23 cellulase genes and 19 hemicellulase genes in its genome (Jianping et al., 2012). The substrate specific extracellular enzyme synthesis in *N. crassa* showed that expression of 8 hemicellulase genes induced by xylan but no predicted cellulase genes, whereas cellulase and hemicellulase gene expression induced when exposed to Avicel. Thus secretions of extracellular enzymes are highly regulated at gene level where substrate and transcription factor plays critical role. The choice of particular type of fungi for bioconversion of lignocelluloses is based of understanding of regulatory mechanisms of these enzymes and efficient genetic manipulation techniques can be applied for in situ production.

As far as cellulases are concerned, three main enzymatic activities are involved in cellulose hydrolysis: Cellulase is a synergetic enzyme which is used to split cellulose into first precursor molecule glucose and/or different oligosaccharide compounds. Cellulase enzymes may be divided into 3 types: endoglucanase (endo-1, 4-β-D-glucanase, EG, EC 3.2.1.4); cellobiohydrolase or exoglucanase (exo-1, 4-β-D-glucanase, CBH, EC 3.2.1.91) and β-glucosidase (1, 4-β-D-glucosidase, BG, EC 3.2.1.21), whereas EGs being the foremost economical enzyme (Muhammad Imran et al., 2016).

The regulation of (hemi) cellulolytic genes appears to be basically the same among filamentous fungi such as *T. reesei, N. crassa, Asperigillus* spp., even though their regulatory mechanisms are quite complex and present some differences. In case of *Trichoderma reesei*, synthesis of cellulase and hemicellulase is variable. It produces lower levels of β-glucosidase, whereas, *Aspergillus niger* fungi have limited levels of the endoglucanase component. Therefore, co-cultivation methods are applied recently by using two cultures of fungi where they act synergistically for production of cellulase enzymes and achieving an increased rate of lignocellulosic bioconversion.

Jagavati et al. (2012) showed that the co-culture of the *Trichoderma sp.* and *Aspergillus sp.* when taken in equal ratio (1:1) produced high amount of cellulose. Similarly Gupte and Madamwar cultivated *Aspergillus ellipticus* and *Aspergillus fumigatus* and found improved hydrolytic activities as compared to separate cultures in a solid-state fermentation system. Thus these studies showed that one organism complement with other organism for production of cellulose enzymes.

The mutagenesis is also having its own potential for producing mutant cells. Thus this approach can be used for finding efficient cellulose enzyme producing strain. Basically, cellulase production coupled by genetic regulation and biochemical control. In this scenario feedback inhibition occurred by product generation limited yields of the enzymatic constituents. Thus mutant strain which lack synthesis of repressor protein. Veen et al. (1995) showed improved levels of D-glucose metabolism in mutant strain creAd30, which was constructed from *Aspergillus nidulans.*

Genetic Manipulation Techniques

The efficient utilization of mixtures of various sugars is critical for attaining the complete conversion of lignocellulosic sugars. The carbon catabolite repression mechanisms are the key regulatory mechanism responsible for most energy-efficient carbohydrate being utilized first. The xylose is considered to be utilized late during fermentation procedure while using lignocelluloses waste by microorganisms. To triumph over this; Kawaguchi et al. (2006) made an attempt to construct two recombinant nonmedical *Corynebacteria glutamicum.* For the construct of these *C. glutamicum* recombinants, the *E. coli* xylA and xylB genes were isolated by PCR and specifically sub cloned under the control of the strong constitutive trc promoter that is present in vector pCRA1.The resulting transformant showed utilization of xylose with differ in their efficiency. Thus, collectively the introduction of both *E. coli xylA* and *xylB* under the control of a constitutive promoter is an efficient strategy to engineer in *Corynebacteria* a functional xylose catabolism pathway that is not subject to catabolite repression, at least at the *xylA-xylB* gene product level.

A number of designer enzymes, also called glycosynthases, including cellulases and hemicellulases, have been engineered by replacing nucleophilic residues resulting inhigher yields of different oligosaccharides. Mostly, genetic engineering techniques have been established by using either *Agrobacterium tumefaciens* or gene-gun mediated gene transfer in plants include rice, corn, wheat, barley, sorghum, poplar, willow and switch grass (Somleva et al., 2002). Several genes and transcription factors have been identified to increase the cellulose content of plant biomass. Transcriptional factors involved in regulation of cellulase and hemicellulase genes' expression in *T. reesei* are the positive regulators XYR1, ACE2 and theHAP2/3/5 complex, the repressor ACE1 and the carbon catabolite repressor CRE1 (Kubicek et al., 2009), numerous studies report that XYR1 is responsible for positive cellulase transcriptional activators, as its removal not only suppresses cellulase but also xylanase and β-mannanase formation (Stricker et al., 2008). Although XYR1 regulate the expression of different transporters in *T. reesei* which shows a role in signaling pathways involved in cellulase induction. Taken together, these data suggest that the transcription factor XYR1 in *T. reesei* specifically controls cellulase and xylanases gene expression. Furthermore XlnR is the main transcriptional factor observed in *Aspergillus* species where has been demonstrated that XlnR regulates the transcription of the *xlnB*, *xlnC* and *xlnD* genes encoding endoxylanases B, endoxylanase C and β-xylosidase, respectively (van Peiji et al., 1998). However, two major gene superfamilies, i.e., *ces*A and *csl* which is involved in cellulose biosynthesis have been identified that considerably improved the cellulose content in rice, maize and energy crops (Xie and Peng, 2011).

CONCLUSION

In the last two decades, global research programs on alternative energy have been directed towards discovering new and sustainable energy sources as global economy cannot longer depend on fossil fuels or dead carbon. The cloning and sequencing of the various cellulolytic genes will help in characterizing the potential systems for economizing the process of lignocellulosic conversion in future. Thus, energy recovery from food waste is an additional attractive option to pursue, particularly from the energy security viewpoint. This understanding has motivated fundamental research on technologies that help to recover some valuable fuels from food waste to reduce the environmental burden of its disposal, avoid depletion of natural resources, minimize risk to human health and maintain an overall balance in the ecosystem.

REFERENCES

Abdel-Rahman, M. A., Xiao, Y., Tashiro, Y., Wang, Y., Zendo, T., Sakai, K., & Sonomoto, K. (2015). Fed-batch fermentation for enhanced lactic acid production from glucose/xylose mixture without carbon catabolite repression. *Journal of Bioscience and Bioengineering*, *119*(2), 153–158. doi:10.1016/j.jbiosc.2014.07.007

Amore, A., Giacobbe, S., & Faraco, V. (2013). Regulation of cellulase and hemicellulase gene expression in fungi. *Current Genomics*, *14*(4), 230–249. doi:10.2174/1389202911314040002

Brown, M. E., & Chang, M. C. (2014). Exploring bacterial lignin degradation. *Curr. opin. Inchem. biol.*, *19*, 1-7.

Duval, A., & Lawoko, M. (2014). A review on lignin-based polymeric, micro- and nano-structured materials. *Reactive & Functional Polymers*, *85*(0), 78–96. doi:10.1016/j.reactfunctpolym.2014.09.017

Imran, M., Anwar, Z., Irshad, M., Asad, M. J., & Ashfaq, H. (2016). Cellulase, production from species of fungi and bacteria from agricultural wastes and its utilization in industry: A review. *Advances in Enzyme Research*, *4*(02), 44–55. doi:10.4236/aer.2016.42005

Jagavati, S., Adivikatla, V., Paritala, N., & Linga, V. (2012). Cellulase Production by co-culture of *Trichoderma* sp. and *Aspergillus* sp. under submerged fermentation. *Dynamic Biochemistry, Process Biotechnology and Molecular Biology*, *6*(S1), 79-83.

Kawaguchi, H., Vertes, A. A., Okino, S., Inui, M., & Yukawa, H. (2006). Engineering of a xylose metabolic pathway in *Corynebacterium glutamicum*. *Applied and Environmental Microbiology*, *72*(5), 3418–3428. doi:10.1128/AEM.72.5.3418-3428.2006

Kubicek, C. P., Mikus, M., Schuster, A., Schmoll, M., & Seiboth, B. (2009). Metabolic engineering strategies for the improvement of cellulose production by Hypocreajecorina. *Biotechnology for Biofuels*, *2*(1), 19. doi:10.1186/1754-6834-2-19

Kulasinski, K., Keten, S., Churakov, S. V., Derome, D., & Carmeliet, J. (2014). A comparative molecular dynamics study of crystalline, paracrystalline and amorphous states of cellulose. *Cellulose (London, England)*, *21*(3), 1103–1116. doi:10.100710570-014-0213-7

Kumar, R., Singh, S., & Singh, O. V. (2008). Bioconversion of lignocellulosic biomass: Biochemical and molecular perspectives. *Journal of Industrial Microbiology & Biotechnology*, *35*(5), 377–391. doi:10.100710295-008-0327-8

Ma, L., Cui, Y., Cai, R., Liu, X., Zhang, C., & Xiao, D. (2015). Optimization and evaluation of alkaline potassium permanganate pretreatment of corncob. *Bioresource Technology, 180*(0), 1–6. doi:10.1016/j.biortech.2014.12.078

Madamwar, D., & Patel, S. (1992). Formation of cellulases by co-culturing of *Trichoderma reesei* and *Aspergillus niger* on cellulosic wastes. In V. S. Malik & P. Sridhar (Eds.), *Industrial biotechnology* (pp. 471–478). Oxford, UK: IBH.

Martín, C., García, A., Schreiber, A., Puls, J., & Saake, B. (2015). Combination of water extraction with dilute sulphuric acid pretreatment for enhancing the enzymatic hydrolysis of *Jatrophacurcas* shells. *Industrial Crops and Products, 64*(0), 233–241. doi:10.1016/j.indcrop.2014.09.040

Nakayama, R., & Imai, M. (2013). Promising ultrasonic irradiation pretreatment for enzymatic hydrolysis of Kenaf. *Journal of Environmental Chemical Engineering, 1*(4), 1131–1136. doi:10.1016/j.jece.2013.08.030

Qin, W., Wu, L., Zheng, Z., Dong, C., & Yang, Y. (2014). Lignin hydrolysis and phosphorylation mechanism during phosphoric acid–acetone pretreatment: A DFT study. *Molecules (Basel, Switzerland), 19*(12), 21335–21349. doi:10.3390/molecules191221335

Ravindran, R., & Jaiswal, A.K. (2015). A comprehensive review on pre-treatment strategy for lignocellulosic food industry waste: Challenges and opportunities. *Bioresource Technology.*

Silva, G. G. D., Couturier, M., Berrin, J.-G., Buléon, A., & Rouau, X. (2012). Effects of grinding processes on enzymatic degradation of wheat straw. *Bioresource Technology, 103*(1), 192–200. doi:10.1016/j.biortech.2011.09.073

Silva, G. G. D., & Xavier, R. S. G. (2011). Successive centrifugal grinding and sieving of wheat straw. *Powder Technology, 208*(2), 266–270. doi:10.1016/j.powtec.2010.08.015

Singh, J., Suhag, M., & Dhaka, A. (2015). Augmented digestion of lignocellulose by steam explosion, acid and alkaline pretreatment methods: A review. *Carbohydrate Polymers, 117*(0), 624–631. doi:10.1016/j.carbpol.2014.10.012

Sun, Tian, Diamond, & Glass. (2012). Deciphering transcriptional regulatory mechanisms associated with hemicellulose degradation in *Neurospora crassa. Eukaryot Cell., 11*(4), 482–493.

Sun, S.-N., Cao, X.-F., Xu, F., Sun, R.-C., Jones, G. L., & Baird, M. (2014). Structure and thermal property of alkaline hemicelluloses from steam exploded *Phyllostachys pubescens*. *Carbohydrate Polymers*, *101*(0), 1191–1197. doi:10.1016/j.carbpol.2013.09.109

Sun, Y., & Cheng, J. (2002). Hydrolysis of lignocellulosic materials for ethanol production: A review. *Bioresource Technology*, *83*(1), 1–11. doi:10.1016/S0960-8524(01)00212-7

Van Peij, N., Gielkens, M. M. C., de Vries, R. P., Visser, J., & deGraaff, L. H. (1998). The transcriptional activator XlnR regulates both xylanolytic and endoglucanase gene expression in *Aspergillus niger. Applied and Environmental Microbiology*, *64*, 3615–3619.

Veen, P. W. D., Ruijter, G. J. G., & Visser, J. (1995). An extreme cre A mutation in *Aspergillus nidulans* has severe eVects on D-glucose utilization. *Microbiol*, *141*(9), 2301–2306. doi:10.1099/13500872-141-9-2301

Vicuña, R. (2000). Ligninolysis. *Molecular Biotechnology*, *14*(2), 173–176. doi:10.1385/MB:14:2:173

Wiselogel, A., Tyson, S., & Johnson, D. (1996). *Biomass feedstock resources and composition*. Washington, DC: Taylor & Francis.

Wyman, C. E. (1999). Production of low cost sugars from biomass: progress, opportunities, and challenges. In R. P. Overend & E. Chornet (Eds.), *Biomass: a growth opportunity in green energy and value added products* (Vol. 1, pp. 867–872). Oxford, UK: Pergamon Press.

Xie, G., & Peng, L. (2011). Genetic engineering of energy crops: A strategy for biofuel production in China. *Journal of Integrative Environmental Sciences*, *53*, 143–150.

Yoo, J., Alavi, S., Vadlani, P., & Amanor-Boadu, V. (2011). Thermo-mechanical extrusion pretreatment for conversion of soybean hulls to fermentable sugars. *Bioresource Technology*, *102*(16), 7583–7590. doi:10.1016/j.biortech.2011.04.092

Zhang, Q., He, J., Tian, M., Mao, Z., Tang, L., Zhang, J., & Zhang, H. (2011). Enhancement of methane production from cassava residues by biological pretreatment using a constructed microbial consortium. *Bioresource Technology*, *102*(19), 8899–8906. doi:10.1016/j.biortech.2011.06.061

Chapter 5
Fermented Foods and Their Production

Suresh Sopanrao Thorat
Mahatma Phule Krishi Vidyapeeth, India

ABSTRACT

Fermented foods have been produced throughout recorded history. Recently, fermented foods have experienced renewed interest stemming from concerns with nutrition and the increase in vegetarian and natural foods diets. This chapter explores fermented foods and their production.

INTRODUCTION

The traditional fermented foods made in the Orient which includes Japan, Indonesia, India, Pakistan, Thailand, Philippines, Taiwan, China, Korea, and the encompassing areas were produced long before written history. There was an extensive research dealing with fermented foods and drinks from 1878. The studies between 1881 and 1914 were devoted to isolation and description of the new microorganisms associated with the fermentation and also the fermented products. The renewed interest in the fermented foods now is concerned with nutrition, natural foods, expand export markets and add products to convenience foods to add zest and flavor. Traditional fermented foods are popularly consumed and form an integral part of our diet since early history. These can be prepared in the household or in cottage industry using relatively simple techniques and equipments (Aidoo et al., 2006). It is one of the oldest and most economical methods for producing and preserving foods. In addition to preservation, fermented foods can also have added benefits of enhancing flavour, increased digestibility, and improving nutritional and pharmacological values

DOI: 10.4018/978-1-5225-7706-5.ch005

(Jeyaram et al., 2009). Lactic acid bacteria (LAB) perform an essential role in the preservation and production of wholesome fermented foods. Homo-fermentative and hetero-fermentative LAB are generally fastidious on artificial media, but they grow readily in most food substrates and lower the pH rapidly to a point where other competing organisms are no longer able to grow. *Leuconostocs* and *Lactococcus* generally lower the pH to 4.0 –4.5 and some of the *Lactobacilli* and *Pediococci* lower it up to 3.5 (Steinkraus, 1983). India is traditionally rich in fermented foods. In the Indian sub-continent, fermented foods are very common. The nature of the products and base material varies from region to region (Sekar and Mariappan, 2007). At present, there are hundreds of fermented foods with different base materials and preparation methodology. Each fermented food is associated with a unique group of microbiota, which increases the level of proteins, vitamins, essential amino acids and fatty acids. However, fermented foods are still produced traditionally by spontaneous fermentation and only limited knowledge has been obtained regarding the microbiota of these products (Jeyaram et al., 2009).

India, being a huge country has been the home of innumerable religious order of human population and diversity in climatic conditions which has resulted in a large number of fermented foods.

Since the beginning of human civilization there has been an intimate companionship between the human being and the fermentative activities of microorganisms. These fermentative activities have been utilized in the production of fermented foods and beverages, which are defined as those products that have been subordinated to the effect of microorganisms or enzymes to cause desirable biochemical changes. The microorganisms responsible for the fermentation may be the microflora indigenously present on the substrate, or they may be added as starter cultures (Harlander, 1992).

Fermentation is one of the oldest and most economical methods of producing and preserving food (Billings, 1998 and Chavan and Kadam, 1989). Ever since, the technologies for the industrial production of fermented products from milk, meat, fruits, vegetables and cereals are well developed recently and scientific work is actively carried out all over the world (Hirahara, 1998 and Pagni, 1998).

Fermented foods are produced using various manufacturing techniques and microbes. However, there are only four main fermentation processes: alcoholic, lactic acid, acetic acid and alkali fermentation (Soni and Sandhu, 1990). Alcohol fermentation results in the production of ethanol, and yeasts are the predominant organisms. Lactic acid fermentation (e.g. fermented milks and cereals) is mainly carried out by lactic acid bacteria. The other group of bacteria of importance in food fermentations are *Acetobacter* species, which convert alcohol to acetic acid under aerobic condition (McKay and Baldwin, 1990).

The preparation of fermented foods remains as a house art. They are produced in houses, villages and small-scale industries. On the contrary, the preparation of others, such as soy sauce, has evolved to a biotechnological state and is carried out on a large commercial scale (Bol and de Vos, 1997). In the distant past there was no verified data on the economic, nutritional, technical and quality control implications of the indigenous fermented food.

TYPES OF CULTURES USED IN FERMENTED FOODS

The microorganisms involved in a fermented food can be divided into the following categories.

1. **Monoculture:** Fermentations in which only one species of microorganism is necessary to produce the product. The Indonesian tempeh fermentation is an example wherein only *Rhizopus* is necessary. Natto is a food made by fermentation of soybeans with *Bacillus natto*. Ang-kak is a fermented rice product fermented by *Monascus purpurea* and used for coloring of other foods.
2. **Multiculture:** In these fermentations, more than one microorganism is required, belonging taxonomically to different species.
3. **Unimulticulture:** These are fermentations in which two or more strains of the same species are used together. Soybean yogurt is an example in which two strains of *Lactobacillus acidophilus* are employed, with one strain contributing to the flavor and another to acid from the other. In this fermentation, as in the multiculture fermentations two or more strains may be used.
4. **Polyculture:** These are food fermentations in which different microorganisms are many and the species specifically required to make the product are unknown. An example is the mixture of microorganisms found in silage fermentation and in Indochinese fish fermentations. Except for the examples described above, in which only a single microorganism is employed even though there are usually several contaminating microorganisms present. Most Asian food fermentations are carried out by more than one microorganism. The essential microbial cultures may be introduced simultaneously or they may be inoculated in sequence. Sequential inoculation of microorganisms is exemplified in the shoyu fermentation, with koji first is prepared using *Aspergillus oryzae* cultures, followed by a yeast bacterial inoculation and then fermentation by the latter organisms.

TYPES OF FERMENTED FOODS

Milk Based Fermented Foods

Milk and milk-based products are consumed most popularly due to their nutritive value. LAB convert milk sugar lactose into lactic acid and selective strains produce antibacterial substance bacteriocin to suppress milk curdling bacteria. Most of the foods in this category are prepared by simply adding LAB to milk of either cow, buffalo or yak and allowed to ferment. One such fermented food is dahi, its description is found in texts as old as 700 BC.

- **Dahi:** Traditionally, dahi is a naturally fermented milk product obtained from boiled cow or buffalo milk and soured using lactic cultures as *Lactococcus lactis* spp. *lactis*, *Lactococcus lactis* spp. *cremoris*, and *L. diacetylactis* is used separately or in combination. It is used in daily diet as a potential source of B-complex vitamins, folic acid, and riboflavin (Sharma and Lal, 1997 and Sarkar et al., 2015). Dahi is rich in lactic acid bacteria and demonstrates the probiotic effect, which helps in intestinal health as it helps in controlling diarrhea in children (Agarwal and Bhasin, 2002). Lactic acid bacteria produce bioactive compounds such as diacetyl, hydrogen peroxide, and reuterin suppress the normal growth of undesirable flora, especially *E. coli, Bacillus subtilis*, and *Staphylococcus aureus* (Sarkar and Misra, 2001 and Sarkar et al., 2015).
- **Rabdi:** Rabdi is a lactic acid-fermented milk product with pearl millet (Mani et al., 1993 and Sarkar et al., 2015).
- **Shrikhand:** Shrikhand is sweetened dewatered dahi and is famous in western and southern India (Steinkraus, 1996). Dahi is suspended in a muslin cloth until all the free water is drained off to prepare shrikhand. Shrikhand vadi, which is essentially a desiccated shrikhand with extended shelf life.
- **Lassi:** Lassi is prepared by mixing dahi with water, salt, sugar and spices. Lassi is a probiotic product from milk fermented in presence of *L. acidophilus* and *S. thermophilus* as active cultures (Patidar and Prajapati, 1998 and Sarkar et al., 2015). Based on the use of ingredients in Lassi, it is classified as salty Lassi or sweet Lassi. Bhang lassi, a type of lassi is prepared using the extract of cannabis plant leaves (Backes, 2014).
- **Buttermilk:** Buttermilk is the liquid which remains when butter is churned out of cream. Buttermilk is rich source of calcium and protein and easy to digest (Pushpangadan et al., 2012). Kadi is prepared by simmering a mixture of chaach/ buttermilk, besan/gram flour and spices (Sharma and Singh, 2012).

Buttermilk is boiled, then water is discarded and solids are dried hard which later used to prepare soups called churpa or churpe and the predominant microflora is *Lactobacillus plantarum, L. curvatus, L. fermentum, L.paracasei* subsp. *pseudoplantarum* and *Leuconostoc mesenteroides* (Tamang et al., 2005 and Tamang et al., 2009).

- **Cheese:** It is a dairy product derived from milk that is produced in a wide range of flavors, textures, and forms by coagulation of the milk protein casein. It is source of proteins and fat.

Their styles, textures and flavors depend on the origin of the milk (including the animal's diet), whether they have been pasteurised, the butterfat content, the bacteria and mold, the processing, and aging. Cheese is valued for its portability, long life, and high content of fat, protein, calcium and phosphorus.

Traditional Fermented Milk Products of Eastern States of India

Dahi or curd is most popular and commonly used traditional Indian fermented product. A principal flavor-inducing metabolite is diacetyl, which is appreciated more by people of South Asian origin compared to the acetaldehyde flavor in yogurt (Yadav et al., 2007b). Yak milk can be processed into different dairy products. The chemical composition of yak cheese contained around 68.2% of total solid, 49.4% of butterfat on a dry matter basis and 1.37% of salt. It is largely consumed in the Himalayan highland (Prashant et al., 2009 and Thapa, 2002). Chhurpi has a white, soft with a mild to strong flavored taste and is consumed as curry mix with wild edible ferns (Diplazium spp.), pickel and condiments along with boiled rice in meals (Tamang et al., 2000).

Average per capita consumption of Chhurpi is 6.9 g/day in Darjeeling hills, and 9.9 g/day in Sikkim (Yonzan and Tamang, 1998). All LAB strains except *L. mesenteroides* BFE1637 showed high degree of hydrophobicity. This is a significant property of LAB that assists in colonization of epithelial cells. Enzymes such as peptidases and esterase-lipases of LAB strains may play an important role in the improvement of cheese quality. Similarly, Chhu (Sheden) is a strong flavored traditional cheese-like product in Sikkim, Darjeeling hills, Arunachal Pradesh and Ladakh. It is consumed as curry by cooking it in butter along with onions, tomatoes and chilies and is mixed with beef or yak meat. Dewan and Tamang (2007b) explored the microbiota present in Chhu. LAB were predominantly present at 8.1 –8.8 log CFU/g. They isolated 120 strains of LAB. All strains showed high degree of hydrophobicity and did not produce any biogenic amines, which indicates the strains could be used as a starter culture. They had coagulation and acidification property.

Shyow is a thick gel curd like product, prepared from yak milk. Mohi is buttermilk, prepared by churning dahi. It is consumed as a refreshing beverage. Somar is a soft paste, strong flavored with bitter taste and is consumed as soup along with cooked rice or finger-millets by the Sherpas of Sikkim. LAB strains produced various enzymes such as esterase, phosphatase, leucine-arylamidase β-galactosidase and peptidase. These bacterial strains inhibited pathogens such as *Enterobacter agglomerans, Enterobacter cloacae* and *Klebsiella pneumoniae*. LAB strains did not produce any undesirable biogenic amines. Some of the LAB strains showed high degree of relative hydrophobicity (0.75%), as judged by bacterial adherence to hydrocarbon, indicating their hydrophobic nature. Khadi is a delicacy of Gujarat. This fermented food is prepared from sour dahi (curd) or buttermilk mixed with spices and dhal and warmed. The sour curd is made from cow's milk, inoculated with a culture of curd and allowed to ferment for a day or even longer at room temperature. The sour curd is either used directly for further processing or churned into buttermilk and then processed. This food is often served with rice or the naan (Sukumar and Ghosh, 2010). LAB isolated from khadi exhibited antibacterial activity against *S. aureus, Escherichia coli* and *P. aeruginosa* and also showed basic probiotic properties (Sukumar and Ghosh, 2010). LAB species isolated from fermented milk products include *Streptococcus cremoris, S. lactis, S. thermophilus, Lactobacillus bulgaricus, L. acidophilus, L. helveticus, L. cremoris, L. plantarum, L. curvatus, L. fermentum, L. paracasei* subsp. *pseudoplantarum, L. alimentarius, L. kefir, L. hilgardii, Enterococcus faecium, L. mesenteroides, L. farciminis, L. brevis, L. lactis subsp. cremoris, L. casei* subsp. *casei* and *L. bifermentans*. Mitra et al. (2007) isolated *L. lactis* from dahi which produced nisin-like (Nisin Z) bacteriocin that inhibited important food pathogens *L. monocytogenes* and *S. aureus*.

In Himachal Pradesh, traditional milk based products are prepared from the milk of several species of indigenous cattle, buffalo, sheep, goats, and Churu (hybrid of cow and yak) (Savitri and Bhalla, 2007). Nudu, a functional food is prepared by cooking wheat flour in milk with little salt and is consumed with ghee (Savitri and Bhalla, 2007).

Cereal, Cereal and Legume, Legume Based Fermented Foods

Cereals such as rice (*Oryza sativum*), ragi flour (*Eleusine coracana*), wheat flour (*Triticum spp.*), barley flour (*Hordeum vulgare*), and pulses such as black gram dhal, red gram, green gram dhals are predominantly used in the preparation. These cereals and legumes are considered as effective substrates for the production of probiotic-incorporated functional food, as they can be used as a source of non-digestible carbohydrates, which stimulate the growth of *Lactobacilli* and *Bifidobacteria*. They contain water-soluble fibers such as β-glucan, arabinoxylan, galacto-oligosaccharides and fructo-oligosaccharides, which are digested by selective groups of LAB (Swennen et al., 2006). Cereals and legumes on fermentation by bacteria yield short-chain fatty acids (SCFAs). The SCFA contributes to acidic condition in the large intestine, which favors the growth of lactic acid bacteria (Macfarlane et al., 2006 and Roopashri and Vardaraj, 2009). In the preparation of fermented foods such as idli, dosa, adai dosa, kallappam, ambali and dhokla, the batter is prepared from the basic ingredients, and this batter is left overnight at room temperature for fermentation, occasionally sodium bicarbonate is added to provide anaerobic conditions for the growth of yeast and LAB. The fermented batter is prepared into either as steamed cakes (idli) or as pan cakes (dosa, appam) before it gets too soured.

- **Idli:** Idli, is white colored and spongy fermented product made from milled rice and dehulled black gram batter (Agarwal and Bhasin, 2002). Rice and pulses are necessary for its preparation and also mixed natural microflora needed for quality idli (Achaya, 1994). Black gram (*Phaseolus mungo* L.) is the primary ingredient having natural fermentation microflora and acts as the substrate for the fermentation of batter. Sour buttermilk is also used as a starter (Radhakrishnamurty, 1961). For batter making, the ratio of dehulled black gram and milled rice is 1:2. Fermentation of idli batter improves the nutrition and protein efficiency value (Reddy et al., 1982). Leavening is important in the process and the microorganisms involved are *Lactobacillus mesenteroides*, *Streptococcus faecalis* (Mukherjee et al., 1965) and probiotic microorganisms such as *Lactobacillus plantarum* and *Lactobacillus lactis*. These produce vitamin B12 and β-galactosidase enzyme, which promotes health (Iyer et al., 2013). Idli is a natural fermented food, both bacteria and yeasts are generally introduced by the ingredients. *Leuconostoc mesenteroides* is the most commonly encountered bacterium (Nout and Sarkar, 1999).
- **Koozhu** is the Tamil name for porridge made from millet. Finger millet, a traditional South Indian weaning food, also consumed in the fermented form, as koozhu in rural and urban households (Antony et al., 1998). Koozhu is made from Kezhvaragu or Cumbu flour and broken rice (called noyee in Tamil) in a mud pot. Koozhu is consumed as a breakfast and during festival time by the rural population. Koozhu is easily digested and it cools the body; therefore, during summer street vendors sell it as a cool drink in South India.
- **Fermented rice or Pazhaiya soru** is prepared by adding water to cooked rice and by incubating the mixture overnight, and finally adding buttermilk and salt and directly consumed (Sekar and Mariappan, 2007). It is an early morning diet for farmers prior to going to the field. Predominant microbiota isolated from these foods include: *Weissella paramesenteroides, Lactobacillus*

fermentum, L. plantarum, Enterococcus faecalis, Pediococcus acidilactici, P. cerevisiae, L. mesenteroides.

- **Dosa:** Dosa is a fermented dish like idli, mainly consumed in the parts of south India (Steinkraus, 1996). It contains milled rice and dehulled black gram. During fermentation, the volume of the batter doubles and as fermentation time increases, the protein content of batter also increases (Soni et al., 1985). The batter is prepared by mixing wet rice and black gram with water, which undergoes fermentation for 8-20 h. The fermented suspension is spread on a heated plate with little oil (Battacharya and Bhat, 1997). Dosa is consumed with chutney and sambar (Purushothaman et al., 1977). LAB isolated from dosa inhibited pathogens such as *Bacillus cereus, Staphylococcus aureus, Listeria monocytogenes, Pseudomonas aeruginosa, V. parahaemolyticus* and *Aeromonas hydrophila* through their antibacterial protein bacteriocin (Pal et al., 2005).
- **Ambali:** It is a finger millet-based fermented semi-liquid product of south India. It prepared by mixing of finger millet flour with water to make thick batter followed by cooking and fermentation. The fermentation of finger millet is carried by microorganisms viz., *Leuconostoc mesenteroides, Lactobacillus fermentum* and *Streptococcus faecalis* (Ramakrishnan, 1980). Ambali contain high amount of calcium and low resistance starch because of use of finger millet (Mangala et al., 1999).
- **Dhokla:** Dhokla is similar to idli except that the dal used is Bengal gram (*Cicer arietinum*). Lactic acid and acetoin gives sour taste and good flavor (Aidoo et al., 2006). It is prepared by fermentation of bengal gram and rice. The method of preparation is the same as idli, but it is steamed openly (Steinkraus, 1995). The antioxidant property of Dhokla helps in curing age related diseases and oxidative stress-induced degenerative diseases (Moktan et al., 2011). So, dhokla is good food product for people suffering from diabetes.
- **Bhatooru, Marchu and Chilra:** These are leavened breads or roti's constituting the staple diet of rural people in Himachal Pradesh (Savitri and Bhalla, 2007). Bhatooru, marchu and chilra are prepared by using wheat/ barley/ buckwheat flour as substrate and inoculum 'Malera' and 'Treh' respectively which is rich in lactic acid bacteria as *Lactobacillus plantarum, L. acidophilus, Leuconostoc, Bacillus* spp., *Lactococcus lactis, Sacchromyces cerevisiae, Debaromyces hansenii,* and *Lactococcus lactis* (Kanwar et al., 2007). The fermented dough of bhatooru and chilra is baked on hot plates while marchu is deep fried in mustard oil. Marchu is prepared during functions and other ceremonies and is taken alongwith tea. Chilra is a

favorite snack of the tribe served with coriander chutney, potato and mutton soup (Thakur et al., 2004).

- **Kinema:** It is a soybean based fermented food (Deka, 2012) and consumed in eastern Himalayan regions of Darjeeling hills and Sikkim. It is source of proteins (Tamang and Nikkuni, 1998) and is prepared from soybeans. For its preparation soybean seeds are soaked overnight and then cracked slightly in a mortar and pestle. Grits are put in bamboo basket lined with fern frond and kept for fermentation at 25-40°C for 2-3 days (Tamang, 2001). The microorganisms involved in fermentation for preparation of Kinema are *Bacillus subtilis, Enterococcus faecium, Candida parapsilosis* and *Geotrichum candidum* (Sarkar et al, 1994). The product of fermented soybean is salted, deep fat fried and consumed as soup along with rice and vegetables (Sarkar et al., 1993). A remarkable increase in free amino acids, mineral contents, vitamin-B complex and antioxidant activity was reported during kinema fermentation (Sarkar et al., 1997, Tamang and Nikkuni, 1998 and Tamang et al., 2009).
- **Tungrymbai:** It is a traditional fermented food product prepared from soybean seeds used in Meghalaya. In its preparation the soybean seeds are cleaned, washed and soaked in double quantity of water for about 4-6 hours, outer skin is removed and then cooked in the same water till all water is absorbed. The cooked soybeans are pressed in a bamboo basket, the inner surface lined by leaves of *Clinogyne dichotoma*, cooled and then left to ferment at temperature 25–40ºC for 3-4 days. The fermented product is brown mass with typical odor (Sohliya et al., 2009, Jeyaram et al., 2009 and Agrahar-Murugkar and Subbulakshmi, 2006). The microflora of Tungrymbai is *Bacillus subtilis* and *Enterococcus faecium, Candida parapsilosis, Saccharomyces bayanus, Saccharomycopsis fibuligera* and *Geotrichum candidum* (Sohliya et al., 2009 and Sarkar et al., 1993). It is a popular fermented soybean based sticky food which serves as a cheap source of high protein food in local diet (Sohliya et al., 2009). Increase in carotene and folic acid has been reported in tungrymbai (Murungkar and Subbulakshmi 2006)
- **Hawaijar:** It is an alkaline-fermented soybean product consumed in Manipur. It is the main source of soluble proteins (26 to 27%) (Thingom and Chhetry, 2011) and has characteristic flavor and stickiness (Premarani and Chhetry, 2010). The molecular identification of microorganism mainly involved in the fermentation of this product is documented by Jeyaram et al. (2008). In preparation of hawaijar, bamboo baskets, banana or fig leaves are used to help in fermentation. Presence of *Bacillus spp.* gives high fibrinolytic activity to the product (Singh et al., 2014).

- **Kulcha nan and Bhatura:** Fermented snack foods like Bhatura and Kulcha (white wheat flour product) are made in India. For the fermentation of these products, mainly *Saccharomyces cerevisiae* and LAB are used (Sanjeev and Sandhu, 1990).
- **Kurdi and taotjo:** Kurdi are solid fried chips, and taotjo popular in eastern India is a condiment made from fermentation of roasted wheat meal by *Aspergillus oryzae* (Adams, 1998, Chavan and Kadam, 1989, Harlander, 1992, Sankaran, 1998, Soni and Sandhu, 1990 and Blandinob et al., 2003).

Cereal/Pulse and Buttermilk-Based Fermented Food

- **Jalebi:** It is also known as **zulbia**, is a sweet popular food in some parts of South Asia, West Asia, North Africa, and East Africa. It is prepared by frying maida flour batter in circular shapes, followed by soaking in sugar syrup. The lactic acid bacteria carry the fermentation process.

Citric acid or lime juice is sometimes added to the syrup, as well as rose. Jalebi is eaten with curd or rabri (North India). The fermented batter is deep fat fried in oil in spiral shapes and immersed in sugar syrup for a few minutes. The pH decreases from 4.4 to 3.3 and there is a 9% increase in the volume of batter.

- **Selroti:** It is a rice-based fermented food and is ring shaped, consumed in Sikkim and Darjeeling. Rice is soaked overnight in cold water and water is removed. The soaked rice is powdered and mixed with wheat flour, sugar, butter, and condiments. The powder is molded into soft dough using water. The batter is incubated to ferment at ambient temperature for 2-4 hours, followed by molding into a ring and fried. It is served as a confectionery product (Yonzan and Tamang, 2009). Selroti is good source of digestible proteins. The microorganisms in the preparation of fermented food Selroti are viz., *Lactobacilli*, *Pediococci*, *Enterococci*, and *Leuconostocs* (Yonzan and Tamang, 2010).
- **Khaman:** It is prepared from Bengal gram dhal. Idli, Dhokla and Khaman are steamed till the batter is leavened and acidified. The acid content inhibits the growth of food poisoning and food spoilage organisms (Sekar and Mariappan, 2007).
- **Sez:** It is semi-fermented food of Uttaranchal and is made from rice. In most cases, sez is extracted while preparation of rice jann (local beer) (Roy et al., 2004).
- **Adai and vada**: These both are cereal legume based breakfast or snack food in India. Microorganisms required for the fermentation of both the products

predominantly are *Pediococcus, Streptococcus, Leuconostoc* (Adams, 1998; Chavan and Kadam, 1989; Harlander, 1992; Sankaran, 1998; Soni and Sandhu, 1990 and Blandinob et al., 2003).

- **Bhattejaanr and anarshe:** Anarshe is rice based sweetened snack food fermented by lactic acid bacteria, whereas bhattejaanr is sweet and sour alcoholic paste fermented by *Hansenula anomala* and *Mucor rouxianus* (Adams, 1998; Chavan and Kadam, 1989; Harlander, 1992; Sankaran, 1998; Soni and Sandhu, 1990 and Blandinob et al., 2003).

Vegetable, Bamboo Shoot (BS) and Unripe Fruits-Based Fermented Foods

The lactic acid fermentation of vegetables for the finished products is important process because it improves the nutritive value, palatability, acceptability, microbial quality and shelf-life of the fermented product (Kingston et al., 2010). Moreover, this is a remarkable procedure to store the perishable vegetable in the absence of cold-storage or refrigeration, where majority of rural people cannot afford canned or frozen foods. Fermented BS products are consumed as a traditional food by ethnic people of North-eastern states of India (Tamang et al., 2009b). In India, BSs are harvested annually in Sikkim (26.2 tons), Meghalaya (435 tons) and Mizoram (426.8 tons). BSs are low in fat and cholesterol, but very high in potassium, carbohydrates and dietary fibers. Vitamins, amino acids and antioxidants can be extracted from BSs (Choudhury et al., 2011). LAB are mainly dominant in fermented vegetables and BS products (Tamang and Tamang, 2009). *Pediococcus pentasaceous, L. cellubiosus, L. plantarum, L. fermentum, L. brevis, L. mesenteroides, L. lactis, E. faecium* and *P. acidilactici* are predominant LAB species found in fermented vegetables. Tamang et al., (2009b) determined the functional properties of LAB isolated from ethnic fermented vegetables (gundruk, sinki, khalpi and inziangsang) of the Himalayas. LAB strains showed strong acidification and coagulation activities. They showed antimicrobial activity, particularly a strain *L. plantarum* isolated from inziangsang, a fermented leafy vegetable product, was inhibitory towards *S. aureus* and *P. aeruginosa*. LAB strains showed various enzymatic activities such as alkaline phosphatase, esterase, esterase lipase, lipase, leucine arylamidase, valine arylamidase, cysteine-arylamidase, acid phosphatase, napthol-AS-B1-phosphohydrolase, a-galactosidase, β-galactosidase, a-glucosidase, b-glucosidase, N-acetyl-b-glucosaminidase and also degraded oligosaccharides. Some strains of *L. plantarum* showed more than 70% hydrophobicity and adherence to the mucus secreting HT-29 MTX cells. Seventeen LAB strains from kanji responded positively when screened for probiotic properties such as acid tolerance, bile salt tolerance, antimicrobial activity against food-borne pathogens (*B. cereus, L. monocytogenes, E. coli, S. aureus, Yersinia enterolytica*), β-galactosidase activity, antibiotic susceptibility and cholesterol assimilation (Reddy et al., 2007). During fermentation of radish taproot product sinki, *L. plantarum* utilizes mannitol to remove the bitter flavour from the finished product (Tamang and Sarkar, 1993). BS-based fermented foods contain *Lactobacillus plantarum, L. brevi, L. corniformis, L. delbrueckii, Leuconostoc fallax, L. lactis, L. mesenteroides, Enterococcus durans, S. lactis, L. casei, Tetragenococcus halophilus* and *L. fermentum* as predominant LAB species, they also showed functional probiotic properties (Tamang et al., 2009b). During fermentation of BSs, *L. lactis* helps in reducing cyanogen glycoside content that causes major diseases of nervous system, goiter and miscarriage (Singh et al., 2007b).

Miscellaneous Bamboo Based Fermented Food

- **Mesu:** It is bamboo shoot derived fermented food (Nehal, 2013). Locally available species of bamboo are used such as choya bans (*Dendrocalamus hamiltonii* Nees and Arnott), bhalu bans (*D. sikkimensis* Gamble) and karati bans (Bambusatulda Roxb) (Tamang and Sarkar, 1996) which are defoliated, chopped and pressed tightly into a green bamboo hollow stem. The stem is covered with leaves and kept for fermentation for about 7-15 days. Mesu is mostly consumed as pickle (Sekar and Mariappan, 2007; Tamang and Sarkar, 1996; Tamang and Tamang, 2009a and Tamang and Tamang, 2009b).
- **Soibum:** It a fermented bamboo shoot product is food of the people in Manipur state. Noney/ kwatha and andro are two types of fermentation procedures adopted. The outer inedible and hard casings of succulent bamboo sprouts are peeled off while the soft portions are chopped and pressed tightly into wooden or earthen pots and left to ferment for 6-12 months in both the methods (Jeyaram et al., 2009). The microflora of Soibum includes *Enterococcus durans, Streptococcus lactis, Bacillus* spp., *Candida* spp., *Saccharomyces* spp., *Torulopsis* spp. It is consumed as regular side dish with steamed rice. The bamboo shoots are rich in potassium, carbohydrates, dietary fibers, vitamins and various antioxidants (Tamang and Tamang, 2009b).

Fermented Vegetable and Fruits Products

- **Gundruk:** Gundruk is a fermented food prepared from leaves of rayo-saag, mustard, or cauliflower. Leaves are cut, crushed, fermented in earthened jars for 7-10 days and dried in sunlight for 3-4 days (Tamang and Tamang, 2009). The predominant microorganisms found as Lactic acid bacteria comprise *Lactobacillus, Leuconostoc,* and *Pediococcus. Lactobacillus fermentum, L. plantarum, L. casei, Pseudoplantarum* and *Pediococcus pentosaceus*

(Tamang et al., 2005b). The Gundruk soup can be used as appetizer (Tampang et al., 2005). Gundruk has a good amount of ascorbic acid, carotene, and lactic acid, and acts as an anticarcinogen (Tamang, 2010). It is consumed as soup or pickle (Tamang et al., 2012).

- **Pickled Cucumber:** A pickled cucumber is prepared where cucumber is fermented by immersing in solutions of brine or vinegar.
- **Dill Pickle:** A gherkin is a variety of cucumber: (Kathryn Hawkins, 2007). Gherkins are pickled in jars with vinegar or brine alongwith addition of sugar.

Brined Pickles

Brined pickles are made using the traditional fermentation process in a brine. The brine concentration can vary between 20 to more than 40 g of salt per lit of water. The fermentation process depends on *Lactobacillus* bacteria.

Lime

Lime pickles are soaked in pickling lime rather than in a salt brine. The lime is washed off the pickles followed by addition of vinegar, sugar and spices.

Nutrition

Like pickled vegetables such as sauerkraut, sour pickled cucumbers are low in calories. They also contain a moderate amount of vitamin K, Sweet pickled cucumbers, including bread-and-butter pickles, are higher in calories due to their sugar content.

Lactobacilli species such as *L. plantarum* and *L. brevis* have been shown to add to the nutritional value of pickles (Tokatli et al., 2015).

- **Sinki:** It is a non-salted fermented radish tap root of the Gorkha (Tamang et al., 2012) and Gurung tribe of Sikkim prepared by fermentation in a 1m deep pit which is plastered with mud and warmed by burning (Tamang et al., 2012, and Tamang and Tamang, 2009b). It is prepared by washing of radish, cutting into small pieces and drying under sunlight in naaglo (local utensil made of bamboo for winnowing the grains) for 3-4 days. Dried pieces are added in a pit. The pit is covered with cow dung and soil paste and left for over 15 days for fermentation (Singh et al., 2007). Microorganisms observed as *Lactobacillus fermentum, L. brevis* and *L. plantarum* (Tamang, 1993). It has an acidic flavor, mostly used as soup and pickle (Tamang and Tamang, 2009b). It contains 14.5% protein, 2.5% fat and 11.3% ash on dry weight. It is an effective appetizer, cures diarrhea, stomach pain and consumed mostly during the lean period (Tamang and Tamang, 2009b).

- **Anishi:** It is prepared from edible *Colocasia* sp. leaf (Deka, 2012).The fresh mature green leaves are washed, staked one above the other and wrapped in banana leaf. It is then kept for about a week till the leaves turn yellow and then ground into paste and cakes are made out of it which are dried. The addition of chilly, salt and ginger is made and cooked with dry meat. It is sour in taste and used as a condiment (Tamang and Tamang, 2009b).

Meat and Fish-Based Fermented Foods

Meat is highly susceptible to microbial spoilage. Drying, smoking and fermentation of meat are critical steps in the traditional processing of meat (Oki et al., 2011). In India, people of the North-eastern region ferment meat of yak, goat, pig, fish and crab for preservation for longer period. **Kargyong** is an ethnic sausage-like fermented product prepared from yak, beef and pork. Three varieties of Kargyong are prepared and consumed: yak kargyong (prepared from yak meat), lang kargyong (prepared from beef) and faak kargyong (prepared from pork). Yak kargyong is a popular fermented sausage in Sikkim, Ladakh, Tibet, Arunachal Pradesh and Bhutan in the Himalayas (Rai et al., 2010).

Fermented fish products are important dietary components in the protein deficient South-East Asia. Preservation of fish by salt is an age-old technology. This method of preservation still enjoys popularity in many developing countries (Singh et al., 2007b). The LAB present in are mainly responsible for acidification. The LAB bring about a significant increase in soluble nitrogen, free amino acids, proteolytic activity and B-vitamins including thiamine, riboflavin and cyanocobalamine.

- **Ngari:** The fermented fish product Ngari is the food of people in Manipur and eaten with rice (Deka, 2012). *Phoubu*, a fish species is used for the preparation of Ngari. *Phoubu* is sun dried and washed with water (Jeyaram et al, 2009). It is then covered with gunny bags, pressed and packed in an earthen pot of 45-50 kg capacity and sealed. Finally, the solid state fermentation takes place (Thapa et al.,2004 and Nehal, 2013).
- **Hentak:** Hentak is a ball-like thick paste prepared by fermentation of a mixture of sun-dried fish (*Esomusdanricus*) powder and petioles of aroid plants (*Alocasimacrorhiza*) in Manipur. The mixture is allowed to ferment for 7–9 days. It is to be eaten only after 2 weeks of complete fermentation giving proper texture and aroma to the dish. The microflora of Hentek is as *Bacillus cereus, B. subtilis, Staphylococcus aureus*, *Enterococcus faecium, Candida* spp. (Thapa et al., 2004). However, on being stored for a few months the balls

harden which can then be used as a reserve food by its propounding it to a paste with a little water and stored as balls (Thapa et al., 2004; Thapa, 2002 and Jeyaram et al., 2009). Hentak is eaten as curry with boiled rice (Thapa, 2002). Sometimes it is given to women in the final stages of their pregnancy and patients recovering from sickness or injury (Sarojnalini and Singh, 1988 and Sarkar et al., 2015).

Health Benefits of Fermented Foods

Fermentation enhances digestibility, flavor and aroma of food and exerts health promoting benefits through biological enrichment of food substrates with protein, essential amino acids, essential fatty acids, and vitamins. It also helps in destroying undesirable compounds present in raw foods. Food contains many antinutritive factors that not digested by people which are reduced during fermentation. It can make food pleasantly sour or tangy, and develops flavor. Anti-nutrients are the natural or synthetic compounds that interfere with the absorption of nutrients, which can be minimized/ destroyed by fermentation. Foods that are tough, difficult to digest or unpalatable raw can be improved by fermentation, and reducing the need for cooking. Many of the fermented foods were observed to have beneficial effect during ailment by the local people and they are used as a special diet or medicine for ages. Fermented foods are easily digested and often used as food for infant and invalids. It is a prescribed diet in the hospitals for patients undergoing treatment (Steinkraus, 1996). Koozhu is included in the daily diet of rural and agricultural workers and is claimed to be a nourishing health food (Antony and Chandra, 1997). Fermented milk dahi can be used to cure intestinal disease such as diarrhea (Agarwal and Bhasin, 2002) intake of dahi has anti-cholesteremic (Sinha and Sinha, 2000), anticarcinogenic (Arvind et al., 2010), anti-diabetic (Yadav et al., 2007a), angiotensin-converting enzyme inhibition effect (Harun-ur-Rashid et al., 2007) and anti-atopic dermatitis effect (Watanabe et al., 2009). Certain fermented vegetable products (gundruk, sinki and iniziangsang) are said to be good appetizers and the ethnic people use these foods for remedies from indigestion (Tamang, and Tamang 2009). Gundruk soup is considered as a tonic for old age people (Singh et al., 2007b). Fermented radish root pieces are called sinki. It is very effective in curing diarrhea and stomach pain, and is consumed mostly during lean period. Iromba is prepared from tree bean (*Perkia roxburgii*) and is considered an appetizer (Singh et al., 2007b). Fermented rai helps in curing stomach pain and gas trouble, and significantly improves digestion (Singh et al., 2007b). Kanjika or kanji, a lactic acid fermented rice product is prescribed for a number of chronic diseases (Reddy et al., 2007). Carrot Kanji is considered to have high nutritional value and cooling and soothing properties (Sura et al., 2001). Beetroot kanji is considered to have potential to prevent infection and malignant disease (Kingston et al., 2010). Handua or Kardi is used as medicine against any digestive problems, particularly against constipation. The Meitei community in Northeastern India use fermented BSs (Soibum) along with fermented fish against the plague disease (Singh et al., 2007b). Ngari-dry fish of Puntus breed *Phutunis* or *Eromus dandricus* is mixed with Iromba (fermented BSs) and this dish called Yongchak Iromba is considered as an appetizer and is well known to the old age women of the Meitei community (Singh et al., 2007b). Hentak is given to the mothers during confinement and patients during convalescence (Sarojnalini and Singh, 1988). Most of the claims are based on traditional belief, hence there is a scope to conduct experiment-based research to prove these claims.

Preservation, Food Security and Cultural Importance

Preservation of foods by fermentation is a widely practiced from ancient time (Parvez et al., 2006). Fermentation ensures not only increased shelf life and microbiological safety of a food but also make some foods more digestible and in some cases reduces toxicity of the substrate.

Fermentation technologies play an important role in ensuring the food security of millions of people around the world, particularly marginalized and vulnerable groups [Marshall and Mejia, 2011]. This is achieved through improved food preservation, increasing the range of raw materials that can be used to produce fermented food products and removing anti-nutritional factors to make food safe to eat. Moreover, there exist many examples of fermentation by-products which can be safely fed to nutritionally supplement livestock, thereby further strengthening the livelihood system. These provide a good source of undegradable protein and water soluble vitamins, but need to be stored cool and fed within a week, or otherwise ensiled, to prolong their shelf-life (FAO, 1999). Fermentation is a cheap and energy efficient means of preserving perishable raw materials, which is accessible to even the most marginalized, landless, physically incapacitated rural, peri-urban and urban poor. Fermentation requires very little sophisticated equipment's, either to undertake or subsequently store the fermented products, and has had a major impact on nutritional habits, traditions, and culture. As such, traditional fermentation still serves as a substitute for refrigeration or otherwise safe keeping of food, and is also directly utilized to make good of edible leftovers.

- **Flavor Enhancement:** Fermentation makes the food palatable by enhancing its aroma and flavor. These organoleptic properties make fermented food more popular than the unfermented one in terms of consumer acceptance (Blandino et al., 2003).

- **Improvement of Nutritional Quality:** Fermented foods can be more nutritious than their unfermented counterparts. Microorganisms help in both catabolic and anabolic processes. Fermentation can enhance nutritional value especially in plant foods, which involves enzymatic splitting of cellulose, hemicellulose, and related polymers that are not digestible by humans into simpler sugars and sugar derivatives. Cellulosic materials in fermented foods can be nutritionally improved for humans by the action of microbial enzymes [Potter and Hotchkiss, 2006]. In addition to improving digestibility by enhancing the activity of enzymes, fermentation also reduces the levels of antinutrients such as phytic acid and tannins in food thereby increasing bioavailability of minerals.
- **Biodegradation of Phytate:** Fermented food has the ability to biodegrade the phytic acid. Phytic acid or phytate (myoinositol hexakisphosphate, IP6) is the primary storage form of phosphorus in mature seeds of plants and it is particularly abundant in many cereal grains, oilseeds, legumes, flours and brans. Phytate has a strong chelating capacity and forms insoluble complexes with divalent minerals of nutritional importance such as iron, zinc, calcium and magnesium. Phytases are widespread in various microorganisms including filamentous fungi, Gram positive and Gram-negative bacteria and yeasts [Lopez et al,. 2002]. The phytase activities of yeast during bread making for reduction of phytate content of bread have been examined [Lopez et al,. 2002].
- **Improvement of Immunity System:** The immune system acts to protect the host from infectious agents and a variety of noxious agents existing in the environment [Schoen et al., 2009].
- **Prevention of Toxic Effects of Mycotoxins:** Mycotoxins are secondary metabolites produced by fungi belonging mainly to the *Aspergillus, Penicillium* and *Fusarium* genera. Contamination of agricultural products by mycotoxins is a worldwide dilemma. The most important mycotoxins are the aflatoxins, ochratoxins, fumonisins, deoxynivalenol, zearalenone and trichothecenes [Schatzmayr et al,. 2006]. Various fermented food microorganisms are able to some extent and with varied efficiency to degrade mycotoxins to less- or non-toxic products. Inhibition of mycotoxin absorption in the gastrointestinal tract is another way to prevent the toxic effects of mycotoxins. The mechanism of detoxification by yeast is due to the adhesion of mycotoxins to cell-wall components.
- **Bioavailability of Nutrient:** Beneficial functions of probiotic microorganism like yeasts are improvement of bioavailability of minerals through the hydrolysis of phytate, folate biofortification and detoxification of mycotoxins due to surface binding capacity of the yeast cell wall.

- **Folate Biofortification:** Folates (vitamin B9) are the essential cofactors in the biosynthesis of nucleotides and therefore crucial for the cellular replication and growth. *S. cerevisiae* is a rich dietary source of native folate and produces high levels of folate per weight [Patring et al,. 2005].
- **Intestinal pH Balance:** A healthy large intestine (or colon) has a slightly acidic pH, which tends to inhibit or destroy putrefactive bacteria. Putrefactive bacteria can produce foul smelling gases and are damaging to health when present in large numbers in the intestine. An acidic pH and a healthy population of friendly bacteria will inhibit the growth of undesirable bacteria, moulds, mould spores and yeasts, particularly *Candida* [Lopez et al., 2002].
- **Improvement of Digestion and the Digestibility of Foods:** Healthy bacteria found in naturally fermented foods produce enzymes that can break down foods present in the intestines, making easier absorption of the nutrients.

The inability to digest lactose in lactase-deficient individuals, or milk sugar, is prevalent worldwide. Consumption of lactose by those lacking adequate levels of lactase produced in the small intestine can result in symptoms of diarrhea, bloating, abdominal pain and flatulence [Panesar et al., 2006]. It has been documented that many lactose intolerant individuals are better able to consume fermented dairy foods, such as yoghurt, with fewer symptoms than the same amount of unfermented counterpart. Fermented food yoghurt was found to be helpful in the digestion of lactose because the lactic acid bacteria used to make yoghurt produce lactase and digest the lactose [Ebringer et al., 2008].

- **Probiotics:** Probiotics are defined as live microorganisms which when administered in adequate amounts confer a health benefit on the host (Heller, 2001, FAO/WHO, 2001). Dairy foods are well suited to promoting the positive health impact in lactose intolerance, urinary tract infections in woman, gut function, traveler's diarrhea, infantile diarrhea, antibiotic associated diarrhea, helicobacter pylori gastritis, inflammatory bowel disease (IBD), irritable bowel syndrome (IBS) and colorectal cancer (CRC), immune function, infant health, atopic disease and atopic dermatitis for probiotics (Heller, 2001). Probiotics are the live microorganisms which when administered in adequate amount it will give health benefits to the consumer (FAO, 2002), such as *Lactobacillus plantarum, L. casei, L. acidophilus,* and *Streptococcus lactis,* which are supplemented by food that beneficially affect the host by improving its intestinal balance (Tamang, 2009).

- **Protection Against Infection:** Gastrointestinal infections including diarrhea result from a change in the gut microflora caused by an invading pathogen. It is suggested that viable lactic acid bacteria interfere with the colonization and subsequent proliferation of food borne pathogens, thus preventing the manifestation of infection [Gandhi, 2000]. The beneficial effects of lactic acid bacteria and cultured milk products have also been attributed to their ability to suppress the growth of pathogens either directly or through production of antibacterial substances. Fermented food have been reported to effective in prevention of various gastrointestinal infections [Panesar, 2011].
- **Anticarcinogenic Effect:** It has been reported that fermented food products can work against certain types of cancers [Hosono et al,. 1986].
- **Antihypertensive Activity:** Two antihypertensive peptides have been purified from sour milk fermented with *L. helveticus* and *S. cerevisiae* starter cultures. These two peptides inhibit angiotensin-converting enzyme that converts angiotensinogen I to angiotensinogen II, which is a potent vasoconstrictor [Maeno et al,. 1996].
- **Lowering of Serum Cholesterol:** Reports indicate that fermented food products to have hypo cholesteraemic effect. It has been found that *L. acidophilus* has exhibited the ability to lower serum cholesterol levels [Grunewald, 1992].

CONCLUSION

The large diversity in traditional fermented foods of India is because of the regional health foods which evolved according to the climate, culture, and cropping practices of a particular region. These fermented products are having various benefits in processing, production and health. Some traditional fermented foods are not explored geographically, hence the study of these fermented products is required in formulating the product with defined micro-flora to maintain the characteristics and quality of food. There is still more scope for research on the large-scale production of fermented foods and to look for new applications of fermented foods.

REFERENCES

Achaya, K. T. (1994). *Indian food: a historical companion*. Delhi: Oxford University Press.

Adams, M. R. (1998). Fermented weaning foods. In J. B. Wood (Ed.), *Microbiology of fermented foods* (pp. 790–811). London: Blackie Academic. doi:10.1007/978-1-4613-0309-1_25

Agarwal, K. N., & Bhasin, S. K. (2002). Feasibility studies to control acute diarrhea in children by feeding fermented milk preparation Actimel and Indian Dahi. *European Journal of Clinical Nutrition*, *56*(4), S56–S59. doi:10.1038j.ejcn.1601664 PMID:12556949

Agrahar-Murugkar, D., & Subbulakshmi, G. (2006). Preparation techniques and nutritive value of fermented foods from the Khasi tribes of Meghalaya. *Ecology of Food and Nutrition*, *45*(1), 27–38. doi:10.1080/03670240500408336

Aidoo, K. E., Nout, N. J. R., & Sarkar, P. K. (2006). Occurrence and function of yeasts in Asian indigenous fermented foods. *FEMS Yeast Research*, *6*(1), 30–39. doi:10.1111/j.1567-1364.2005.00015.x PMID:16423068

Antony, U., & Chandra, T. S. (1997). Microbial population and biochemical changes in fermenting finger millet (*Eleusine coracana*). *World Journal of Microbiology & Biotechnology*, *13*(5), 533–537. doi:10.1023/A:1018561224777

Antony, U., George, M. L., & Chandra, T. S. (1998). Inhibition of *Salmonella typhimurium* and *Escherichia coli* by fermented Finger millet (*Eleusine coracana*). *World Journal of Microbiology & Biotechnology*, *14*(6), 883–886. doi:10.1023/A:1008871412183

Arvind, K., Nikhlesh, K. S., & Pushpalata, R. S. (2010). Inhibition of 1,2-dimethylhydrazine induced colon genotoxicity in rats by the administration of probiotic curd. *Molecular Biology Reports*, *37*(3), 1373–1376. doi:10.100711033-009-9519-1 PMID:19330535

Backes, M. (2014). *Cannabis Pharmacy: The Practical Guide to Medical Marijuana*. New York: Black Dog & Leventhal.

Battacharya, S., & Bhat, K. K. (1997). Steady shear rheology of rice blackgram suspensions and suitability of rheological models. *Journal of Food Engineering*, *32*(3), 241–250. doi:10.1016/S0260-8774(97)00027-7

Billings, T. (1998). *On fermented foods*. Available: http://www.livingfoods.com

Blandinob, A., Al-Aseeria, M. E., Pandiellaa, S. S., Canterob, D., & Webba, C. (2003). Review: Cereal-based fermented foods and beverages. *Food Research International*, *36*(6), 527–543. doi:10.1016/S0963-9969(03)00009-7

Bol, J., & de Vos, W. M. (1997). Fermented foods: an overview. In J. Green (Ed.), *Biotechnological innovations in food processing* (pp. 45–76). Oxford, UK: Butterworth-Heinemann.

Chavan, J. K., Kadam, S. S., & Beuchat, L. R. (1989). Nutritional improvement of cereal fermentation. *CRC Critical Reviews in Food Science and Nutrition*, *28*(5), 349–400. doi:10.1080/10408398909527507 PMID:2692608

Choudhury, D., Sahu, J. K., & Sharma, G. D. (2011). Bamboo shoot based fermented food products: A review. *Journal of Scientific and Industrial Research (New Delhi, India)*, *70*, 199–203.

Dewan, S., & Tamang, J. P. (2007b). Microbial and analytical characterization of Chhu-A traditional fermented milk product of the Sikkim Himalayas. *Journal of Scientific and Industrial Research (New Delhi, India)*, *65*, 747–752.

Ebringer, L., Ferenčík, M., & Krajčovič, J. (2008). Beneficial health effects of milk and fermented dairy products--review. *Folia Microbiologica*, *53*(5), 378–394. doi:10.100712223-008-0059-1 PMID:19085072

FAO. (1998). *Fermented Fruits and Vegetables-A Global Perspective* (Vol. 134). Rome, Italy: FAO Agricultural Services Bulletin.

FAO. (1999). Fermented Cereals—A Global Perspective. FAO Agricultural Services Bulletin, 138.

FAO. (2002). *Guidelines for evaluation of probiotics in food.* Report of a Joint FAO/ WHO Working Group on Drafting Guidelines for the Evaluation of Probiotics in Food.

FAO/WHO Expert Consultation Report. (2001). *Evaluation of health and nutritional properties of powder milk and live lactic acid bacteria*. Author.

Gandhi, D. N., Marwaha, S. S., & Arora, J. K. (Eds.). (2000). *Food Processing: Biotechnological Applications*. New Delhi: Asiatech Publishers Inc.

Grunewald, K. K. (1992). Serum Cholesterol levels in rats fed skim milk fermented by *Lactobacillus acidophilus*. *Journal of Food Science*, *47*(6), 2078–2079. doi:10.1111/j.1365-2621.1982.tb12955.x

Harlander, S. (1992). Food biotechnology. In J. Lederberg (Ed.), *Encyclopedia of microbiology* (pp. 191–207). New York: Academic Press.

Harun-ur-Rashid, M., Togo, K., Useda, M., & Miyamoto, T. (2007). Probiotic characteristics of lactic acid bacteria isolated from traditional fermented milk 'Dahi' in Bangladesh. *Pakistan Journal of Nutrition*, *6*(6), 647–652. doi:10.3923/pjn.2007.647.652

Hawkins, K. (2007). *Allotment Cookbook*. New Holland Publishers.

Heller, K. J. (2001). Probiotic bacteria in fermented foods: Product characteristics and starter organisms. *The American Journal of Clinical Nutrition*, *73*(2), 374s–379s. doi:10.1093/ajcn/73.2.374s PMID:11157344

Hirahara, T. (1998). Functional food science in Japan. In T. Mattila-Sandholm & K. Kauppila (Eds.), Functional food research in Europe (pp. 19–20). Julkaisija-Utgivare.

Hosono, K. T., Kashina, T., & Kada, T. (1986). Antimutagenic properties of lactic acid-cultured milk on chemical and fecal mutagens. *Journal of Dairy Science*, *69*(9), 2237–2242. doi:10.3168/jds.S0022-0302(86)80662-2 PMID:3097092

Iyer, B. K., Singhal, R. S., & Ananthanarayan, L. (2013). Characterization and in-vitro probiotic evaluation of lactic acid bacteria isolated from idli batter. *Food Sci. Technol.*, *50*, 1114–1121. PMID:24426023

Jeyaram, K., Mohendro Singh, W., Premarani, T., Devi, A. R., Chanu, K. S., Talukdar, N. C., & Singh, M. R. (2008). Molecular identification of dominant microflora associated with 'Hawaijar'da traditional fermented soybean (*Glycine max* (L.)) food of Manipur. India. *International Journal of Food Microbiology*, *22*(3), 259–268. doi:10.1016/j.ijfoodmicro.2007.12.026 PMID:18281117

Jeyaram, K., Singh, A., Romi, W., Devi, A. R., Singh, W. M., & Dayanithi, H. (2009). Traditional fermented foods of Manipur. *Indian Journal of Traditional Knowledge*, *8*(1), 115–121.

Kanwar, S. S., Gupta, M. K., Katoch, C., Kumar, R., & Kanwar, P. (2007). Traditional fermented foods of Lahaul and Spiti area of Himachal Pradesh. *Indian Journal of Traditional Knowledge*, *6*, 42–45.

Kingston, J. J., Radhika, M., Roshini, P. T., Raksha, M. A., Murali, H. S., & Batra, H. V. (2010). Molecular characterization of lactic acid bacteria recovered from natural fermentation of beet root and carrot Kanji. *Indian Journal of Microbiology*, *50*(3), 292–298. doi:10.100712088-010-0022-0 PMID:23100843

Lopez, H. W., Leenhardt, F., Coudray, C., & Remesy, C. (2002). Minerals and phytic acid interactions: Is it a real problem for human nutrition? *International Journal of Food Science & Technology*, *37*(7), 727–739. doi:10.1046/j.1365-2621.2002.00618.x

Macfarlane, S., Macfarlane, G. T., & Cummings, J. H. (2006). Prebiotics in the gastrointestinal tract. *Alimentary Pharmacology & Therapeutics*, *24*(5), 701–714. doi:10.1111/j.1365-2036.2006.03042.x PMID:16918875

Maeno, M., Yamamoto, N., & Takano, T. (1996). Identification of an antihypertensive peptide rom casein hydrolysate produced by a proteinase from *Lactobacillus helveticus* CP790. *Journal of Dairy Science*, *79*(8), 1316–1321. doi:10.3168/jds.S0022-0302(96)76487-1 PMID:8880454

Mangala, S. L., Malleshi, N. G., Tharanathan, R. N., & Not Available, N. A. (1999). Resistant starch from differently processed rice and ragi (finger millet). *European Food Research and Technology*, *209*(1), 32–37. doi:10.1007002170050452

Mani, U., Prabhu, S., Damie, S., & Mani, I. (1993). Glycemic index of some commonly consumed foods in Western India. *Asia Pacific Journal of Clinical Nutrition*, *2*, 111–114. PMID:24352140

Marshall E. and Mejia D., (2011) Traditional fermented food and beverages for improved livelihoods. *FAO Diversification Booklets, 21.*

McKay, L. L., & Baldwin, K. A. (1990). Applications for biotechnology: Present and future improvements in lactic acid bacteria. *FEMS Microbiology Reviews*, *87*(1-2), 3–14. doi:10.1111/j.1574-6968.1990.tb04876.x PMID:2271224

Mitra, S., Chakrabartty, P. K., & Biswas, S. R. (2007). Production of Nisin Z by *Lactococcus lactis* Isolated from Dahi. *Applied Biochemistry and Biotechnology*, *143*(1), 41–53. doi:10.100712010-007-0032-5 PMID:18025595

Moktan, B., Roy, A., & Sarkar, P. K. (2011). Antioxidant activities of cereal-legume mixed batters as influenced by process parameters during preparation of dhokla and idli, traditional steamed pancakes. *International Journal of Food Sciences and Nutrition*, *62*(4), 360–369. doi:10.3109/09637486.2010.532116 PMID:21142877

Mukherjee, S. K., Alburry, M. N., Pederson, C. S., Vanveen, A. G., & Steinkraus, K. H. (1965). Role of *Leuconostoc mesenteroides* in leavening the batter of idli, a fermented food of India. *Applied Microbiology*, *13*, 227–231. PMID:14325884

Murughar, D. A., & Subbulakshmi, G. (2006). Preparation techniques and nutritive value of fermented foods from the khasi tribes of Meghalaya. *Ecology of Food and Nutrition*, *45*(1), 27–38. doi:10.1080/03670240500408336

Nehal, N. (2013). Knowledge of traditional fermented food products harbored by the Tribal Folks of the Indian Himalayan Belt. *Int. J.Agri. Food Sci. Technol.*, *4*, 401–414.

Nout, M. J. R., & Sarkar, P. K. (1999). Lactic acid food fermentation in tropical climates. *Antonie van Leeuwenhoek, 76*(1/4), 395–401. doi:10.1023/A:1002066306013 PMID:10532396

Oki, K., Rai, A. K., Sato, S., Watanbe, K., & Tamang, J. P. (2011). Lactic acid bacteria isolated from Ethnic preserved meat products of the Western Himalayas. *Food Microbiology, 28*(7), 1308–1315. doi:10.1016/j.fm.2011.06.001 PMID:21839380

Pagni, J. (1998). *Demonstrating bacteria for health.* VTT Biotechnology and Food Research (Catalogue 5).

Pal, V., Jamuna, M., & Jeevaratnam, K. (2005). *Isolation and characterization of bacteriocin producing lactic acid bacteria from a south Indian special dosa (appam) batter.* J Cult Collect.

Panesar, P. S. (2011). Fermented dairy products: Starter cultures and potential nutritional benefits. *Food Nutrit. Sci., 2*, 47.

Panesar, P. S., Panesar, R., Singh, R. S., Kennedy, J. F., & Kumar, H. (2006). Review Microbial production, immobilization and applications of β-D-galactosidase J. *Chem. Tech. Biotech., 81*(4), 530–543. doi:10.1002/jctb.1453

Patidar, S., & Prajapati, J. (1998). Standardization and evaluation of lassi prepared using *Lactobacillus acidophilus* and *Streptococcus thermophilus. Journal of Food Science and Technology, 35*, 428–431.

Patring, J. D., Jastrebova, J. A., Hjortmo, S. B., Andlid, T. A., & Jagerstad, I. M. (2005, April). Development of a simplified method for the determination of folates in baker's yeast by HPLC with ultraviolet and fluorescence detection. *Journal of Agricultural and Food Chemistry, 53*(7), 2406–2411. doi:10.1021/jf048083g PMID:15796570

Prashant, Tomar, Singh, Gupta, Arora, & Joshi. (2009). Phenotypic and genotypic characterization of *Lactobacilli* from Churpi cheese. *Dairy Sci Technol, 89*, 531–540.

Premarani, T., & Chhetry, G. (2010). Evaluation of traditional fermentation technology for the preparation of hawaijar in Manipur. *Assam Univ. J. Sci. Technol., 6*, 82–88.

Purushothaman, D., Dhanapal, N., & Rangaswami, G. (1977). *Microbiology and biochemistry of idli fermentation.* Symposium on Indigenous Fermented Food, Bangkok, Thailand.

Pushpangadan, P., Dan, V. M., Ijinu, T., & George, V. (2012). Food, nutrition and beverage. *Indian Journal of Traditional Knowledge, 11*, 26–34.

Radhakrishnamurty, R., Desikachar, H. S., Srinivasan, M., & Subrahmanyan, V. (1961). Studies on idli fermentation. II. Relative participation of black gram flour and rice semolina in the fermentation. *Journal of Scientific and Industrial Research (New Delhi, India)*, *20*, 342–345. PMID:14038559

Rai, A. K., Palni, U., & Tamang, J. P. (2009). Traditional knowledge of the Himalayan people on the production of indigenous meat products. *Indian Journal of Traditional Knowledge*, *8*(1), 104–109.

Ramakrishnan, C. (1980). Studies on Indian fermented foods. *Baroda J. Nutr.*, *6*, 1–57.

Reddy, K. B. P. K., Raghavendra, P., Kumar, B. G., Misra, M. C., & Prapulla, S. G. (2007). Screening of probiotic properties of lactic acid bacteria isolated from Kanjika, an ayruvedic lactic acid fermented product: An *in-vitro* evaluation. *The Journal of General and Applied Microbiology*, *53*(3), 207–213. doi:10.2323/jgam.53.207 PMID:17726302

Reddy, N. R., Sathe, S. K., Pierson, M. D., & Salunkhe, D. K. (1982). Idli, an Indian fermented food: A review. *Journal of Food Quality*, *5*(2), 89–101. doi:10.1111/j.1745-4557.1982.tb00736.x

Roopashri, A. N., & Varadaraj, M. C. (2009). Molecular characterization of native isolates of lactic acid bacteria, *Bifidobacteria* and yeasts for beneficial attributes. *Applied Microbiology and Biotechnology*, *83*(6), 1115–1126. doi:10.100700253-009-1991-y PMID:19407995

Roy, B., Kala, C. P., Nehal, A. F., & Majila, B. S. (2004). Indigenous fermented food and beverages: A potential for economic development of the high altitude societies in Uttaranchal. *Journal of Human Ecology (Delhi, India)*, *15*(1), 45–49. doi:10.1080/09709274.2004.11905665

Sanjeev, K. S., & Sandhu, K. D. (1990). Indian fermented foods; microbiological and biochemical aspects. *Indian Journal of Microbiology*, *30*, 135–157.

Sankaran, R. (1998). Fermented food of the Indian subcontinent. In Microbiology of fermented foods. Blackie Academic and Professional. doi:10.1007/978-1-4613-0309-1_24

Sarkar, P. K., Cook, P. E., & Owens, J. D. (1993). *Bacillus* fermentation of soybeans. *World J. Microb.Biot.*, *9*(3), 295–299. doi:10.1007/BF00383066 PMID:24420029

Sarkar, P. K., Jones, I. J., Craven, G. S., Somerset, S. M., & Palmer, C. (1997). Amino acid profiles of Kinema, a soyabean-fermented food. *Food Chemistry*, *24*, 337–339.

Sarkar, P. K., Kumar, L. D. H., Dhumal, C., Panigrahi, S. S., & Choudhary, R. (2015). Traditional and ayurvedic foods of Indian origin. *J Ethn Foods*, *2*(3), 97–109. doi:10.1016/j.jef.2015.08.003

Sarkar, P. K., Tamang, J. P., Cook, P. E., & Owens, J. D. (1994). Kinema – a traditional soybean fermented food: Proximate composition and microflora. *Food Microbiology*, *11*(1), 47–55. doi:10.1006/fmic.1994.1007

Sarkar, S., & Misra, A. (2001). Bio-preservation of milk and milk products. *Indian Food Indian*, *20*, 74–77.

Sarojnalini, C., & Singh, W. V. (1988). Composition and digestibility of fermented fish foods of Manipur. *Journal of Food Science and Technology*, *25*, 349–351.

Savitri, B. T. C. (2007). Traditional foods and beverages of Himachal Pradesh. *Indian Journal of Traditional Knowledge*, *6*(1), 17–24.

Schatzmayr, G., Zehner, F., Taubel, M., Schatzmayr, D., Klimitsch, A., Loibner, A. P., & Binder, E. M. (2006). Microbiological for deactivating mycotoxins. *Molecular Nutrition & Food Research*, *50*(6), 543–551. doi:10.1002/mnfr.200500181 PMID:16715543

Schoen, C., Ernwiss, D., Schulz, A., Chema, D., Schweikart, J., Ernwiss, D., ... Baehr, V. (2009). Regulatory effects of a fermented food concentrate on immune functioning parameters in healthy volunteers. *Nutrition (Burbank, Los Angeles County, Calif.)*, *25*(5), 499–505. doi:10.1016/j.nut.2008.10.022 PMID:19121921

Sekar, S., & Mariappan, S. (2007). Usage of traditional fermented products by Indian rural folks and IPR. *Indian Journal of Traditional Knowledge*, *6*, 111–120.

Sharma, N., & Singh, A. (2012). An insight into traditional foods of Northwestern area of Himachal Pradesh. *Indian Journal of Traditional Knowledge*, *11*, 58–65.

Sharma, R., & Lal, D. (1997). Effect of dahi preparation on some water-soluble vitamins. *Indian Journal of Dairy Science*, *50*, 318–320.

Simango, C. (1997). Potential use of traditional fermented foods for weaning in Zimbabwe. *Journal of Social Science and Medicine*, *44*(7), 1065–1068. doi:10.1016/S0277-9536(96)00261-4 PMID:9089926

Singh, A., Singh, R. K., & Sureja, A. K. (2007). Cultural significance and diversities of ethnic foods of Northeast India. *Indian Journal of Traditional Knowledge*, *6*(1), 79–94.

Singh, A., Singh, R. K., & Sureja, A. K. (2007b). Cultural significance and diversities of ethnic foods of Northeast India. *Indian Journal of Traditional Knowledge*, *6*(1), 79–94.

Singh, T. A., Devi, K. R., Ahmed, G., & Jeyaram, K. (2014). Microbial and endogenous origin of fibrinolytic activity in traditional fermented foods of Northeast India. *Food Research International*, *55*, 356–362. doi:10.1016/j.foodres.2013.11.028

Sinha, P. R., & Sinha, R. N. (2000). Importance of good quality dahi in food. *Indian Dairyman*, *52*, 45–47.

Sohliya, I., Joshi, S.R., Bhagobaty, R.K., & Kumar, R. (2009). Tungrymbai- A traditional fermented soybean food of the ethnic tribes of Meghalaya. *Indian J. Trad. Knowl.*, *8*, 559-561.

Soni, S. K., Sandhu, D., & Vilkhu, K. (1985). Studies on dosa and indigenous Indian fermented food: Some biochemical changes accompanying fermentation. *Food Microbiology*, *2*(3), 175–181. doi:10.1016/0740-0020(85)90032-2

Soni, S. K., & Sandhu, D. K. (1990). Indian fermented foods: Microbiological and biochemical aspects. *Indian Journal of Microbiology*, *30*, 135–157.

Steinkraus, K. H. (1983). Lactic acid fermentation in the production of foods from vegetables, cereals and legumes. *Antonie van Leeuwenhoek*, *49*(3), 337–348. doi:10.1007/BF00399508 PMID:6354083

Steinkraus, K. H. (1995). *Handbook of Indigenous Fermented Foods* (2nd ed.). CRC Press.

Steinkraus, K. H. (1996). *Handbook of Indigenous Fermented Foods*. Marcel Dekker.

Sukumar, G., & Ghosh, A. R. (2010). *Pediococcus* spp. – A potential probiotic isolated from Khadi (an Indian fermented food) and identified by 16S rDNA sequence analysis. *African Journal of Food Science*, *4*(9), 597–602.

Sura, K., Garg, S., & Garg, F. C. (2001). Microbiological and biochemical changes during fermentation of Kanji. *Journal of Food Science and Technology*, *38*, 165–167.

Swennen, K., Courtin, C. M., & Delcour, J. A. (2006). Non-digestible oligosaccharides with prebiotic properties. *Critical Reviews in Food Science and Nutrition*, *46*(6), 459–471. doi:10.1080/10408390500215746 PMID:16864139

Tamang, B., & Tamang, J. P. (2009a). Lactic acid bacteria isolated from indigenous fermented bamboo products of Arunachal Pradesh in India and their functionality. *Food Biotechnology, 23*(2), 133–147. doi:10.1080/08905430902875945

Tamang, B., & Tamang, J. P. (2009b). Traditional knowledge of biopreservation of perishable vegetables and bamboo shoots in Northeast India as food resources. *Indian Journal of Traditional Knowledge, 8*, 81–95.

Tamang, J. P. (2001). *Kinema, Feature: Fermented soybean foods in daily life*. Food Culture.

Tamang, J.P. (2010). Benefits of traditional fermented foods. *Fermented-Foods*.

Tamang, J. P., Chettri, R., & Sharma, R. M. (2009). Indigenous knowledge of Northeast women on production of ethnic fermented soybean foods. *Indian Journal of Traditional Knowledge, 8*(1), 122–126.

Tamang, J. P., Dewan, S., Olasupo, N. A., Schillinger, V., & Holzapful, H. (2000). Identification and enzymatic profiles of predominant lactic acid bacteria isolated from soft variety chhurpi, a traditional cheese typical of the Sikkim Himalayas. *Food Biotechnology, 14*(1&2), 99–112. doi:10.1080/08905430009549982

Tamang, J. P., & Nikkuni, S. (1998). Effect of temperatures during pure culture fermentation of kinema. *World Journal of Microbiology & Biotechnology, 14*(6), 847–850. doi:10.1023/A:1008867511369

Tamang, J. P., & Sarkar, P. K. (1993). Sinki: A traditional lactic acid fermented radish tap root product. *The Journal of General and Applied Microbiology, 39*(4), 395–408. doi:10.2323/jgam.39.395

Tamang, J. P., & Sarkar, P. K. (1996). Microbiology of mesu, a traditionally fermented bamboo shoot product. *International Journal of Food Microbiology, 29*(1), 49–58. doi:10.1016/0168-1605(95)00021-6 PMID:8722186

Tamang, J. P., Tamang, B., Schillinger, U., Franz, C. M., Gores, M., & Holzapfel, W. H. (2005). Identification of predominant lactic acid bacteria isolated from traditionally fermented vegetable products of the Eastern Himalayas. *International Journal of Food Microbiology, 105*(3), 347–356. doi:10.1016/j.ijfoodmicro.2005.04.024 PMID:16055218

Tamang, J. P., Tamang, B., Schillinger, U., Franz, C. M. A. P., Gores, M., & Holzapfel, W. H. (2005b). Identification of predominant lactic acid bacteria isolated from traditional fermented vegetable products of the Eastern Himalayas. *International Journal of Food Microbiology*, *105*(3), 47–356. doi:10.1016/j.ijfoodmicro.2005.04.024 PMID:16055218

Tamang, J. P., Tamang, B., Schillinger, U., Guigas, C., & Holzapfel, W. H. (2009b). Functional properties of lactic acid bacteria isolated from the ethnic fermented vegetables of the Himalayas. *International Journal of Food Microbiology*, *135*(1), 28–33. doi:10.1016/j.ijfoodmicro.2009.07.016 PMID:19666197

Tamang, J. P., Tamang, N., Thapa, S., Dewan, S., Tamang, B., Yonzan, H., ... Kharel, N. (2012). Microorganisms and nutritional value of ethnic fermented foods and alcoholic beverages of North East India. Indian J. Trad. Knowl., 11:7-Thakur N., Savitri and Bhalla T. C. (2004). Characterization of some traditional fermented food and beverages in Himachal Pradesh. *Indian Journal of Traditional Knowledge*, *3*, 325–335.

Thapa, N., Pal, J., & Tamang, J. P. (2004). Microbial diversity in ngari, hentak and tungtap, fermented fish products of North East India. *World Journal of Microbiology & Biotechnology*, *20*, 599–607.

Thapa, T.B. (2002). Diversification in processing and marketing of yak milk based products, *TAAAS/IYIC/Yakfoundation/FAOROAP/ICIMOD/ILR/Workshop*, 484–489.

Thingom, P., & Chhetry, G. (2011). Nutritional analysis of fermented soybean (Hawaijar). *Assam Univ J Sci Technol*, *7*(1), 96–100.

Tokatli, M., Gulgor, G., Elmaci, S. B., & Isleyen, N. A. (2015). *In-vitro* properties of potential probiotic indigenous lactic acid bacteria originating from Traditional Pickles. *BioMed Research International*, *2015*, 1–8. doi:10.1155/2015/315819

Watanabe, T., Hamada, K., Tategaki, A., Kishida, H., Tanaka, H., & Kitano, M. (2009). *Oral administration of lactic acid bacteria isolated from traditional south Asian fermented milk*. Academic Press.

Yadav, H., Jain, S., & Sinha, P. R. (2007a). Antidiabetic effect of probiotic dahi containing *Lactobacillus acidophilus* and *Lactobacillus casei* in high fructose fed rats. *Nutrition (Burbank, Los Angeles County, Calif.)*, *23*(1), 62–68. doi:10.1016/j.nut.2006.09.002 PMID:17084593

Yadav, H., Jain, S., & Sinha, P. R. (2007b). Evaluation of changes during storage of probiotic dahi at 78C. *International Journal of Dairy Technology*, *60*(3), 205–210. doi:10.1111/j.1471-0307.2007.00325.x

Yonzan, H., & Tamang, J. P. (1998). Consumption pattern of traditional fermented foods in the Sikkim Himalaya. *Journal of Hill Research*, *11*, 112–115.

Yonzan, H., & Tamang, J. P. (2009). Traditional processing of Selroti-A cereal based ethnic fermented food of the Nepalis. *Indian Journal of Traditional Knowledge*, *8*, 110–114.

Yonzan, H., & Tamang, J. P. (2010). Microbiology and nutritional value of Selroti, an ethnic fermented cereal food of the himalayas. *J. Food Biotechnology*, *24*(3), 227–247. doi:10.1080/08905436.2010.507133

Chapter 6
Recent Insight Into Fermented Foods and Production

Dixit V. Bhalani
CSIR-CSMCRI, India

Arvind Kumar Singh Chandel
CSIR-CSMCRI, India

Poonam Singh Thakur
RTM University Nagpur, India

ABSTRACT

The fermented beverages and foods either of plant or animal source play a vital role in the food of society in several parts of the world. The fermented of foods not only afford vital sources of nutrients but also have abundant potential in maintaining health and also preventing various diseases. The bacteria and yeasts are the major groups of microorganisms related to traditional fermented of the foods. Numerous diverse types of traditional fermented beverages and foods are formed at domestic level in the various countries. The advancement of fermentation technology provides value addition to waste food by their complete conversion into the different value-added products. The recent research suggests that the biological functions of fermented foods affect the health due to functional microbes involved during fermentation which provides several health-promoting benefits to the consumers. The emphasis of this chapter is to describe the fermentation technology and their potential to minimize the wastage of foods by conversion of value-added products and their benefits.

DOI: 10.4018/978-1-5225-7706-5.ch006

INTRODUCTION

The fermented foods are an essential part of human life from food to medicine and many more, Different types of fermented food used worldwide from morning to evening people are using such as curd, butter, cheese, bread, pickles, wine, beer, fermented vegetables, antibiotics, food supplements fermented meats etc. The scope of food fermentation extended from producing alcoholic beverages, fermented milk, fermented meat and vegetable products to genetically engineered superbugs to carry out efficient fermentation to treatment and utilization of waste and overall producing nutritious and safe products with attractive qualities. (Simango, 1997)

The fermented foods are that food which produced by alteration of the raw material of either animal or vegetable origin by the activities of microorganisms in the absence of oxygen. Bacteria, yeast and moulds can be used to produce a diverse range of products that differ in flavour, texture and stability from the original raw material. The Fermented foods are those foods which are subjected to microorganisms or enzymes to get desirable biochemical changes and cause significant modification to food materials. (Nout, 2003) (Figure 1: Simple representation of fermentation) The concept of fermentation in a biochemical sense the term fermentation refers to the metabolic process in which organic compounds (particularly carbohydrates or sugars) are broken down to release energy without the involvement of terminal electron acceptor such as oxygen. Incomplete oxidation of the substrate occurs so that only a relatively small amount of ATP energy is released compared with the energy generated if a terminal electron acceptor is involved. Partial oxidation of carbohydrate or sugars can give rise to a variety of organic compounds. The compounds formed by microorganisms vary from organism to organism and are produced via different metabolic pathways (Gerardi, 2003).

Figure 1. Yeast mediated glucose fermentation

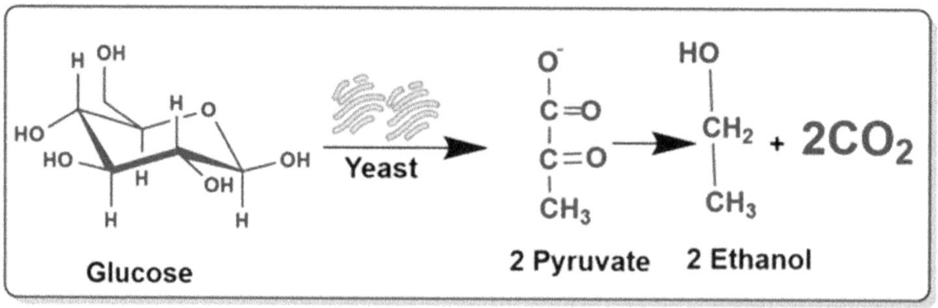

Table 1. Pathways of bacterial fermentation

Pathway	Microbes	Main Product	End Products
Propionic acid	Propionibacterium, Bifidobacterium	Swiss cheese	Acetic acid, propionic acid, CO_2
Acetone-butanol-ethanol	Clostridium acetobutylicum	Commercial solvents, gasoline alternative	Acetone, butanol, ethanol, CO_2
Alcohol	Candida, Saccharomyces	Beer, bread	Ethanol, CO_2
Butanediol	Klebsiella, Enterobacter	Chardonnay wine	Formic and lactic acid; ethanol; acetoin; 2,3 butanediol; CO_2; hydrogen gas
Mixed acid	Escherichia, Shigella	Vinegar, cosmetics, pharmaceuticals	Acetic, formic, lactic, and succinic acids; ethanol, CO_2, hydrogen gas
Lactic acid	Streptococcus, Lactobacillus	Sauerkraut, yogurt, cheese	Lactic acid
Butyric acid	Clostridium butyricum	Butter	Butyric acid, CO_2, hydrogen gas

The term fermentation can also be applied to any industrial process that produces a material that is beneficial to humans and if the process depends on the activity of one or more microorganisms. These processes, known as industrial fermentations, are usually carried out on a large scale and in vessels in which organism are normally grown in liquid media (Caplice et al., 1999). Some industrial fermentation in a biochemical sense but the majority is aerobic processes in which the microorganisms use oxygen and metabolize carbohydrates completely. Benefits of fermented foods: Fermented foods are an extremely valuable addition to the human diet for a variety of reasons (Stanbury et al., 2013).

Fermented foods increase the variety of foods that are available, adding to our diet a group of highly nutritious products with unique characteristics. There are, for example, about 1000 different types of cheeses. Use as ingredients: Fermented foods form important ingredients for a wide variety of dishes and are often used to impact special flavours, e.g. pepperoni in pizzas, yoghurts in curries, cheeses in a whole range of food, including soups, and soy sauce in stir-fry dishes. Enhancement in nutritional quality: The fermentation process may improve the nutritional values of raw material. Here are some examples: (Stanbury et al., 2013)

The tape fermentation doubles the protein content of cassava and increases the level of essential amino acids. The existence of yeasts in fermented food will increase the vitamin B content. Antinutritional factors such as glucosinolates, phytase, and lectins may be removed by the fermentation process. Fermentation may produce an increase in the accessibility of minerals (Liener, 1994). (Figure 2: simple fermentation of a glucose molecule) Fermentation often preserves a raw material, improving safety with regard to foodborne pathogens and enhancing shelf life; compare the shelf of raw milk (only a few days) with the shelf life of yoghurt, curd (many weeks). And some fermented foods are claim to have definite health benefits while the scientific proofs for this are very few. (Cleveland, 2001) Reports suggested that fermented milk products such as yoghurt can reduce serum cholesterol levels and help to avoid obesity, vitamin D deficiency, and cancers, mainly those associated with the colon. Bio-yoghurts (AB and ABT yoghurts) are supposed to have supportive effects on the gastrointestinal tract (GIT) microflora, those assisting recovery of a normal balanced flora after intake oral antibiotic therapy (Nagpal et al., 2012). Some of the fermented food is effortlessly digested than the original raw material. People who not able to digest lactose properly (show lactose intolerance) can normally consume some types of fermented dairy products (particularly yoghurts) without harmful effects. Human body shoes Lactose intolerance is due to the deficiency of the enzyme galactosidase in digestive juices, which converts lactose into glucose and galactose (Scrimshaw et al., 1988). Consumption of dairy products leaves unabsorbed lactose in the gut which is fermented by normal gastrointestinal tract (GIT) flora giving flatulence, abdominal pain diarrhoea and gastric distress. The fermentation of milk converts the harmful lactose to more easily digestable lactate, and the β galactosidase in liver. The starter culture organisms appears to support in the digestion of any remaining lactose. Legume (such as pea, soybean) contains oligosaccharides such as stachyose which is fermented in the gut to produces gas and the associated socially embarrassing flatus (Saavedra et al., 1989). The oligosaccharides are broken down to readily digestible monosaccharides and disaccharides during mould fermentations of legumes, therefore removing the problem.

Figure 2. Value added products from glucose by fermentation

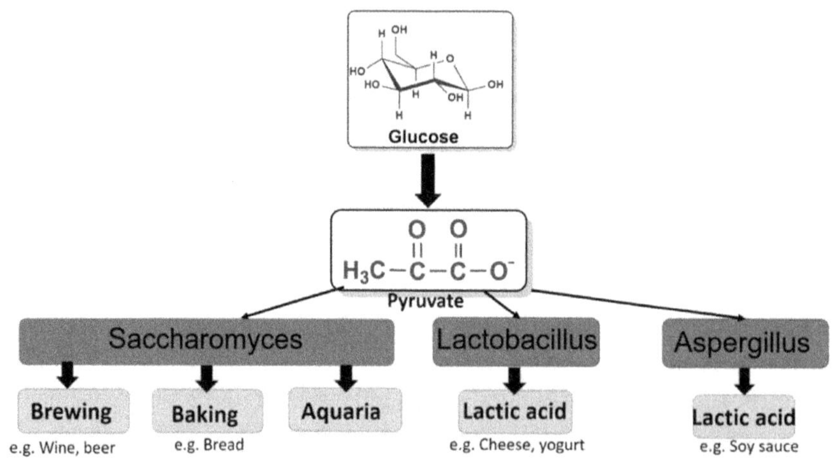

- **Detoxification of Raw Materials:** The fermentation process may remove toxic chemicals which present in the fermentation raw material. for example, Cassava fermentation, removes a cyanogenic (cyanide producing) glycoside; cassava is toxic if eaten raw and by tradition, lactic acid bacteria are the generally used microorganism for the preservation of foods (Chaves-Lopez et al., 2014). Modern methods of gene technology make it possible for the microbiologist to design and develop starter cultures with specific desired qualities. Many microbiologists studies deal with identification of organism's isolated from various fermented foods. The lactic acid bacteria are isolated from tomatoes that are naturally fermented under partial anaerobic conditions. These are *Lactobacillus brevis Leuonostoc mesenteroides*, and *Streptococcus sp*. In Asia mainly moulds of the genera *Aspergillus, Amyomyces, Rhizopus, Mucor, Neurospora Actinomucor*, and *Monascus*, are used in the manufacture of fermented foods. In Europe, mould ripened foods are primarily cheese and meats, usually using a *Penicillium* species. Gari which was prepared by fermenting cassava slurry contains *Bacillus, Penicillium spp.* and *Aspergillus* as major organisms. Lactic acid bacteria: the predominant group of fermentative microorganisms (Leroy et al., 2004). This group comprised of 11 genera of gram-positive bacteria: *Pediococcus, Carnobacterium, Oenococcus, Enterococcus, Lactococcus, Streptococcus, Wessells Lactosphaera, Vagococcus, Lecconostoc,* and *Lactobacillus.* Related to this group are genera such as *Propionibacterium, Aerococcus,* and *Microbacterium*, while this is a roughly defined group with no exact boundaries, as for fermenting organisms, they deficient from electron transport systems or cytochromes with functional heme-linked; they do not have a functional Krebs cycle. This chapter also covers the various recent techniques used for quality and quantity improvement of the value added final fermented products. Most common groups of microorganisms involved in food substance fermentations (Tomkins, et al., 1988). (Figure 3: Various food products)
- **Bacteria:** The essential bacteria in required for food fermentations are the *Lactobacillaceae* which can produce lactic acid from carbohydrates. Other important bacteria, are *acetobacter* species which produces acetic acid particularly in the fermentation of fruits and vegetables.
- **Yeasts:** Similar to moulds and bacteria, yeasts shows advantageous, and non-advantageous effects in foods and the most useful yeasts for food fermentation are *Saccharomyces*, playing roles such as the leavening of bread and the manufacture of alcohol and invert sugar;

Figure 3. Fermentation products

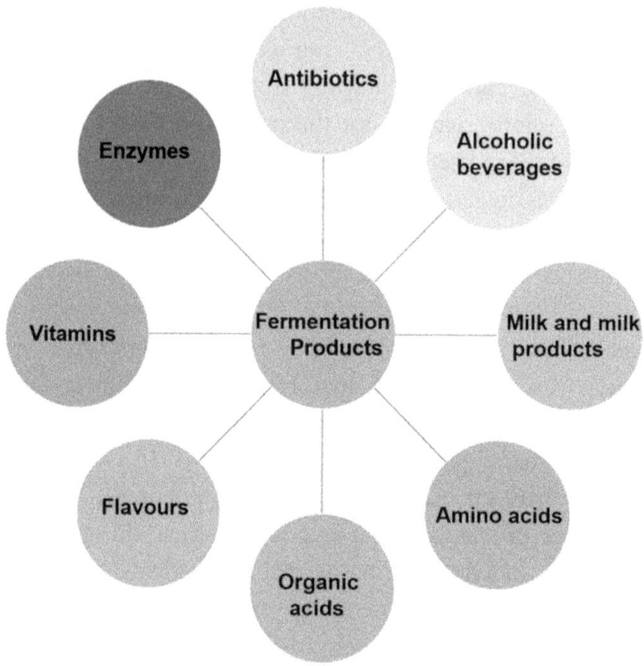

- **Moulds:** Are also involved in the food processing sector, both as preservers and spoilers. Almost all food fermentations are the result of more than one microorganism, either working together or in a sequence, but the growth is usually initiated by bacteria, followed by yeasts and then moulds. (Hutkins et al., 2008)

Classification of Fermented Foods

The fermented foods are food substrates that are attacked or overgrown by edible microorganisms whose enzymes, particularly proteases, amylases and lipases hydrolyse the protein polysaccharides, and lipids to non-toxic products with flavours, aromas and textures pleasant and attractive to the consumer. If the products of enzyme activities have nasty odours or annoying, unattractive flavours or the products are toxic or disease creating, the foods are defined as spoiled. Food fermentations may classified in multiple ways (Dirar, 1993): by categories (Yokotsuka, 1982) –

Table 2. Fermented food products and involved microorganisms

Fermented Food	Involve Microorganism
Fish sauce	*Streptococcus, Micrococcus* and *Bacillus species*
Sauerkraut	*Leuconostoc mesenteroides* and *Lactobacillus plantarum*
Chocolate	*Acetobacter fabarum*
Soy sauce	*Aspergillus oryzae* and *Aspergillus soyae*
Wine	*Acetobacter fabarum*
Coffee	*Saccharomyces cerevisiae var. Ellipsoideus*
Pickels	*L. mesenteroides, L. Brevis, L. Plantarum, P.Cerevisiae,*
Vegetables	*Acetobacter lovaniensis*
Vinegar	*Acetobacter malorum*
Stinky tofu	*Bacillus sphaericus*
Natto	*Bacillus subtilis*
Beer and Ale	*Accharomyces carlsbergensis, Saccharomyces cerevisiae.*
Yogurt	*Bifidobacterium adolescentis*
Dairy	*Bifidobacterium lactis*
Cheese	*Candida colliculosa*
Limburger cheese	*Debaryomyces kloeckeri*
Ham	*Enterococcus faecium*
Beer	*Dekkera bruxellensis*
Sourdough bread	*Lactobacillus acidifarinae*
Manchego cheese	*Enterococcus faecium*
Kefir	*Candida colliculosa*
Olive	*Leuconostoc spp.*
Palm wine	*Zymomonas mobilis*
Kimchi	*Tetragenococcus koreensis*
Fresh cheese	*Streptococcus sp.)*
Semi hard cheese	*Lactobacillus casei, Streptococcus cremoris*
Processed cheese	fungi or fungal spores used during ripening

1. Alcoholic beverages fermented by yeasts
2. Pickles fermented with *Lactobacillus*
3. Milk fermented with *Lactobacillus*
4. Vinegar fermented with Acetobacter
5. Plant proteins fermented with moulds with or without *Lactobacillus*
6. Fish or meat fermented with *Lactobacillus*

As per (Campbell-Platt, 1987) classification –

1. Dairy products
2. Cereal products
3. Beverages
4. Fish products
5. Legumes
6. Fruit and vegetable products
7. Meat products; by commodity

According to (Odunfa, 1988) (1) fermented starchy roots, (2) fermented cereals, (3) alcoholic beverages, (4) fermented vegetable proteins and (5) fermented animal protein; by commodity (Kuboye, 1985) (1) cereal, (2) cassava-based, (3) legumes and (4) beverages. (Dirar, 1993) states that the Sudanese traditionally classify their foods not on the basis of microorganisms or commodity but on a functional basis: (1) Kissar (staples) - porridges and breads such as acedia and kissra, (2) Milhat (sauces and relishes for the staples), (3) marayiss (30 types of opaque beer, clear beer, date wines and meads and other alcoholic drinks) and (4) Akil-munasabat (food for special occasions). Steinkraus classified fermentations according to the following categories that will serve as the basis for this Chapter (Steinkraus, 1997). (Figure 4: Classification of alcoholic and non-alcoholic beverages)

Figure 4. Classification of alcoholic and non-alcoholic beverages

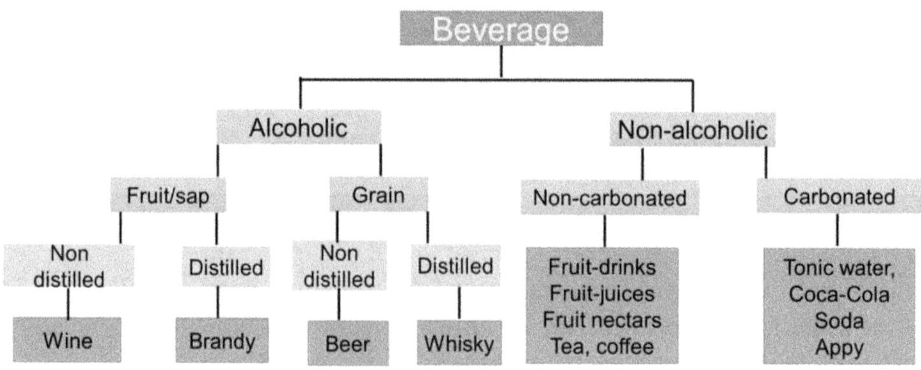

FERMENTED FOOD PRODUCTS

Cheese

Cheese is a concentrated form of two major milk components milk protein (casein) and milkfat. Besides milk, it contains a selected strain of bacteria, a milk clotting agent and sodium chloride. Cheese is a one of the popular dairy food product which was prepared from milk. It was available in numbers of varieties, in different textures, forms and flavours. It is prepared by the coagulation process of milk protein casein. It contains fat and milk proteins. For the preparation of cheese, first of all the milk is acidified and enzyme is added which causes the coagulation (Fankhauser, 2007).

Yoghurt

Yoghurt is formed by the controlled fermentation of milk by two species of bacteria *Lactotococcus thermophilus Streptococcus thermophiles* and *Lactobacillus bulgaricus.* The lactose is fermented to lactic acid, and it is the one which causes the characteristic curd to form. *Streptococcus thermophilus* brings the pH of the milk down to 5.5, *Lactobacillus bulgaricus* converts lactose into lactic acid, and proteolytic enzymes from *L. bulgaricus* break down milk proteins into peptides. These peptides stimulate the growth of *L. themophilus* which in turn produces formic acid and carbon dioxide. These are the growth stimulants for *L. Bulgaricus.* Yogurt prepared from milk contains 5% fat, 81% water, 4% carbohydrates, 9% protein, and 4% sugars. It provides 406 kilojoules energy per 100 gm. Yogurt is an enriched source of vitamins, it contains riboflavin (23% DV) and vitamin B12 (31% DV).

Kefir

Kefir is a kind of beverage formed by the action of lactic acid bacteria, yeast and acetic acid bacteria on milk, which generates a unique fermented milk product with unique properties. It is formed by adding a starter culture called "kefir grains" directly to milk. The kefir grains are a mass of several different bacteria and yeasts fixed in a complex matrix of protein & carbohydrate. Traditionally prepared kefir contained 1-2% alcohol, due to the complex fermentative process of kefir grains.

Acidophilus Milk

The skimmed milk is fermented by *Lactobacillus acidophilus* which has therapeutic importance in the GIT. Also useful in the treatment of people especially those suffering from frequent diarrhoea and intestinal gas. It has a very harsh acid taste and thus faces rejection by consumers to overcome this; sweet acidophilus milk was developed. In this, the bacteria are grown separately and added to pasteurised milk and the inoculated milk is kept at 4°C. When this milk is consumed, the beneficial bacteria are activated in the warm stomach and intestinal tract.From its first report in 1922, *Lactobacillus acidophilus* plays an important role in curing disorders of the gastrointestinal tract. Nowadays, several types of Acidophilus milk are available which belong to class of probiotic foods (Kneifel and Bonaparte, 2003).

Cultured Buttermilk

The Cultured buttermilk is the liquid remaining after-ripened, or sour cream is churned into butter using mesophilic starters. Various products are produced by using diverse strains of lactic acid bacteria as starter cultures and different fractions of whole milk as the starting substrate. The cream is starting substrate; butter is normally made by churning cream that has been soured by lactic acid bacteria. Sour cream uses S. cremoris or S. lactis for producing lactic acid and *Leuconostoc cremoris* for characteristic flavour. Ripened cream uses *S. cremoris* or S. lactis to produce lactic acid rapidly, and *L. citrovorum* produces the essential flavours. The cultured buttermilk production process involves pasteurization, homogenization and culturing system (Aryana et al., 2017).

Lassi

Lassi is a by-product which obtained in the preparation of country butter (ghee) from curd by indigenous methods. The curd is churned with the recurrent addition of water up to butter granules are formed. The product obtained by manual removal of the butter granules is called lassi. (Steinkraus, 1997).

Kombucha

Kombucha is a kind of fermented beverage made from a combination of black tea and sugar. It contains yeast and bacteria that are accountable for initiating the fermentation process once combined with sugar. After being fermented, kombucha becomes carbonated and contains vinegar, B-vitamins, enzymes, probiotics and a high concentration of acid (acetic, gluconic and lactic). There are reasons to drink kombucha every day because it improves digestion, helps with weight loss, increases energy, detoxes the body, supports the immune system, reduces joint pain and prevents cancer (Teoh et al., 2004).

Shrikhand

Shrikhand is sweetened dewatered curd. It is very widely held in western and some parts of southern India (Steinkraus, 1997). It has a unique rich flavour and a fairly long shelf life. To prepare shrikhand, the curd is suspended in a muslin cloth until all the free water has drained off. The semisolid mass in then whisked with sugar through a fine cloth, coloured and scented with saffron or rose-water, and flavoured with cardamom if desired. The composition of shrikhand is moisture 34.48 - 35.66%, fat 1.93 - 5.56%, protein 5.33 - 6.13%, reducing sugar 1.56 - 2.18% and nonreducing sugar (sucrose) 55.55- 53.76% (Mital et al., 1977).

Pickles

Pickles contain tannins, vitamins, minerals, and antioxidants along with that gut-friendly bacteria. Its use full in the vitamin K deficiency, as one small pickle, possess about 18-20% vitamins of our daily requirements. this vitamin that's an essential fat-soluble vitamin which plays a vital role in bone and heart health. Pickles are prepared by pickling process. It is the method of life expanding or preserving of food products via anaerobic fermentation process in brine or by food immersion in vinegar. The end product is termed as a pickle. This process will change the taste, flavour and texture of food. In the part of east asia, vinegar and vegetable oil (vinaigrette) is also employed as a media for pickling. Variety of foods can be pickled such as eggs, vegetables, meats, and fruits (Chou, 2012).

Miso

Miso is produced by fermentation of barley, soybean, or brown rice with koji fungus (*Aspergillus oryzae*). The miso fermentation process includes two steps: koji generation, and combination of the koji with other ingredients and then that mixture is kept for the enzymatic digestion, fermentation and ageing. This has anti-ageing substances which help to maintain healthy skin. It also boots up the immune system and lowers the risk of cancer, improves bone health and promotes a healthy nervous system (Shibasaki et al., 1962).

Sauerkraut

Sauerkraut is one of the ancient traditional fermented foods. Prepared from the fermentation of cabbage, contains a high amount of in dietary fiber, vitamin A, vitamin K and vitamins B. vitamin C, It's also an abundant source of calcium, iron, sodium, copper, magnesium and manganese. Sauerkraut has a range of advantageous effects on human health; it boosts digestive health, aids circulation, fights inflammation, strengthens bones and reduces cholesterol levels. Some of the sauerkraut based phytochemicals exhibit anti-inflammatory, antioxidant, and chemopreventive activities in counter to some particular types of cancer (Penas et al., 2016).

Tempeh

The tempeh is a fermented food a which made from soybean by addition of a tempeh starter (live mold). When it sits for a one day or two days, it converts into a cake-like product.Tempeh decreases cholesterol, rises bone density, decreases menopausal symptoms, encourages muscle recovery, and It contains high levels of vitamins B2, B3, B5,B6, and B12. As it contains large amounts of protein and calories, it is utilized by the poor population of several countries. In general, its a costeffective and highly nutritious food which can be utilized by each group of the society (Babu et al., 2009).

Natto

Natto is a fermented food produced from fermented soybeans. Natto is a prevalent dish in Japan containing fermented soybeans. It encompasses the extremely powerful probiotic *Bacillus subtilis*, which has been modulating the immune system and cardiovascular system; it also improves the digestion of vitamin K2. In addition to these natto benefits, it contains a powerful anti-inflammatory enzyme called nattokinase that has been proven to fight cancer (Sumi et al., 1987).

Raw Cheese

Raw cheeses are prepared from milk that hasn't been pasteurized. Goat milk, sheep milk and A2 cows soft cheeses are particularly high in probiotics, including *bulgaricus*, *bifudus*, *thermophillus*, and *acidophilus*. Probiotics benefits include healing digestive issues, neurological disorders and mental health problems; plus, probiotics boost the immune system and destroy harmful bacteria (Schoder et al., 2011).

Apple Cider Vinegar (ACV)

Vinegar is supposed to initiate as far back as 400 B.C. in Greece when Apple cider vinegar has subtle fruit notes that can add flavour to salad dressings and sauces. The procedure begins by crushing the apples into juice, adding yeast, and letting it ferment in barrels. physicians suggest apple cider vinegar can benefit into the control of blood sugars after a sugary meal. Also, apple cider vinegar has polyphenols, a potential cancer-fighting nutrient. Apple cider vinegar also shows good antibactrial activity, it kills 80% of germs, molds, bacteria, and viruses. It is also employed as deordarizing and cleaning agent for domestic application (Bragg et al., 2003).

Nutritional Value of Fermented Foods

Usually, it observed that the soluble fraction of food is increases during fermentation. The quantity as well as the quality of the food proteins as denoted by biological value, and frequently the content of water-soluble vitamins is usually increased, but the antinutritional factors were decline during the fermentation process (Boye et al., 2010). Fermentation fallouts in a minor proportion of dry matter in the food. And the concentrations of minerals, protein and vitamins appear to be increased when it was calculated on the dry weight basis (Adams, 1990). Single and mixed culture fermentation of pearl millet flour with lactobacilli and yeast, expressively enhances the total quantity of soluble sugars, reducing and non-reducing sugar content, with a synchronized decrease in its starch content (Khetarpaul & Chauhan, 1990). Grouping of fermentation and cooking improved the nutrition value of all tested sorghum seeds and minimized the content of antinutritional factors to a safety level in comparison with other processing methods (Obizoba & Atii, 1991). Combined culture fermentation of pearl millet flour with Saccharomyces cerevisiae, Saccharomyces diastaticus, Lactobacillus fermentum and Lactobacillus brevis was found to raise its biological utilisation in rats (Khetarpaul & Chauhan, 1991). Fermentation caused a significant reduction in lipid and lignin contents of okara, which is an insoluble residue recovered as a by-product in the production of soybean milk. On the other hand fermented okara not raising PER and also not raising the weight gain in rats (Guermani et al., 1992) related to then-fermented samples.

Proteins

The fermentation process enhances the protein efficiency ratio (PER) of wheat, that is due to the rise in the availability of lysine. Equal amount mixture of soybeans and wheat would produce an improved pattern of amino acids. After fermentation, the PER value of the mixture was raised to such a level that can be comparable to that of casein (Hesseltine & Wang, 1980). Fermentation process may not enhance the protein content, and amino acids except urea or ammonia are added as a source of nitrogen to the fermentation media (Reed, 1981).

Germination enhances the relative nutritional value (RNV) of maize from 65% to 81%. The fermentation increases the RNV to 87% of that germinated maze flour (Lay & Fields, 1981). Fermentation of legumes for cooking dhokla and fermentation of millet for making ambali did not confirm any improvement in the values recorded for TD, PER, BV and NPU in comparison to the unfermented food products (Chavan 1989). Steamed and soaked washed seeds of *Lathyrus sativus* had a score of 14 with methionine and cystine as the limiting amino acids. On tempe fermentation the score was increased to 16 and autoclaving obeyed by tempe fermentation boosted the score up to 21 (Moslehuddin & Hang, 1987). Solid substrate type fermentation of cassava with added urea enhanced the protein content from 1% to 10,7% (Daubresse et al., 1987). Fermentation of cassava promoted the utilisation of the diets, estimated as protein efficiency ratio and biological value (Friedman, 1996). The protein content of cassava deducted from 2,36 g/100g to 1,61 g/100g while fermentation (Padmaja et al., 1994). Natural fermentation process provided a higher yield of leaf protein concentrate related to heat coagulation.

Fermentation process significantly enhances the biological value of the leaf protein. Leaf protein concentrate which coagulated by fermentation was free from grassy odour and is usually more agreeable by human users as related with leaf protein concentrates (LPCs) coagulated by heat at natural pH or isoelectric pH of leaf extract (Pandey & Srivastava, 1993).

Vitamins

While the fermentation process, some micro-organisms generates vitamins at a higher rate than others do. The content of riboflavin and thiamine in ambali and dhokla was around 50% higher after fermentation. Fermented milk products, in general, displayed an increment in folic acid content and a slight reduction in vitamin B12 while other B vitamins were affected slightly (Patel et. al., 2013) in relation to raw milk. The levels of vitamin B12, folacin and riboflavin were raised by lactic acid fermentation of maize flour, while the level of pyridoxine was reduced (Murdock & Fields, 1984).

Kefir produced from ten distinct kefir grain cultures exhibited significant (>20%) rise for cobalamin, pyridoxine, folic acid and biotin and decline exceeding 20% for riboflavin, thiamine, pantothenic acid and nicotinic acid depending on the culture used. There was a 40% rise in thiamine content in two of the cultures. While riboflavin showed a small rise in two cultures, pyridoxine raised more than 120% in 3 cultures (Kneifel & Mayer, 1991). During tempe fermentation, the rhizopus strains were found to generate nicotinic acid, riboflavin, nicotinamide and vitamin B6, but not vitamin B12. The addition of cobalt and 5,6- dimethyl benzimidazole was found to enhance the vitamin B12 content of tempe (Keuth & Bisping, 1994).

Minerals

Fermentation process cant affects the mineral content unless some salts are added to the product during fermentation or by leaching when liquid part is isolated from the fermented food. Sometimes, when fermentation is conducted in metal containers, some minerals are get solubilized by the fermented product, which may cause an increment in mineral content. Phytate content in bread was reduced when the quantity of yeast or the fermentation time was increased (Harland & Harland, 1980). Phytate content in locust bean seeds was reduced from 0,51 mg/g to 0,31 mg/g by fermentation (Eka, 1980). Natural lactic fermentation of maize meal reduced phytate phosphorus by 78% (Chompreeda & Fields, 1984). The decline of phytate content while dough fermentation for whole grain flour was about 50% (Roos et al., 1990). Fermentation by *Saccharomyces diastaticus* obeyed by *Lactobacillus brevis* completely excreted phytic acid from pearl millet flour (Khetarpaul & Chauhan, 1991). Tannin content in the bambara nut milk could be minimized by fermentation (Obizoba & Egbuna, 1992). There was an impressive enhancement in protein availability and concentration when fermentation of siljo, which is a traditional Ethiopian fermented food. A study on the impact of fermentation of cowpea (*Vigna unguiculata*) on the nutritional quality of the cowpea meal revealed that 72h fermentation enhanced the content of protein, ash and lipid levels while lowering the levels of tannin and phytate (Nnam, 1995). Trypsin inhibitors, riboflavin and thiamine, were decreased significantly during fermentation. A decrease in protein content was seen during the first 2 days of fermentation, and after that, the reduction was not notable. (Gupta et al., 1998). Vaishali investigated the effect of natural fermentation on in vitro zinc bioavailability in cereal-legume mixtures determined that fermentation enhanced the zinc solubility (2-28%) and the zinc uptake by intestinal segment (1-16%) to a notable level (Vaishali et al., 1997).

HEALTH EFFECTS OF FERMENTED FOODS

Probiotic Effect

One of the important reasons for the increasing interest in fermented foods is due to its ability to improve the functions of the human digestive system in some positive ways. This particular benefaction is called probiotic effect. Previously, in 1900, Metchnikoff pointed out the use of fermented milk in the diet for stopping of particular diseases of the gastrointestinal tract and the progression of healthy day to day life. Since then some studies have now revealed that the fermented food products do have a positive effect on health status in multiple ways. The human intestinal microbial

Table 3. World-wide fermented foods products

Country	Products
Indian sub-continent	Acar, Achar, Tandal achar, Garam nimboo achar Pickled fruit and vegetables Gundruk Fermented, dried vegetable Lemon pickle, Lime pickle, Mango pickle
Europe and World	Mushrooms, Yeast Moulds Olives, Sauerkohl, Sauerruben Pickled fruit and vegetables Grape vinegar, Vinegar,Wines, Citron Fermented fruits
Americas	Cucumber pickles, Dill pickles, Olives,Sauerkraut Pickled fruit and vegetables Lupin seed, Oilseeds Pickled oilseed Vanilla, Wines Fermented fruit and vegetable
Africa	Fruit vinegar Vinegar Hot pepper sauce Lamoun makbouss, Mauoloh, Msir, Mslalla, Olive Pickled fruit and vegetables, Oilseeds, Ogili, Ogiri, Hibiscus seed Fermented fruit, vegetable seeds, Wines Fermented fruits
East Asia	Bossam-kimchi, Chonggak-kimchi, Danmoogi, Dongchimi, Kachdoo kigactuki,Kakduggi, Kimchi, Mootsanji, Muchungkimchi, Oigee, Oiji, Oiso baegi, Tongbaechukimchi, Tongkimchi, Totkal kimchi. Cha-ts'ai, Hiroshimana, Jangagee, Nara, senkei, Narazuke, Nozawana, Nukamisozuke, Omizuke, Pow tsai, Red in snow, Seokbakji, Shiozuke, Szechwan cabbage, Tai-tan tsoi, Takana, Takuan, Tsa Tzai, Tsu, Umeboshi, Wasabi-zuke, Yen tsai
South East Asia	Bai-ming, Leppet-so, Miang Fermented tea leaves, Nata de coco, Nata de pina Fermented fruit juice, Asinan, Burong mangga, Dalok, Jeruk, Kiam-chai, Kiam-cheyi, Kong-chai, Naw-maidong,Pak-siam-dong, Paw-tsay, Phak-dong, Phonlami-dong, Sajur asin, Sambal tempojak,Santol, Si-sek-chai, Sunki, Tang-chai, Tempoyak, Vanilla,

flora is supposed to weigh about 1000 grams and may contain 1016 – 1017 Colony forming units are expressing more than 500 strains. For physiological desires, it can be estimated to be a specialised organ of the body with a broad variety of functions in nutrition, immunology and metabolism (Gustafsson, 1983).

Flatulence Reducing Effect

In the fermentation of the beans for the development of tempe, the trypsin inhibitor is inactivated, and the amount of several oligosaccharides which generally induce flatulence are significantly diminished (Hesseltine, 1983). When *Lactobacillus* innoculated bean flour fermented with 20% moisture content, revealed a decline of the stachyose content (Duszkiewicz-Reinhard et al., 1994).

Anticholesterolemic Effect

Hepner stated the hypercholesteremic effect of yoghurt in human subjects suffering a one-week dietary supplement (Hepner et al., 1979). Investigation on supplementation of infant formula with Lb. acidophilus determined that the serum cholesterol in infants was decreased from 147 mg/ml to 119mg/100 ml (Harrison & Peat, 1975). In in vitro study, the ability of 23 strains of lactic acid bacteria isolated from different fermented milk products, and binding of bacterial cell to cholesterol was investigated. No cholesterol was detected inside the cells (Taranto et al., 1997). Poppel and Schafsma have also reported the ability of yoghurt to reduce the cholesterol in the serum by controlled human examinations (Poppel and Schafsma, 1996). The probable role of lactic acid bacteria in lowering amount cholesterol and various mechanisms by which it may be possible has been reviewed by Haberer (Haberer et al., 1997). Brigidi have cloned a gene encoding cholesterol oxidase from Streptomyces lividans into *Bacillus*, *Lactobacillus* and *E. coli* (Brigidi et al., 1993). Impact on transit time, bowel function and glycemic index: The transit time for 50% (t50) of the gastric content was significantly decreased for normal unfermented milk (42 ±10 min) in association with a fermented milk product indigenous to Sweden called "långfil" or ropy milk (62±14 min). In another study, (Wilhelm, 1993) states an increment in transport time and recovered bowel function in patients with habitual constipation.

Enhance Nutrient Absorption

As per various reports, the fermented foods help in the digested food transformation into their most bioavailable form. The bioavailable nutrients are able to easily and quickly absorbed from the GIT and perform many functions. The calcium bioavailability is highly influenced by the presence of an acid which produced by *lactobacillus* bacteria which helps into maintain the bone density of the post-menopause women (Parvez et al., 2006).

Anticancerogenic Effect

The fermented foods were exhibits anticarcinogenic effects. *Lactobacilli* plays an important role in diminishing or removing carcinogens and procarcinogens in the alimentary canal (Mital & Garg, 1995). The enzymes b-glucuronidase, nitroreductase and azoreductase, which are present in the intestinal canal, are perceived to convert procarcinogens to carcinogens (Goldin & Gorbach, 1984). Oral administration of *Lb rhamnosus GG* was confirmed to reduce the faecal concentration of b-glucuronidase in humans implying a reduction in the reformation of procarcinogens to carcinogens. Fermented milk having *Lactobacillus acidophilus* fed together with fried meat patties significantly reduced the excretion of mutagenic substances related to ordinary fermented milk with *Lactococcus* given together with fried meat patties (Lidbeck et al., 1992). The food fermentation processes are also stated to lessen the mutagenicity of foods by degrading the mutagenic substances while the fermentation process.

Immunoactive Effects

In fermented milk products, some lactic acid bacteria are present, which are observed to play an essential role in the immune system of the host after colonisation in the gut (De Simone, 1986). Oral administration of *Lactobacillus casei* induced enhancement of the function of the peritoneal macrophages and raised the generation of IgA (Sato et al., 1988). The mechanism behind this effect is not precisely understood, but it is considered that the *Lactobacillus*, their enzymes or the metabolic products being in the fermented food product may function as antigens, stimulating generation of antibodies.

Marin have examined the impact of *Lactobacillus* employed in fermented dairy products on the production of cytokines by macrophages. The results showed that for most strains, direct synergy with macrophages began a concentration-dependent raise in tumour necrosis factor and interleukin (Marin et al., 1997). A study by perdigon determined that the *Lactobacillus casei* could inhibit enteric infections and excite the IgA in undernourished animals but also translocate in to the bacteria, while yoghurt could prevent germination of intestinal carcinoma via the enhanced activity of IgA, T cells and macrophages (Perdigon et al., 1995). Marteau & Rambaud resolved that there is a potential of applying lactic acid bacteria for therapy and immunomodulation in mucosal diseases, particularly in the gastrointestinal tract (Marteau and Rambaud, 1993).

Gastrointestinal Health

Fermented foods can promote the function of the gastrointestinal system. This function enhancement is because of the capacity of fermented foods to raise the number of essential bacteria in the gut. These bacteria are capable of digesting food, fight off deadly bacteria, and reduce indications of diarrhoea and constipation (Rolfe, 2000).

Integrative Medicine

Integrative medicine is a holistic strategy to healthcare, which considered the whole person, including body and mind in the restorative (healing) process. The significant element of integrative medicine is gut health and its association with diseases. Registered Dietitians working with an integrative medicine team usually inspire consuming probiotics, preferably through food, to reintroduce essential bacteria and thus promote optimal digestive function (Parvez et al., 2006). The relationship between fermented foods and health can be determined back to both ancient Rome and China.

Prebiotics

Prebiotics are not bacteria themselves, but natural, non-digestible food ingredients that essential bacteria can utilise as an energy source. Prebiotics thus promote digestive health and also may improve calcium absorption. Sources of prebiotics in foods include whole wheat foods, bananas and vegetables such as leeks, artichokes and asparagus (Rolfe, 2000).

Effect of Fermentation on Pathogenic Organisms

A group of children served with lactic acid fermented gruel for nine months. They had a mean number of 2,1 diarrhoea episodes while the group of children which fed with unfermented gruel had 3,5 diarrhoea episodes (Lorri and Svanberg, 1994). Although *Shigella, Salmonella, Campylobacter, Vibrio, Escherichia* and *Yersinia* are the most frequent organisms connected with bacterial diarrhoea, other enterotoxigenic genera, including *Pseudomonas, Klebsiella, Enterobacter, Serratia, Aeromonas, Proteus, Providencia, Flavobacterium* and *Achromobacter*, have also been described (Nout et al., 1989). In addition, it was noticed that there was no notable difference between the function of the pathogens in fermented porridge or nonfermented porridge (acid-supplemented), which indicates that the anti-microbial effect is because of presence of lactic and acetic acids at lowered pH and that other anti-microbial materials do not perform a detectable role (Nout et al., 1989).

Benefits of Fermentations

The fermented foods are beneficial to the human health by various different ways which listed in table 5.

Fermented Food Production

Generally, there are two types of fermentation processes: solid state and submerged fermentation.

Solid State Process

In solid state fermentation (SSF), the fermentation use solids as a support in the absence water. But, some minimum amount of moisture is needed in the solid support for the micro-organism growth and their metabolism. SSF process provides prospects in the processing and utilization of agro-industrial solid waste. The process

Table 4. Worldwide fermented beverages products

Name of Distilled Beverage	Name of Fermented	Source
Scotch whisky	Beer, ale	Barley
Rye whisky	Rye beer	Rye
Bourbon whiskey	Corn beer	Corn
Wheat whisky, Korn	Wheat beer	Wheat
Shochu (Japan), soja (Korea)	Sake sonti	Rice
Brandy, cognac (France), Branntwein (Germany), pisco (Peru/Chile)	Wine (most commonly thought of from grapes)	Juice of fruits, other than apples or pears
Slivovitz, tzuica, palinca	Plum wine	Juice of plums
Pear brandy	Perry, or pear cider	Juice of pears
Tequila, mezcal	Pulque	Juice of agave
Grappa (Italy), Trester (Germany), marc (France)	Pomace wine	Pomace
Applejack (or apple brandy), Calvados	Cider, Apfelwein	Juice of apples
Brandy, cognac (France), Branntwein (Germany), pisco (Peru/Chile)	Wine (most commonly thought of from grapes)	Juice of fruits, other than apples or pears
Rum, cachaça, aguardiente, Guaro	Basi, betsa-betsa (regional)	Juice of sugar Cane, or molasses
Distilled mead ("mead brandy" or "honey brandy")	Mead	Honey
Vodka: potato mostly used in Ukraine, otherwise grain	Potato beer	Potato and/or grain
Araka	Kumis	Milk

is eco-friendly (low water utilization), low in energy consumption and also solve the issue of solid waste disposal.

SSF can be a potential technology in the generation of microbial-fermentation based products such as fermented foods, bioethanol (fuels), pharmaceuticals, and chemical products. Nowadays, commercial scale design and development of fermentation reactors is possible with the help of mathematical modelling and biochemical engineering. Currently solid state fermentation is design and developed for the bioprocesses such as biological detoxification of agro-industry based waste, biodegradation and bioremediation of hazardous waste, biopulping, biological nutritional enhancement, generation of value added products, e.g., secondary metabolites (enzymes, alkaloids, antibiotics, plant growth promotors), biosurfactants, biopesticides, and bioethanol (biofuel) (Pandey, 2003).

Table 5. Benefits of fermented foods

Benefits of Fermentation	
Preservation	➢ Enrich and preserves food
Manufacture of crucial nutrients and feed additives	➢ Antioxidants, vitamins, emulsifiers, pigments, trace elements, zeolite, minerals, salt, chelates, sterols, nucleotides, amino acids, feeding attractants, prebiotics, enzymes, organic acids, gut modifiers, immune enhancers, anti-viral, anti-parasitical, anti-fungal, essential oils, growth promotors, antibiotics, binders, hormones.
Improvement of safety	➢ Acid production ➢ Removal of toxic compounds and harmful bacteria ➢ Acid and alcohol production
Agriculture based food industry	➢ Mushroom production ➢ Food additives ➢ Conventional food fermentations ➢ Bio conservation by-products
Improvement of nutritional value/ Health benefits	➢ Improved digestibility and bowl health ➢ Improved intestinal function ➢ Retention of Micronutrients ➢ Increased fiber content ➢ Synthesis of probiotic compounds ➢ ➢ Fermentation introduces helpful bacteria (probiotics) which balance our natural bacterial colonies. ➢ Improved Ca absorption and help body for more absorption of the live nutrients in our food. ➢ Improves overall immunity.
Industrial fermentation	➢ Antibiotic production ➢ Fungal metabolism ➢ Enzyme production ➢ Organic acid production ➢ Alcohol (Ethanol) production
Enhancements of flavor	➢ Enhances flavor and taste of food
Agriculture	➢ Plant growth hormones ➢ Biocontrol ➢ Bio insecticide

Submerged Fermentation Process

Submerged Fermentation (SmF) or Liquid Fermentation (LF) utilizes liquid as media of fermentation. In most of the industrial scale fermentation processes, the submerged fermentation process is adopted. In SmF, the microorganisms are spread in the liquid nutrient media with the help of bioreactor. The bioreactor is either a stirred or non-stirred vessel which can be run in continuous or batch mode. SmF is the most suitable technique for bacterial culture. End product purification is easy with Smf technique compare to others. In early days, it was employed for the isolation of secondary metabolites (Subramaniyam et al., 2012).

Global Initiatives for Waste Food: Treatment and Utilization

In coming decades, chemicals and fuel demands will be increase rapidly. So, there is a need of green and sustainable approaches for chemical sector. Today process developers are searching for the innovative utilization of waste products. Waste food products can be a potential source as a starting raw material for the generation of chemicals and fuels. In current scenario, the quantity of food waste produced worldwide is increases day by day. Mostly they are transported to land filling. Some different ways are also employed such as: advanced valorization, and composting. But, very minute quantity is employed for that (Lin et al., 2013 and 2014).

Current Food Waste Management Practices for Supply Chain Food Waste: composting, animal feeding, landfilling and incineration. Animal feeding is one of the easy routes for supply chain food waste. But, it was restricted by the regulatory. Land spreading and composting of food waste is environmentally acceptable, because it will satisfy the farmer's necessity of fertilizers (Lin et al., 2013).

Waste Framework Directive and United Kingdom on Waste Management: In waste frame work directive, the valorization of any type of waste is not clarified properly. However the main aim of Directive's policy is to convert waste in to energy and valuable products. On 18th May 2010, European commission decided that anaerobic digestion (AD) and composting of food waste are the most capable steps in bio-waste management. Presently, major part of waste food in the UK is composted, landfilled, or incinerated. Anaerobic digestion is not widely accepted at commercial scale. In UK, there are only three AD plants working currently for the treatment of food waste from retailers, caterers, and producers. The commercial level anaerobic digestion units will produces power, heat and fertilizers (Lin et al., 2013).

Management of Citrus Peel Waste: Grape fruits, oranges, lime, and lemons are the most common citrus fruits. These are generally cultivated in India, Brazil, Japan, US, Spain, South Africa, Italy, China, and Turkey. Around 31.2 million tons of citrus fruits are processed per year. Out of that, 15.6 million metric tons waste was generated (Finley 2013; Lin et al., 2013).

Various approaches for valorization of citrus peel waste (Lin et al., 2013):

- Production of pectic (Angel Siles Lopez et al., 2010)
- Production of biogas (Methane) (Aslanzadeh and Ozmen, 2009)
- Isolation of pectin by acid hydrolysis (Ma et al., 1993)
- Activated carbon production (Ma et al., 1993)
- Isolation of dietary fibers (Marin et al., 2007)
- Generation of bioethanol (Angel Siles Lopez et al., 2010)

Food Waste Treatment in Spain: It is estimated that around 15 million tones food waste per annum are produced in Spain. Composting of food waste is the main approach which was adopted by the Spain. Domestic and restaurant based waste oils are the main food waste component in Spain. Spain, Italy and Greece are generating more than 30% of olive generation worldwide. The trans-esterification technology is employed for the generation of value added products from waste oils (Lin et al., 2013).

Strategies for Food Waste Treatment in the UK: Very large quantity of grease and animal fat are generated in Europe. They are conventionally used for manufacturing of cosmetic precursor, and supplement in pet food. Nowadays, fats based biofuel production is find a great importance in UK and European Union. UK based Brocklesby Ltd company has invented a procedure for the recovers oil and fats from various waste foods (Dobbs et al., 2011; Lin et al., 2013).

Food Waste Utilization in Greece: Greece based food industries are mainly involved in the production of dairy products, meat, sugar based foods, and wine-beers type of beverages. In Greece, anaerobic digestion (AD) is the most promising technology for the transformation of waste organic matter into biogas. AD is not adopted for the energy generation, but, its byproduct (digestate) is employed as a soil conditioner (Lin et al., 2013).

Food Waste Crisis in Hong Kong: With an average population density of around 6480/km^2 in 2009, The Hong Kong is the most dense places in the world. Due to this compactness, they are facing a critical waste management issues. Around 3237 tons of food waste is generated on daily basis. In early days, landfilling and composting of food waste was practiced. But, nowadays the development of conversion process and integrated management is aimed for the food waste management (Lin et al., 2013).

CONCLUSION

The fermentation technology provides a potential tool to reduce the wastage of foods and utilization of wastage foods to form value added products from the waste foods. The advancement of this technology cuts potentially cuts the wastage of the food material. The several traditional foods and beverages in the world are formed by fermentation process including either bacteria including or yeasts (fungus) or combinations of both. The Lactic acid fermentation plays a significant and predominant role in the production of traditional foods and beverages. The fermentation of the dairy products such as curd, yoghurt has been widely studied. Over the last century, and the production processes involved have been highly standardized and industrialized.

Numerous of these products are performed either following natural, spontaneous fermentations or by employing the back-slopping method. While the back-slopping procedure results in a higher initial number of microorganisms that are beneficial to the fermentation that would be present in the raw material itself- being introduced, thereby ensuring a faster fermentation than the spontaneous fermentation, nonetheless, given that the specific microflora involved will vary markedly from area to area and even between families within the similar geographical regions. Consequently, a very high degree of variability exists between each of these fermentations. This coupled with the different processing parameters which are also being employed between the different fermentation regimes, and geographical region makes the production of a uniform product(s) extremely difficult to achieve. For this intention, further research is immediately needed to determine the microbiology and biochemistry of these traditional foods before the major microorganisms in these fermentations can be isolated and subsequently characterized for use in the development of starter cultures for the production of these foods in a consistent manner, both from a quality- and food-safety perspective. With increased industrial development and development, there is a need for larger-scale manufacture of fermented foods and beverages of frequently high quality. The successful transfer of the production of traditional fermented products from the domestic level to the industrial scale production level requires several steps. These include the isolation and identification of microorganisms associated with fermentation and the determination of their precise role in the fermentation process, using both standards- and molecular-based microbiological approaches. Improvement in process controls for the manufacture of fermented foods is also required, as is an advance in the overall quality from a microbiological standpoint of the raw materials used in the production of these fermented foods and in processes which simulate those currently employed in the traditional production schemes.

ACKNOWLEDGMENT

Dr P S Thakur thanks to Department of Zoology S. A. Sci. College, RTM University Nagpur, for providing research facility, and Dixit V. Bhalani thanks DST for providing DST-INSPIRE Fellowship and Arvind K. Singh Chandel thanks, CSIR, India for providing a research fellowship (CSIR-JRF). DV and AK also thanks to AcSIR, CSIR-CSMCRI for providing research facility.

REFERENCES

Adams, M. R. (1990). Topical aspects of fermented foods. *Trends in Food Science & Technology*, *1*, 140–144. doi:10.1016/0924-2244(90)90111-B

Afrc, R. F. (1989). Probiotics in man and animals. *The Journal of Applied Bacteriology*, *66*(5), 365–378. doi:10.1111/j.1365-2672.1989.tb05105.x PMID:2666378

Agte, V. V., Gokhale, M. K., & Chiplonkar, S. A. (1997). Effect of natural fermentation on in vitro zinc bioavailability in cereal–legume mixtures. *International Journal of Food Science & Technology*, *32*(1), 29–32. doi:10.1046/j.1365-2621.1997.00372.x

Ángel Siles López, J., Li, Q., & Thompson, I. P. (2010). Biorefinery of waste orange peel. *Critical Reviews in Biotechnology*, *30*(1), 63–69. doi:10.3109/07388550903425201 PMID:20148755

Aryana, K. J., & Olson, D. W. (2017). A 100-Year Review: Yogurt and other cultured dairy products. *Journal of Dairy Science*, *100*(12), 9987–10013. doi:10.3168/jds.2017-12981 PMID:29153184

Aslanzadeh, S., & Özmen, P. (2009). *Biogas production from municipal waste mixed with different portions of orange peel*. Academic Press.

Babu, P. D., Bhakyaraj, R., & Vidhyalakshmi, R. (2009). A low cost nutritious food “tempeh”-a review. *World J Dairy Food Sci*, *4*(1), 22–27.

Boye, J., Zare, F., & Pletch, A. (2010). Pulse proteins: Processing, characterization, functional properties and applications in food and feed. *Food Research International*, *43*(2), 414–431. doi:10.1016/j.foodres.2009.09.003

Bragg, P. C., Patricia Bragg, N. D., & Bragg, P. C. (2003). *Apple Cider Vinegar Miracle Health System*. Health Science Publications, Inc.

Brigidi, P., Bolognani, F., Rossi, M., Cerre, C., & Matteuzzi, D. (1993). Cloning of the gene for cholesterol oxidase in Bacillus spp., Lactobacillus reuteri and its expression in Escherichia coli. *Letters in Applied Microbiology*, *17*(2), 61–64. doi:10.1111/j.1472-765X.1993.tb00371.x PMID:7763933

Campbell-Platt, G. (1987). *Fermented foods of the world. A dictionary and guide*. Butterworths.

Caplice, E., & Fitzgerald, G. F. (1999). Food fermentations: Role of microorganisms in food production and preservation. *International Journal of Food Microbiology*, *50*(1-2), 131–149. doi:10.1016/S0168-1605(99)00082-3 PMID:10488849

Chavan, J. K., Kadam, S. S., & Beuchat, L. R. (1989). Nutritional improvement of cereals by fermentation. *Critical Reviews in Food Science and Nutrition*, *28*(5), 349–400. doi:10.1080/10408398909527507 PMID:2692608

Chaves-López, C., Serio, A., Paparella, A., Martuscelli, M., Corsetti, A., Tofalo, R., & Suzzi, G. (2014). Impact of microbial cultures on proteolysis and release of bioactive peptides in fermented milk. *Food Microbiology*, *42*, 117–121. doi:10.1016/j.fm.2014.03.005 PMID:24929726

Chompreeda, P. T., & Fields, M. L. (1984). Effects of heat and fermentation on the extractability of minerals from soybean meal and corn meal blends. *Journal of Food Science*, *49*(2), 566–568. doi:10.1111/j.1365-2621.1984.tb12469.x

Chou, L. (2003, Fall). Chinese and Other Asian Pickles. *Flavor and Fortune*.

Cleveland, J., Montville, T. J., Nes, I. F., & Chikindas, M. L. (2001). Bacteriocins: Safe, natural antimicrobials for food preservation. *International Journal of Food Microbiology*, *71*(1), 1–20. doi:10.1016/S0168-1605(01)00560-8 PMID:11764886

Daubresse, P., Ntibashirwa, S., Gheysen, A., & Meyer, J. A. (1987). A process for protein enrichment of cassava by solid substrate fermentation in rural conditions. *Biotechnology and Bioengineering*, *29*(8), 962–968. doi:10.1002/bit.260290807 PMID:18576545

De Simone, C. (1986). Microflora yogurt and the immune system. *International Journal of Immunotherapy*, 19–23.

Dirar, H. A. (1993). *The indigenous fermented foods of the Sudan: a study in African food and nutrition*. CAB international.

Dobbs, R., Oppenheim, J., Thompson, F., Brinkman, M., & Zornes, M. (2011). *Resource Revolution: Meeting the world's energy, materials, food, and water needs*. Academic Press.

Duszkiewicz-Reinhard, W., Gujska, E., & Khan, K. (1994). Reduction of stachyose in legume flours by lactic acid bacteria. *Journal of Food Science*, *59*(1), 115–117. doi:10.1111/j.1365-2621.1994.tb06911.x

Eka, O. U. (1980). Effect of fermentation on the nutrient status of locust beans. *Food Chemistry*, *5*(4), 303–308. doi:10.1016/0308-8146(80)90051-5

Fankhauser, D. B. (2007). Fankhauser's Cheese Page. *Taken To, 9*(23), 2007.

Friedman, M. (1996). Nutritional value of proteins from different food sources. A review. *Journal of Agricultural and Food Chemistry*, *44*(1), 6–29. doi:10.1021/jf9400167

Gerardi, M. H. (2003). *The microbiology of anaerobic digesters*. John Wiley & Sons. doi:10.1002/0471468967

Goldin, B. R. (1983). The effect of oral administration of Lactobacillus and antibiotics on intestinal bacterial activity and chemical induction of large bowel tumors. *Developments in Industrial Microbiology*, *25*, 139–150.

Graf, W. (1983). Studies on the therapeutic properties of acidophilus milk. In *Symposia of the Swedish Nutrition Foundation*. Almqvist och Wiksell International.

Guermani, L., Villaume, C., Bau, H. W., Chandrasiri, V., Nicolas, J. P., & Mejan, L. (1992). *Composition and nutritional value of okara fermented by Rhizopus oligosporus*. Sciences des Aliments.

Gupta, U., Rudramma, Rati, E. R., & Joseph, R. (1998). Nutritional quality of lactic fermented bitter gourd and fenugreek leaves. *International Journal of Food Sciences and Nutrition*, *49*(2), 101–108. doi:10.3109/09637489809089389 PMID:9713580

Gustafsson, B. E. (1983). Introduction to the ecology of the intestinal microflora and its general characteristics. *Nutrition and the Intestinal Flora*, 11-16.

Haberer, P., Holzapfel, W. H., & Wagner, H. (1997). Moegliche Rolle von Milchsaeurebakterien bei der Cholesterinsenkung im Blutserum. *Mitteilungsblatt-Bundesanstalt Fur Fleischforschung Kulmbach*, *1*(136), 202–207.

Harland, B. F., & Harland, J. (1980). Fermentative reduction of phytate in rye, white and whole wheat breads. *Cereal Chemistry*, *57*(3), 226–229.

Harrison, V. C., & Peat, G. (1975). Serum cholesterol and bowel flora in the newborn. *The American Journal of Clinical Nutrition*, *28*(12), 1351–1355. doi:10.1093/ajcn/28.12.1351 PMID:45573

Hepner, G., Fried, R., St Jeor, S., Fusetti, L., & Morin, R. (1979). Hypocholesterolemic effect of yogurt and milk. *The American Journal of Clinical Nutrition*, *32*(1), 19–24. doi:10.1093/ajcn/32.1.19 PMID:581636

Hesseltine, C. W., & Wang, H. L. (1980). The importance of traditional fermented foods. *Bioscience*, *30*(6), 402–404. doi:10.2307/1308003

Hutkins, R. W. (2008). *Microbiology and technology of fermented foods* (Vol. 22). John Wiley & Sons.

Keuth, S., & Bisping, B. (1994). Vitamin B12 production by Citrobacter freundii or Klebsiella pneumoniae during tempeh fermentation and proof of enterotoxin absence by PCR. *Applied and Environmental Microbiology, 60*(5), 1495–1499. PMID:8017933

Khetarpaul, N., & Chauhan, B. M. (1990). Fermentation of pearl millet flour with yeasts and lactobacilli: In vitro digestibility and utilisation of fermented flour for weaning mixtures. *Plant Foods for Human Nutrition (Dordrecht, Netherlands), 40*(3), 167–173. doi:10.1007/BF01104139 PMID:2217082

Khetarpaul, N., & Chauhan, B. M. (1991). Biological utilisation of pearl millet fluor fermented with yeasts and lactobacilli. *Plant Foods for Human Nutrition (Dordrecht, Netherlands), 41*(4), 309–319. doi:10.1007/BF02310625 PMID:1796088

Khetarpaul, N., & Chauhan, B. M. (1991). Sequential fermentation of pearl millet by yeasts and lactobacilli—effect on the antinutrients andin vitro digestibility. *Plant Foods for Human Nutrition (Dordrecht, Netherlands), 41*(4), 321–327. doi:10.1007/BF02310626 PMID:1796089

Kneifel, W., & Mayer, H. K. (1991). Vitamin profiles of kefirs made from milks of different species. *International Journal of Food Science & Technology, 26*(4), 423–428. doi:10.1111/j.1365-2621.1991.tb01985.x

Kuboye, A. O. (1985, October). Traditional fermented foods and beverages of Nigeria. In *Proceedings of the IFA/UNU workshop on the development of indigenous fermented foods and food technology in Africa, Douala, Cameroon. Stockholm: International Foundation for Science* (pp. 224-36). Academic Press.

Lay, M. M. G., & Fields, M. L. (1981). Nutritive value of germinated corn and corn fermented after germination. *Journal of Food Science, 46*(4), 1069–1073. doi:10.1111/j.1365-2621.1981.tb02993.x

Leroy, F., & De Vuyst, L. (2004). Lactic acid bacteria as functional starter cultures for the food fermentation industry. *Trends in Food Science & Technology, 15*(2), 67–78. doi:10.1016/j.tifs.2003.09.004

Lidbeck, A., Övervik, E., Rafter, J., Nord, C. E., & Gustafsson, J. Å. (1992). Effect of Lactobacillus acidophilus supplements on mutagen excretion in faeces and urine in humans. *Microbial Ecology in Health and Disease, 5*(1), 59–67. doi:10.3109/08910609209141305

Liener, I. E. (1994). Implications of antinutritional components in soybean foods. *Critical Reviews in Food Science and Nutrition, 34*(1), 31–67. doi:10.1080/10408399409527649 PMID:8142044

Lin, C. S. K., Koutinas, A. A., Stamatelatou, K., Mubofu, E. B., Matharu, A. S., Kopsahelis, N., & Luque, R. (2014). Current and future trends in food waste valorization for the production of chemicals, materials and fuels: A global perspective. *Biofuels, Bioproducts & Biorefining*, *8*(5), 686–715. doi:10.1002/bbb.1506

Lin, C. S. K., Pfaltzgraff, L. A., Herrero-Davila, L., Mubofu, E. B., Abderrahim, S., Clark, J. H., & Thankappan, S. (2013). Food waste as a valuable resource for the production of chemicals, materials and fuels. Current situation and global perspective. *Energy & Environmental Science*, *6*(2), 426–464. doi:10.1039/c2ee23440h

Lorri, W., & Svanberg, U. (1994). Lower prevalence of diarrhoea in young children fed lactic acid-fermented cereal gruels. *Food and Nutrition Bulletin-United Nations University*, *15*, 57–57.

Ma, E., Cervera, Q., & Sánchez, G. M. M. (1993). Integrated utilization of orange peel. *Bioresource Technology*, *44*(1), 61–63. doi:10.1016/0960-8524(93)90209-T

Marín, F. R., Soler-Rivas, C., Benavente-García, O., Castillo, J., & Pérez-Alvarez, J. A. (2007). By-products from different citrus processes as a source of customized functional fibres. *Food Chemistry*, *100*(2), 736–741. doi:10.1016/j.foodchem.2005.04.040

Marin, M. L., Tejada-Simon, M. V., Murtha, J., Ustunol, Z., & Pestka, J. J. (1997). Effects of Lactobacillus spp. on cytokine production by RAW 264.7 macrophage and EL-4 thymoma cell lines. *Journal of Food Protection*, *60*(11), 1364–1370. doi:10.4315/0362-028X-60.11.1364

Marteau, P., & Rambaud, J. C. (1993). Potential of using lactic acid bacteria for therapy and immunomodulation in man. *FEMS Microbiology Reviews*, *12*(1-3), 207–220. doi:10.1111/j.1574-6976.1993.tb00019.x PMID:8398215

Mitall, B. K., & Garg, S. K. (1995). Anticarcinogenic, hypocholesterolemic, and antagonistic activities of Lactobacillus acidophilus. *Critical Reviews in Microbiology*, *21*(3), 175–214. doi:10.3109/10408419509113540 PMID:8845062

Mittal, B. K., Prasad, R., & Singh, S. (1977). Effect of carbohydrates and phosphates on acid production by lactic acid bacteria in soy milk [India]. Research note. *Journal of Food Science and Technology*.

Moslehuddin, A. B. M., & Hang, Y. D. (1987). Effect of processing methods on the nutritional value of Lathyrus sativus seeds. Nutrition Reports International (USA).

Murdock, F. A., & Fields, M. L. (1984). B-Vitamin Content of Natural Lactic Acid Fermented Cornmeal. *Journal of Food Science*, *49*(2), 373–375. doi:10.1111/j.1365-2621.1984.tb12425.x

Nagpal, R., Behare, P. V., Kumar, M., Mohania, D., Yadav, M., Jain, S., ... Henry, C. J. K. (2012). Milk, milk products, and disease free health: An updated overview. *Critical Reviews in Food Science and Nutrition*, *52*(4), 321–333. doi:10.1080/10408398.2010.500231 PMID:22332596

Nnam, N. M. (1995). Evaluation of nutritional quality of fermented cowpea (Vigna unguiculata) flours. *Ecology of Food and Nutrition*, *33*(4), 273–279. doi:10.1080/03670244.1995.9991435

Nout, M. J. R., Rombouts, F. M., & Hautvast, J. G. A. J. (1989). Accelerated natural lactic fermentation of infant food formulations. *Food and Nutrition Bulletin*, *11*(1), 65–73. doi:10.1177/156482658901100102

Nout, M. R. (2003). Traditional fermented products from Africa, Latin America and Asia. In *Yeasts in food-beneficial and detrimental aspects* (pp. 451–473). Behr's Verlag. doi:10.1533/9781845698485.451

Nout, M. R., Rombouts, F. M., & Havelaar, A. (1989). Effect of accelerated natural lactic fermentation of infant good ingredients on some pathogenic microorganisms. *International Journal of Food Microbiology*, *8*(4), 351–361. doi:10.1016/0168-1605(89)90006-8 PMID:2701696

Obizoba, I. C., & Atii, J. V. (1991). Effect of soaking, sprouting, fermentation and cooking on nutrient composition and some anti-nutritional factors of sorghum (Guinesia) seeds. *Plant Foods for Human Nutrition (Dordrecht, Netherlands)*, *41*(3), 203–212. doi:10.1007/BF02196388 PMID:1924184

Obizoba, I. C., & Egbuna, H. I. (1992). Effect of germination and fermentation on the nutritional quality of bambara nut (Voandzeia subterranea L. Thouars) and its product (milk). *Plant Foods for Human Nutrition (Dordrecht, Netherlands)*, *42*(1), 13–23. doi:10.1007/BF02196068 PMID:1546053

Odunfa, S. A., & Komolafe, O. B. (1989). Nutritional characteristics of staphylococcus species from fermenting african locust bean (parkia-biglobosa). *Nahrung-food*, *33*(7), 607–615.

Padmaja, G., George, M., Moorthy, S. N., Bainbridge, Z., Plumb, V., Wood, J. F., & Powell, C. J. (1994). Nutritional evaluation of the starchy flour obtained from cassava tubers on fermentation with a mixed-culture inoculum. *Journal of Agricultural and Food Chemistry*, *42*(3), 766–770. doi:10.1021/jf00039a033

Pandey, A. (2003). Solid-state fermentation. *Biochemical Engineering Journal*, *13*(2-3), 81–84. doi:10.1016/S1369-703X(02)00121-3

Pandey, V. N., & Srivastava, A. K. (1993). A simple, low energy requiring method of coagulating leaf proteins for food use. *Plant Foods for Human Nutrition (Dordrecht, Netherlands)*, *43*(3), 241–245. doi:10.1007/BF01886225 PMID:8506239

Parvez, S., Malik, K. A., Ah Kang, S., & Kim, H. Y. (2006). Probiotics and their fermented food products are beneficial for health. *Journal of Applied Microbiology*, *100*(6), 1171–1185. doi:10.1111/j.1365-2672.2006.02963.x PMID:16696665

Patel, A., Shah, N., & Prajapati, J. B. (2013). Biosynthesis of vitamins and enzymes in fermented foods by lactic acid bacteria and related genera-A promising approach. *Croatian Journal of Food Science and Technology*, *5*(2), 85–91.

Peñas, E., Martinez-Villaluenga, C., & Frias, J. (2016). Sauerkraut: production, composition, and health benefits. In Fermented Foods in Health and Disease Prevention (pp. 557-576). Academic Press.

Perdigon, G., Alvarez, S., Rachid, M., Agüero, G., & Gobbato, N. (1995). Immune system stimulation by probiotics. *Journal of Dairy Science*, *78*(7), 1597–1606. doi:10.3168/jds.S0022-0302(95)76784-4 PMID:7593855

Reed, G. (1981). Use of microbial cultures-yeast products. *Food Technology*, *35*(1), 89–94.

Rolfe, R. D. (2000). The role of probiotic cultures in the control of gastrointestinal health. *The Journal of Nutrition*, *130*(2), 396S–402S. doi:10.1093/jn/130.2.396S PMID:10721914

Roos, U., Caneva, P., Escher, F., & Amadò, R. (1990). Influence of fermentation on the phytic acid content of conventionally produced bread. *Processing and Quality of Foods: Food Biotechnology: Avenues to Healthy and Nutritious Products*, *2*, 105–110.

Saavedra, J. M., & Perman, J. A. (1989). Current concepts in lactose malabsorption and intolerance. *Annual Review of Nutrition*, *9*(1), 475–502. doi:10.1146/annurev.nu.09.070189.002355 PMID:2669882

Sato, K., Saito, H., Tomioka, H., & Yokokura, T. (1988). Enhancement of host resistance against Listeria infection by Lactobacillus casei. *Microbiology and Immunology*, *32*(12), 1189–1200. doi:10.1111/j.1348-0421.1988.tb01483.x PMID:2853287

Schoder, D., Melzner, D., Schmalwieser, A., Zangana, A., Winter, P., & Wagner, M. (2011). Important vectors for Listeria monocytogenes transmission at farm dairies manufacturing fresh sheep and goat cheese from raw milk. *Journal of Food Protection*, *74*(6), 919–924. doi:10.4315/0362-028X.JFP-10-534 PMID:21669068

Scrimshaw, N. S., & Murray, E. B. (1988). The acceptability of milk and milk products in populations with a high prevalence of lactose intolerance. *The American Journal of Clinical Nutrition*, *48*(4), 1142–1159. doi:10.1093/ajcn/48.4.1142 PMID:3140651

Shibasaki, K., & Hessbltine, C. W. (1962). Miso fermentation. *Economic Botany*, *16*(3), 180–195. doi:10.1007/BF02860037

Simango, C. (1997). Potential use of traditional fermented foods for weaning in Zimbabwe. *Social Science & Medicine*, *44*(7), 1065–1068. doi:10.1016/S0277-9536(96)00261-4 PMID:9089926

Stanbury, P. F., Whitaker, A., & Hall, S. J. (2013). *Principles of fermentation technology*. Elsevier.

Steinkraus, K. H. (1997). Classification of fermented foods: Worldwide review of household fermentation techniques. *Food Control*, *8*(5-6), 311–317. doi:10.1016/S0956-7135(97)00050-9

Subramaniyam, R., & Vimala, R. (2012). Solid state and submerged fermentation for the production of bioactive substances: A comparative study. *Int J Sci Nat*, *3*(3), 480–486.

Sumi, H., Hamada, H., Tsushima, H., Mihara, H., & Muraki, H. (1987). A novel fibrinolytic enzyme (nattokinase) in the vegetable cheese Natto; a typical and popular soybean food in the Japanese diet. *Experientia*, *43*(10), 1110–1111. doi:10.1007/BF01956052 PMID:3478223

Tamime, A. Y., & Robinson, R. K. (1999). *Yoghurt: science and technology*. Woodhead Publishing.

Taranto, M. P., Sesma, F., de Ruiz Holgado, A. P., & de Valdez, G. F. (1997). Bile salts hydrolase plays a key role on cholesterol removal by Lactobacillus reuteri. *Biotechnology Letters*, *19*(9), 845–847. doi:10.1023/A:1018373217429

Teoh, A. L., Heard, G., & Cox, J. (2004). Yeast ecology of Kombucha fermentation. *International Journal of Food Microbiology*, *95*(2), 119–126. doi:10.1016/j.ijfoodmicro.2003.12.020 PMID:15282124

Tomkins, A. M., Alnwick, D., & Haggerty, P. (1988). Fermented foods for improving child feeding in eastern and southern Africa: a review. In Improving young child feeding in eastern and southern Africa: household level food technology; proceedings of a workshop held in Nairobi, Kenya, 11-16 Oct. 1987. IDRC.

Van Poppel, G., & Schaafsma, G. (1996, November). Cholesterol lowering by a functional yoghurt. In *Fd. Ing. Europe Conf. Proc., Maarssen* (pp. 12-14). Academic Press.

Yokotsuka, K. (1982). *Chemical Compositions of Brandies*. Institute Winery of Yamanashi University.

Chapter 7
Microbiological Monitoring in the Biodegradation of Food Waste

Satish Chandra Pandey
Kumaun University, India

Anupam Pandey
Kumaun University, India

Tushar Joshi
Kumaun University, India

Veni Pande
Kumaun University, India

Diksha Sati
Kumaun University, India

Mukesh Samant
Kumaun University, India

ABSTRACT

Food wastage and its exponential growth emerged as a global environmental issue, and its improper disposal has become a threat to human health and environment. Deterioration of food wastes releases various greenhouse gases that increases global warming and produces large amounts of toxins and foul odors, such as NH3 and H2S. To reduce this burden there is an urgent need to take appropriate measures by adopting standard management strategies. Microorganisms play an important role in food waste recycling, which appears to be cost-effective and causes less harm to the environment. One such process is anaerobic digestion, which has appeared as one of the most promising and eco-friendly approaches for management that converts organic waste into various useful products. Another sustainable approach is composting. Compost generated by food waste improves soil health and regenerates healthier environment. Thus, through the use of microorganisms, the study paves the way for effective management of food waste in order to minimize potential human and environmental risks.

DOI: 10.4018/978-1-5225-7706-5.ch007

INTRODUCTION

In the most recent years, Food wastage and its disposal has appeared as a serious environmental problem, attracting the attention of scientist, activists and consumers. It is the second largest category of solid waste sent to landfills. To lead a healthy life, most of the people in the world don't even getting sufficient food for their daily requirement and are dying from hunger everyday than other deadly diseases like malaria, AIDS and tuberculosis. While on the other hand around one third of the food produced for human consumption in the world is being wasted everyday through different ways. Food wastage, holds both food waste and food loss, refer to the decline of food in later stages of the food supply chain intended for human consumption. It is not only immoral, but also associated with major economical loss and is severely affecting the world around us. Mainly the food loss occurs at the production level from household stream, as well as industrial and agricultural practices and the reasons are improper skills, poor practices, and lack of infrastructure which have been causing serious hazards for the environment. In India approximately 700 million tons of food waste is generated annually. This includes uneaten food, expired food, food preparation from restaurants, residences, industrial and institutional sources like school cafeteria and factory lunchrooms. The quantity of food waste generated worldwide in developed and developing countries is shown in figure 1. Proliferation of food waste leads to challenges for its safe disposal, with the waste being usually either burned or land filled (Nagavallemma KP et al., 2006). Out of which around 70% of waste holds compostable food items, causing major pollution (Bouallagui et al., 2004). Strategy for the mitigation and management of this type of waste is a major concern, causing increased food prices and the resources required and ultimately destroying environment. To reduce food loss and wastage, there should have food waste hierarchy (Figure 2) to (i) reduce food waste, (ii) redistribute it (e.g. to the homeless), (iii) reuse it as animal feed and compost, (iv) energy recovery through anaerobic digestion and at last, (v) dispose the remain. Composting and incineration are the process that produces greenhouse gases, while wastewater produced from anaerobic digestion causes acidification and eutrophication of local ecosystems. However, a huge population of naturally occurring microorganisms and effective microorganisms (EM) are there, with the potential to convert food waste into valuable outputs like plant nutrients, that help in reducing the C:N ratio to support soil productivity and also contribute in maintaining nutrient flows from one system to another and to minimize ecological imbalance (Novinscak, Filion, Surette, & A Allain, 2008; Umsakul, Dissara, & Srimuang, 2010). Thus waste management associated with the application of biodegradation properties of microorganisms are of great importance that can appropriately harness the process in a better way. Various studies reveals that effective microorganisms (EM) play prominent role in bioremediation, composting, agriculture, control household waste and can treat the leachate coming out from the garbage and remove the foul smell from decomposed food waste (Khaliq, Abbasi, & Hussain, 2006). The biological treatment of these wastes appears to be most cost-effective and causes less harm to the environment (Coker, 2006). This biological process of waste treatment is also known as composting. It is a self-heating, aerobic solid phase biodegradative process of organic materials under controlled conditions and has potential to enhance soil biochemical property (Giusquiani, Pagliai, Gigliotti, Businelli, & Benetti, 1995), biological activity (Pfotzer & Schüler, 1997), protect soil from erosion (P. Bazzoffi, S. Pellegrini, A. Rocchini, M. Morandi, & Grasselli, 1973) and protect plants from soil and seed borne pathogens (Schüler, Pikny, Nasir, & Vogtmann, 1993). Thus compost can be considered as soil conditioner that helps in enhancing crop yield. Another practice is anaerobic digestion (AD), a biological process that converts organic waste into several, potentially useful products, using microorganisms (Paritosh et al., 2017). Thus through utilizing metabolic versatility of microorganisms food waste can be easily treated but only 5-10% of microbial community play actual role of degradation for the target compound. So to enhance the power of microorganism for the breakdown of target compounds and to shorten the process, potential microbial consortium actively involved in the degradation of different components of food waste, under natural condition without producing foul odour should be made. The information contributes to the literatures on the efficient use of anaerobic digestion, composting methods, generation of energy (biogas), quality of compost and their possible effects on environment and human health.

Figure 1. Quantity of food waste generated around the world; Adapted from (Paritosh et al., 2017)

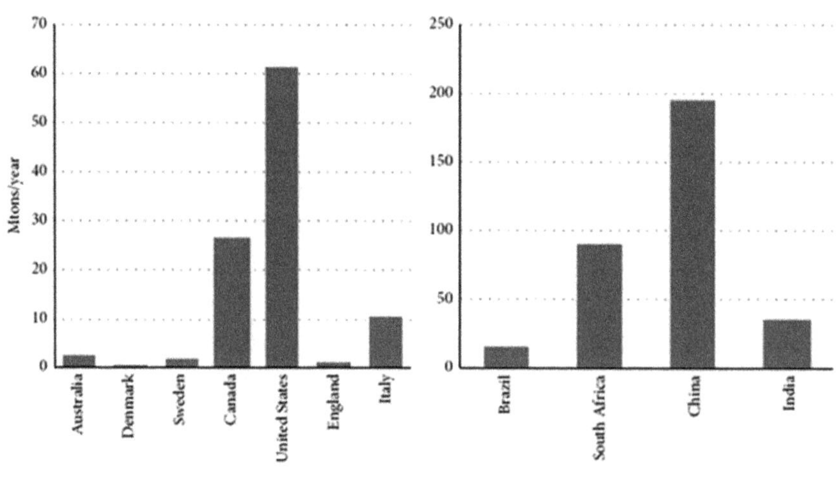

Figure 2. The food waste hierarchy. Adapted from (Ermgassen, Phalan, Green, & Balmford, 2016; Papargyropoulou, Lozano, Steinberger, Wright, & bin Ujang, 2014); AD = anaerobic digestion

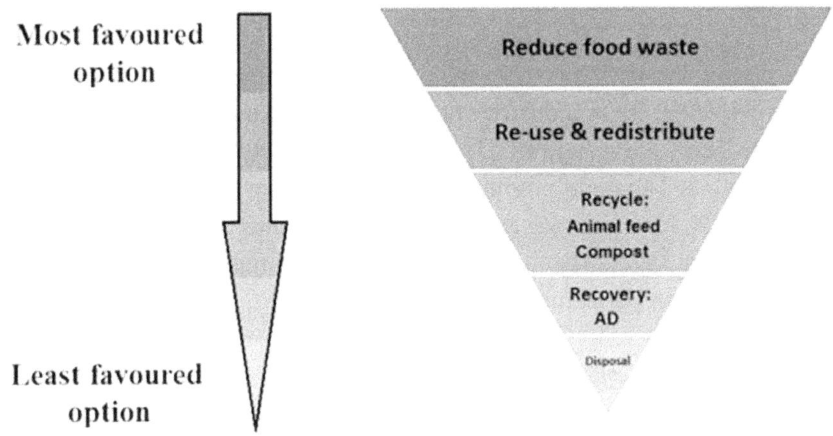

TYPES AND SOURCES OF FOOD WASTE

Every type of food waste (both precooked and leftover) is an organic waste that can be easily biodegraded by composting, aerobic and anaerobic digestion or similar processes into carbon dioxide, water, methane or simple organic molecules. According to FAO, around 1.3 billion tons of food items including fresh vegetables, fruits, and bakery, dairy and meat products are lost. The major composition of food waste is carbohydrates, proteins, fats and traces of inorganic compounds that vary in according to the type of food waste and its constituents. Many organizations are associated with the discharge of food waste production but the most prominently involved are household, hospitality sector and food processing industries. Some majorly associated industries are as follow:

FRUIT AND VEGETABLE PROCESSING INDUSTRY

The wastes generated from fruit and vegetable processing industries involve a huge amount of solid suspensions and a high biochemical oxygen demand (BOD). Some other parameters which are of particular interest to the waste treatment are dissolved oxygen, pH, COD and total solids. Fruit and vegetable waste consist of various by-products with moisture content of 80-90% (Grobe, 1994) and an acidic pH (Riggle, 1989). Chemical composition of the waste depends on the processed fruit or vegetable and chiefly consists of hydrocarbons (sugar, nitrogen and cellulose fibers) and little amount of protein and fat.

Fermentation Industry

The fermentation industry has three main categories: distilling, brewing and wine manufacture. Each of these industries produces a large amount of liquid waste with sharing properties, such as high BOD and COD, but the concentration of the organic compounds varies. The main problem in dealing with fermentation wastewaters is in the flows and loads of the waste. Since the waste water of fermentation industry has high concentration of organic acid, phenols and tannins, so anaerobic treatment is best suited for high performance.

Meat, Poultry and Fish Industry

Meat, poultry and fish industries are responsible for the highest load of food waste production among all the food industries. These industries contains slaughterhouses and processing units where the meat is prepared, cut into pieces and is either cooked, cured, frozen, cured or made into sausages. Waste produced from these sites involve large quantity of blood, residue from intestines, fats, pieces of bones and meat, manure and paunch grass (Cournoyer, 1996). Wastewater generated from slaughterhouse is highly odourous, has high BOD, moisture (90-95%) and nitrogen.

Olive Oil Industry

Olive oil mills are also an important industry, which contributes to waste generation. Olive oil industries produces a dark-colored liquid waste that consist of organic substances such as sugars, organic acids, lipids, polyalcohols, colloids, tannins and pectins. Disposal of oil mill generated wastewaters (OMW) is mainly related to high BOD, COD and high concentration of organic substances like phenols, which makes the process of degradation very difficult and expensive (Saez, Perez, & Martinez, 1992).

Commercial Liquid Food Waste

To minimize the flow of sewer system food waste in the form of waste water originating from kitchen sink, floor drains and dishwashers collects in holding tanks. This putrid smelling waste consist of both organic and inorganic waste (chemical cleaners etc.) and also conatin some hazardous gases such as hydrogen sulphide. This waste is called fats, oils, and grease (FOG) waste also known as brown grease. This is a major problem especially in USA, for the aging sewer systems. According to the US EPA, the reason behind the overflow of sanitary sewer is the improper discharge of FOGs to the collection system. Overflow is associated with annual discharge of 3 billion US gallons (11,000,000 m3) to 10 billion US gallons (38,000,000 m3) of untreated wastewater into local waterways, and responsible for disease and illnesses due to exposure to contamination from overflow of sanitary sewer into recreational waters.

Dairy Industry

The major wastes generated from dairy industries are chemically modified liquid waste that causes pollution of surface water and soil. This waste has dramatic variation in pH (4.2-9.4), high organic load (fatty substances), high load of suspension of solids (SS) around 400-2000mg/l and large variation in waste supply. Main composition of dairy waste is fatty substances, protein, lactose, salts and various kinds of cleaning chemicals. The biggest portions of the chemicals used in dairy industries are detergents. They may be acidic or alkaline. Alkaline detergents are used to remove proteins but through saponification they can also eliminate fats as well. The most widely used alkaline detergent is sodium hydroxide but for specific purposes this can be replaced with other strong bases. Acidic detergents are used to remove inorganic substances, such as milk stone. For this application phosphoric acid or nitric acid are used both alone and in combination. Further the pollution arising from dairy wastewater and cleaning chemicals contains nitrogen and phosphorus from product residues removed by the cleaning processes. The total of these two components alone can accounts to a global pollution load per annum of 3337–5217 tones of nitrogen and 850–1788 tones of phosphorus by the cleaning and disinfection of dairy installations (Romney, 1990).

Food waste produced from agriculture sites and industries are responsible for substantial contamination of the soil and water. After pollution of an area next phase is to find out a feasible option to resolve the problem. Over the past decades various techniques have been, tested, used, approved or rejected. When things get more severe one can either resort to conventional methods, such as prevention and reduction, reuse, employment of degradable materials, recycling, incineration, pyrolysis and landfill, or to modern innovative methods which include composting, biodegradability and bioremediation.

IMPACT OF FOOD WASTE ON ENVIRONMENT

Food waste has a big influence on climate, land, water and biodiversity. Wastage of such a huge amount of food is responsible for a high economic loss. In the recent years, food waste and its management has become serious issue of concern. Now it has been termed as a global paradox regarding the manner which focuses on agriculture for improvising food security and then a third of all the food produced ends up as waste.

Accelerated Carbon Footprint and Climate Change

Wastage of food finally goes to landfills. Where it begins rotting and decomposition and releases around 3.3 billion tons of green house gases (GHG) mainly the methane gas. This is responsible for climate change and global warming. Methane itself accounts for about 20% of greenhouse gas production. It is about 25% more effective than CO2 in trapping heat in the atmosphere. Thus food waste is the third biggest source for emission of green house gases.

Wastage of Fertile Land Areas

Use of land regarding the food is categorized into two main categories: first is for the production of crop and grassland and second is for dumping of uneaten food that has been thrown out. Around 1.4 billion hectare of land constituting almost 1/3 of the planet's arable land is occupied by the unconsumed food. So it can be concluded that more than 30% of the world's fertile land which could be used for agriculture and environmental research is polluted (Pandey et al.,2018). The land if not cared for, loses its ability to yield over time called degradation. Eventually produces far less than can sustain the people living in the region.

Blue Water Footprint

Water is indispensable for life and it's no surprise it is essential for food production as well. An immense volume around 70% of world's fresh water is being used for agriculture. Throwing out 30% of all the food produced means millions of gallons of fresh water that was used in the production and processing of food also goes to waste. This contributes to blue water footprint which refers to the amount of consumed surface and ground water resources that getting waste. It is estimated that food wastage contributes nearly 45 trillion gallons or 24% of all water used for agriculture. It is also affirmed that meat products are the heaviest water users, simply because the animals drink a lot of water—and more importantly, because so much water is needed for the grain that becomes their feed. So throwing out a kilogram of beef causes 50,000 liters of water wastage in meat production. Similarly pouring down of one glass of milk is responsible for wastage of 1000 liters of water.

Biodiversity Loss

Food wastage is also responsible for causing biodiversity loss at a global level. In order to create more land for agriculture, farmers especially in tropical areas are destroying natural flora and fauna. Thus deforestation has led to loss of biodiversity. Some agriculture practices such as mono-cropping is also responsible for diversity loss. Due to increased production of livestock, natural land is converted into pastures. The more livestock graze, the less natural and diverse the area becomes. Another culprit which is responsible for biodiversity loss is marine fishery. This results in overexploitation of areas. Fishes are caught without thought given to how the rapid depletion of fish population will harm their environment. After hunting these fishes then rejected by the stores not meeting their standards, or thrown out by the consumers. Use of pesticides in crop production causes nitrogen, phosphorous and chemical pollution in rivers, streams and coastal waters thus affecting marine life.

IMPACT OF FOOD WASTE ON HUMAN AND ANIMAL HEALTH

General public is not directly exposed for the transmission of infectious diseases originated from food waste. After dumping, the food becomes more prone to be contaminated with various kinds of infectious agents and provide optimum condition for the growth and survival of different infectious agents (Samant et al., 2018). Food waste that can transmit diseases includes food of animal origin or food prepared by products of animal origin, whether raw or cooked e.g. products which contain meat such as meat pies, sausage rolls, pizza, meat scraps, bones, etc. products which have been in contact with meat, for example, sandwiches, bread, pizzas, used cooking oil, any kitchen food waste that may have touched meat, etc.. Food contamination by infected or diseased person during handling is also a reason of food wastage and result in the transmission of various food borne diseases like cholera, hepatitis A, typhoid, jaundice, food poisoning, vomiting, diarrhea, stomach cramps, fever, dark urine, nausea, boils, dizziness etc. Individuals such as waste pickers, drug addicts, children (playing nearby waste) and population residing in the vicinity of municipal site or dumping yard are at particular risk to infectious diseases. On the other hand openly dumped food waste is easily accessible to animals like dog, pig, cow etc roaming around municipal waste in search of food are more susceptible and are directly associated with disease transmission. Scavenging on openly dumped expired and highly contaminated food waste can severely affect animals and sometimes it leads to death of the animal. Thus incorrect disposal of food waste is responsible for development of foot and mouth disease or classical swine fever and several other exotic diseases by feeding food waste to pigs and other animals. Because of disease control concerns feeding of food waste to animal is offence. Some common diseases and their causative agent that transmitted through food waste in are described in Table 1.

RISK MINIMIZATION

Disease transmission through food waste can be minimized at the primary step by vector eradication of protozoans (flies), birds, pigs who act as vector for leishmaniasis, malaria, dengu, psittacosis, foot and mouth disease etc. Use of personal protective equipments such as heavy duty gloves, protective clothing, safety glasses, face protection, respiratory protection (masks) and management controlling system such as training and awareness programmes can help in reducing the risk of disease transmittance from food waste (Samant et al., 2018).

FOOD WASTE HIERARCHY

In developing parts of the world post-harvest loss of food in the initial stage of food supply chain is the major concern. The problem occurs due to socio-economical, financial and structural restriction in the infrastructure of harvesting, storage and transportation techniques. Biological and environmental conditions also play major role in drastic food spoilage. Besides this reduction in the nutritional value, caloric value and edibility of crops, due to extremes temperature, humidity or the action of micro-organisms (Lacey, 1989) also causes significant food waste. Thus for the management of food waste, there are three general levels, the 3R's (Reduce, Reuse, Recycle) where action is needed (Figure 2).

Reduction (Waste Minimization)

High preference should be given for minimizing food wastage at the first level. Beyond improving losses of crops on farms due to poor practices, doing more to better balance production with demand would mean not using natural resources to produce unneeded food in the first place.

Table 1. Major food born diseases caused by microorganisms

S. No.	Microorganism	Disease	Symptoms
1.	*Campylobacter jejuni*	Diarrhea	Fever, blood in stools, abdominal pain, abdominal cramps.
2.	*Salmonella typhi*	Typhoid high fever	Pain in the abdomen or muscles, whole body fatigue, fever, chills, loss of appetite, or malaise
3.	*Staphylococcus aureus*	Infected eczema, psoriasis or any other pus draining lesion	Redness, painful and swollen and the skin
4.	*Escherichia coli*	Diarrhea	Fever, blood in stools, abdominal pain, abdominal cramps.
5.	*Listeria monocytogenes*	Septicemia and meningitis	Shivering, or having cold hands and feet, pale, blotchy complexion, confusion, aching limbs or joints.
6.	*Shigella*	*Shigellosis*	Fever, Abdominal Cramps And Tenesmus, And Frewuent, Small Volume, Bloody Stools Containing Mucous
7.	*Paratyphi*	Paratyphoid fever	Vague chills, sweating, headache, weakness, dry cough, anorexia, sore throat, dizziness, and muscle pains
8.	*Staphylococcus aureus*	Illnesses, skin infections	Pimples, impetigo, boils, cellulitis, folliculitis, carbuncles, scalded skin syndrome, and abscesses
9.	*Clostridium perfringens*	Abdominal pain and stomach cramps	Nausea, Fever and vomiting
10.	*Clostridium botulinum*	Flaccid paralysis of muscles	Double vision, Drooping eyelids, muscle weakness (resulting in a flaccid paralysis)
11.	*Bacillus cereus*	Emetic toxin	Nausea and vomiting
12.	*Yersinia enterocolitica*	Yersiniosis	High body iron levels, watery or bloody diarrhea and fever
13.	*Vibrio parahaemalyticus*	Cholera	Water diarrhea, abdominal cramping, nausea, vomitig
14.	*Norovirus*	Dehydration, malnutrition and even death	Dry mouth and throat, dizziness, decreased urine output

Reuse

In the event of a food excess, re-use within the human food chain, donating surplus food to the people in need or searching secondary market is the best option. In case if the food is not suitable for human consumption, then delivering it to livestock feed is the next best option.

Recycling and Recovery

When food can't be reuse, then the only and best option is recycling and recovery. Composting, Anaerobic digestion, byproduct recycling and incineration are the processes that allow recovery of nutrients and energy from waste food, which is the advantage of dumping food waste in landfills. It reduces greenhouse emissions, which can help in protecting our environment from global warming and climate change overall.

ROLE OF MICROORGANISM IN FOOD WASTE TREATMENT

Since early 1970s the role of environmental microbiology has emerged across the world due to destructive impression of environmentally communicated diseases from various microorganisms on human health and inspired by the comprehensive efficacy of microbes found in the environment. Nowadays, microorganisms are being extensively used for the environmental cleanup of hazardous and infectious waste generated mainly from the organic waste (food waste). The uniqueness, unpredictable nature and biosynthetic potentials of microorganisms made them putative for resolving problems associated with life science and other related fields as well. Microorganisms mainly the bacteria play a prominent role in decomposition of organic waste to more stable less polluting substances and also help in natural cycling. The emerging role of microorganisms in the field of biotechnology has generated new interest in them. Microbes are not only capable of surviving in hostile environment, but are also capable of immobilizing, degrading or detoxifying environmental contaminants. Microorganisms use waste material as food source, thus help in complete elimination or mineralization of toxic compounds by converting them into basic elements such as carbon dioxide and water. Some may result in incomplete or partial degradation of the waste or contaminants to a less complex form. While on the other hand immobilization of the contaminants with the help of microbes is resolved but not eliminated or altered, which is often a potential benefit but rarely a final solution.

The major groups of microbes constituting soil population are bacteria, actinomycetes, fungi, algae, and protozoa. Among them bacteria are the most abundant. They have various advantages over other groups such as high growth rate, diverse metabolic character and potential to degrade varied contaminants. They have been categorized based on their ability to degrade various contaminants. The major factors involved are:

Cell Morphology

Three basic types of cell morphology is found in bacteria, the bacilli or rod shaped bacteria, most abundant in soil microbe, the cocci or spherical shaped bacteria and the spirilla or spirals.

Ability to Grow in Presence or Absence of Oxygen

Ability of the microbes to survive in presence or absence of oxygen is a major biochemical trait, based on this potential bacteria are classified into aerobes, which grow in presence of oxygen, anaerobes, which grow only in absence of oxygen; and facultative anaerobes, which can grow either in absence or presence of oxygen.

Type of Carbon and Energy Sources

Based on carbon and energy source microorganisms are classified into two main classes: Heterotrophs, which require organic substrates as a source of carbon and energy, are the dominant microflora of soil. Autotrophs, obtain carbon by the assimilation of CO_2 and energy from the oxidation of inorganic compounds or sunlight. Autotrophs are further categorized into two types: photoautotrophs which obtain energy from sunlight, and chemoautotrophs, derive the energy from the oxidation of inorganic materials.

Factors Affecting Microbial Activity in Environment

Microorganisms are highly dependent on the physio-chemical properties of the environment. Prominent factors associated with the activity of microbes are temperature, pH, and moisture content.

Acclimatization of Microbes

The growth of the microorganisms in presence of any contaminant mostly observes a starting period, also called acclimatization lag. During this phase, no such biotic change takes place in the level of contaminants. Degradation of biomass can't take place until total microbial population reaches a significant level, due to which they are unable to induce related enzymes. Another reason for the lag in the degradation of contaminants is the preferential depletion of other substrates first. Evaluation of inhibitory or toxic conditions at a site can be done by measuring the activity of indigenous microbes. Bacterial count in ground water ranges from 10^2-10^5 colony forming unit (CFU) per ml of sample, while in soil it is 10^3-10^7 CFU per gram of soil. Higher counts indicate a healthy microbial population, while bacterial count at contaminated site below 10^3 organisms per gram of soil indicates a stresses microbial population.

Temperature

Every microorganism has the ability to survive under specific environmental condition of pH, temperature; salinity etc. for the growth of microorganisms there is an optimum temperature. Beyond that temperature the activity of microorganisms declines. Based on the growth of microbes at different optimum temperature they are classified into three types:

Psychrophiles

They have a growth optima of 15±5°C and minima of 0°C or below. Obligate psychrophiles die if exposed temporarily to room temperature, while the facultative psychrophiles can grow at 0°C.

Mesophiles

Mesophiles grow optimally between 20°C and 45°C and minimally between 15 to 20°C. Majority of the microorganisms and all human pathogen fall under this category. While the microbes effective in bioremediation perform over a temperature range of 10-40°C.

Thermophiles

Thermophilic can grow optimally at temperature of 55-65°C they are differ from mesophiles in having heat stable enzymes and proteins, that functions pretty well in high temperatures. These microbes are majorly involved in composting,

TECHNIQUES OF WASTE MANAGEMENT

The major techniques involved in food waste management are associated with the role of microorganisms or effective microorganisms (EM). Effective Microorganisms (EM) is a group of 80 beneficial coexisting beneficial microorganisms isolated from different environment with a reviving potential on human, animal and the natural environment (Mayer et al., 2010). The prominent microorganisms in EM are yeast, fungi, photosynthetic bacteria, lactic acid bacteria and actinomycetes (Xu, 2001). These Ems are associated with their role such as *Rhodobacter sphaeroides* involve in production of amino acids and amino nucleic, *Streptomyces albus* produce antibacterial matter for pathogenic bacteria and *Lactobacillus plantarum* may accelerate the decomposition and fermentation in composting process (Sreenivasan, 2013; Xu et al., 2000). Since organic acid produced from lactic acid bacteria act as sterilizing agent thus EM can be used to suppress pathogenic microorganisms.

Thus using effective microorganism food waste can be treated through several techniques. Major techniques involved in the decomposition of food waste are as follow:

Anaerobic Digestion

Methane production via anaerobic process is an efficient method for food waste management. This is a cost effective and has low residual production, thus utilizes food waste as renewable source of energy (Morita & Sasaki, 2012; Nasir et al.,2012). The quantity of bioenergy generated via anaerobic digestion were reported earlier (Dung et al., 2014) and is summarized in figure 3. The process of anaerobic digestion (Figure 4) takes three steps (hydrolysis, acid formation, and gas production) for the completion:

Enzymatic Hydrolysis

In the first step of AD, large polymer molecules that cannot be transported to cell membrane are broken down into simpler molecules with the help of hydrolase enzyme secreted by facultative or obligate anaerobic hydrolytic bacteria, e.g. polysaccharides are broken down into oligosaccharides and monosaccharides, production of peptides and amino acids from protein hydrolysis and conversion of lipids or fat into fatty acids and glycerol.

$$nC_6H_{10}O_5 + nH_2O \rightarrow nC_6H_{12}O_6 \tag{1}$$

Figure 3. Worldwide bioenergy potential from FW in developed and developing countries; Adapted from (Paritosh et al., 2017)

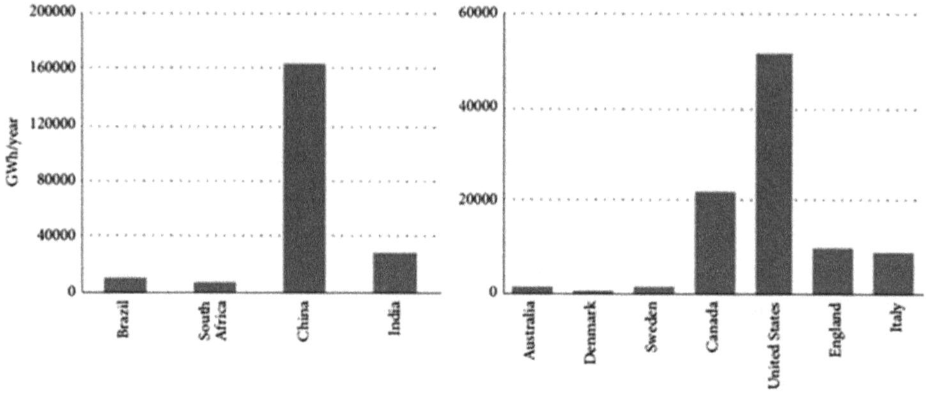

Figure 4. Major steps during anaerobic digestion of food waste

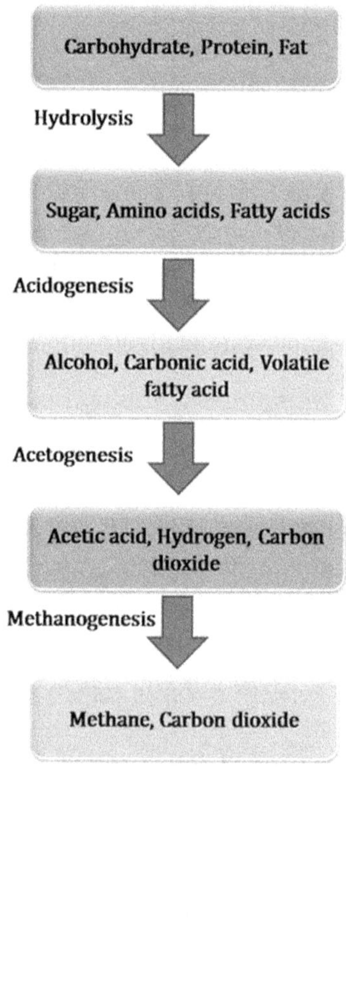

In a study (Mittal, 1997) it is reported that during anaerobic condition, the rate of hydrolysis is somewhat slower than the rate of acid formation, which completely rely on the nature of substrate, concentration of bacteria, pH and temperature of the bioreactor. While some other parameters like size of the substrate, pH, enzyme production, adsorption of enzymes on the substrate particles also influence the rate of hydrolysis. It is reported that *Enterobacter* and *Streptococcus* are the anaerobes that causes hydrolysis of the substrate (Bryant, 1979).

Acidogenesis

In this process, facultative anaerobic bacteria utilize oxygen and carbon, creating an anaerobic condition for methanogenesis. The hydrolytic products (monomers) are subjected to fermentation for the production of volatile fatty acids (acetate, propionate, butyrate, isobutyrate and valerate) alongwith carbon dioxide, hydrogen and ammonia. Acetate, hydrogen and carbon dioxide can be used directly for the production of methane. While a group of syntrophic acetogenic bacteria convert propionate, butyrate, isobutyrate and valerate to formacetate and hydrogen (Bryant, 1979; Mittal, 1997; Schink, 1997).

Acetogenesis

In this process acetogenic bacteria convert acid phase products into acetates and hydrogen. These acetogenic bacteria belong to genera *Syntrophobacter* and *Syntrophomonas* (Schink, 1997). Some acetate molecules are also produce by reduction of carbon dioxide using hydrogen as an electron source. The hydrogen produced in the process causes inhibition of microorganisms. Therefore in anaerobic digesters, acetogenic bacteria show syntrophism with hydrogenetic methanogens that remove hydrogen by utilizing it for methane production. Thus acetogenesis results in the formation of 70% methane while 11% hydrogen is also produced during the process (Schink, 1997) .

$$nC_6H_{12}O_6 \rightarrow 3nCH_3COOH \tag{2}$$

Methanogenesis

Methanogenesis is the last step which is performed by methanogens, belonging to archea. In this phase products of the previous step, that is, acetic acid, hydrogen and carbon dioxide are utilized for methane production. Methane can be produced in two ways by two different types of methanogens: (1) Production of methane from acetic acid using acetoclastic methnogens and (2) utilization of hydrogen to reduce carbon dioxide using hydrogenotrophic methanogens, only 30% of methane is produced during this process (Griffin et al., 1998; Karakashev et al., 2005).

$$CH_3COOH \rightarrow CH_4 + CO_2 \quad (3)$$

$$CO_2 + 4H_2 \rightarrow CH_4 + 3H_2O \quad (4)$$

Microorganisms involved in anaerobic digestion of organic matter are summarized in table 2.

Composting

Another widely used method for the management of food waste is composting. This process is carried out by various kinds of microorganisms such as bacteria, fungi and actinomycetes (Ashraf et al., 2007). The process is responsible for reducing the weight, volume and moisture and also help in minimizing putrid smell of the food waste to the large extent (Bachert et al.,2008). Furthermore composting also play a major role in recovering soil health (Gan et al.,2009; Tandy et al., 2009). By redistributing nutrients and high microbial populations, compost help in reducing soil erosion and water runoff by improving rainfall penetration, which has been shown to reduce the loss of sediment, nutrients, and pesticide losses to streams by 75–95% . This process is simple, cost effective and efficient in reducing organic waste

Table 2. Microorganisms involved in anaerobic digestion of organic matter

Reaction Type	Microorganism	Active Genera	Product
Fermentation	Hydrolytic bacteria	*Bacteriodes, Lactobacillus, Propionibacterium, sphingomonas, Sporobacterium, Megasphaera, Bifidobacterium*	Simple sugars, peptides, fatty acids
Acidogenesis	Syntrophic bacteria	*Ruminococcus, Paenibacillus, Clostridium*	Volatile fatty acids
Acetogenesis	Acetogenic bacteria	*Desulfovibrio, Aminobacterium, Acidominococcus*	Acetic acid
Methanogenesis	Methanogens (Archaea)	*Methanosaeta, Methanolobus, Methanococcoides, Methanohalophilus, Methanosalsus, Methanohalobium, Halomethanococcus, Methanolacinia, Methanogenium, Methanoculleus*	Methane, Carbon dioxide

Adapted from (Lauwers et al., 2013; Ziemi´nski & Frąc, 2012)

(Namkoong et al., 2002). Based on the nature of decomposition process, composting is categorized into aerobic and anaerobic composting (Misra et al.,2003). In aerobic composting, in presence of oxygen, organic food waste is break down into carbon dioxide, ammonia, water and heat. While the anaerobic composting results in the formation of intermediate compound such as methane, organic acid and hydrogen sulphide. The main features of composting are summarized in table 3. Composting can be done in two ways either by traditional (figure 5) or rapid composting methods (table 4) (Misra et al., 2003). In traditional method during anaerobic decomposition, the material are subjected to remain in the pit without turning and watering for three months while aerobic decomposition, takes place by itself without any addition of elements. In large scale passive aeration windrow is used to mix the composting materials, which provide condition for aerobic decomposition. This process takes up to eight weeks. While active composting may takes ten to twelve weeks. On the other hand in rapid composting additional elements such as effective microorganisms and red worms are added to decompose organic food waste. Addition of earth worms to decompose organic waste is called vermicomposting. This process of composting does not require any additional energy to maintain aerobic environment, as this action is already accomplished by the worms. The advantage of the process is that it takes a short duration of less than four weeks for completion. The product of vermicomposting is homogenous and contains low contaminants as compared to traditional composting.

Figure 5. Traditional composting method

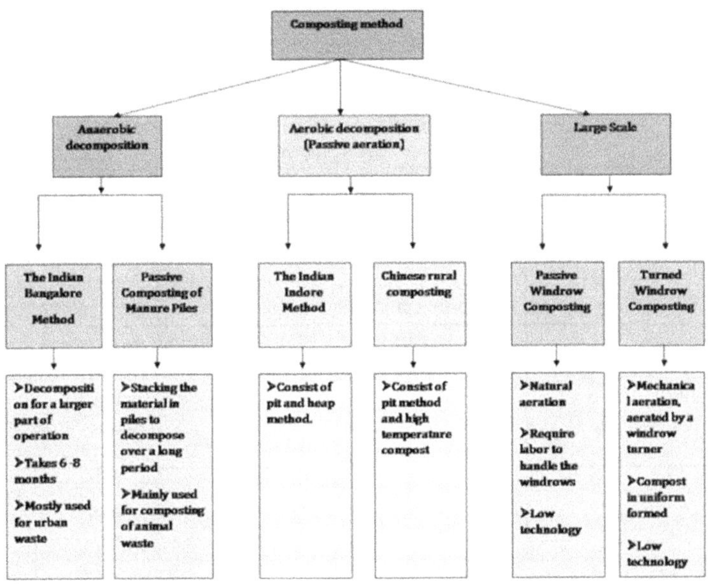

Another advantage of composting is that it helps in decreasing the emission of greenhouse gases to the environment. In landfills, decomposition of food waste takes place anaerobically, thus producing methane gas in to the atmosphere. But in composting process this biodegradable food waste decomposes aerobically and does not produces methane gas, instead generates a large amount of compost that can be utilized in agriculture.

FOODPRINT CAMPAIGNS AND AWARENESS PROGRAMMES

Campaign and promotion programmes for handling and safe disposal of various kinds of food waste is necessary for public health and every member of the public must be aware of the potential health risk. For preventing food wastage some training programs should be developed by the government agencies. For reducing food footprint these campaigns can help farmers, fishers, food processors, supermarkets,

Table 3. Characteristic of aerobic and anaerobic composting; Source

S.NO	Characteristic	Aerobic	Anaerobic
1	Microbes used	Aerobic microbe breaks the organic matter. Stable organic end product produce with Carbon dioxide, ammonia, water, heat and humus.	Anaerobic microbe dominates and develops intermediate compounds including methane, organic acids, hydrogen sulphide and other substances.
2	Oxygen	Process occur in the presence oxygen	Process occur in the absent or in limited of Oxygen supply
3	Temperature	The heat generated accelerates the breakdown of organic mater	Low-temperature process
4	Duration	The processing time is short, depends on the mechanism used (maximum 4 months)	Longer than aerobic composting (6-8 months)
5	Nutrient value	The loses of Nutrients are more from the composting process	Nutrients did not lost during the process of composting
6	Benefits	Greater sludge fertilizer value. Fewer odors Greater sludge fertilizer value	Reduction of pathogens. Decrease in Green house gas emission. Anaerobic digestion reduces reliance on energy imports.
7	Risk	Phytotoxicity risk is Little .	Has strong odors and some present phytotoxicity.

Adapted from (Muttalib et al., 2016)

Table 4. Rapid composting method

S.No	Composting Method	Description
1	Bin composting	• Simplest in-vessel method. • Eliminate weather problems, contain odors • No turning materials and better temperature control
2	Rotating drums	• Uses a horizontal rotary drum to mix, aerate, and move the material through the system. • Air is supplied through the discharge end • The drum can be either open or partitioned
3	Use of Worms (Vermicomposting)	• Used red worms in small up to large scale. • The worm eats all the organic material during the process. • It takes at least 3-4 months to produce compost.
4	Shredding and frequent turnings	• Compost build in pile or bin • Involves inoculating the raw material used for composting with cultures of a cellulose decomposer fungus.
5	Use of Cellulolytic Cultures	• Compost build in pile or bin • Involves inoculating the raw material used for composting with cultures of a cellulose decomposer fungus
6	Aerated Static Pile Composting	• The decomposition was assisted by the air that enters the pile through pulling or pushing mechanisms. • Less odor and consumed small space
7	Silos	• In-vessel technique resembles a bottom-unloading silo. • The aeration system allows air to come up from the base of the silo through the composting materials.
8	Use of Mineral Nitrogen Activator	• Compost build in pile or bin • Nitrogenous fertilizer was added into the center of the pile. • In this high-temperature, bacterially active system, and takes three to four days to produce compost.
9	Use of Effective Microorganisms (EM)	• Compost build in pile or bin • Effective Microorganisms contain culture of bacteria was added to the raw material to enhance composting process.

Adapted from (Muttalib et al., 2016)

individual consumers. To prevent food wastage, UN and FAO have launched such campaigns that mainly emphasize on campaign slogan “Think eat save – reduce your foodprint”. With more such types of campaign, society at large will be informed on ways for reducing foodprint and their impact on environment. Ultimately it will help in resolving the problem of food wastage.

CONCLUSION

Microorganisms help in clean up of food waste in an environment friendly and hygienic manner. They not only decompose or degrade the food waste but also help in producing environment friendly products. They enable us to convert food waste into energy via anaerobic processes in terms of methane, which is very helpful in today's energy crisis. They also help in producing soil manure by decomposing food waste, which is safer to be use than chemical fertilizers. Thus play a prominent role in management of food waste generated from different sectors. Microbes made the process feasible and cost effective, which would be impossible through physical and chemical engineering methods.

REFERENCES

Ashraf, R., Shahid, F., & Ali, T. A. (2007). Association of fungi, bacteria and actinomycetes with different composts. *Pakistan Journal of Botany*, *39*(6), 2141–2151.

Bachert, C., Bidlingmaier, W., & Surapong, W. (2008). Open windrow composting manual. *Orbit (Amsterdam, Netherlands)*, V.

Bazzoffi, P., Pellegrini, S., Rocchini, A., & Morandi, M. (1973). Municipal waste in a plantation of young slash pine: Effects on soil and trees. *Journal of Environmental Quality*, *2*(4), 441–444. doi:10.2134/jeq1973.00472425000200040006x

Bouallagui, H., Torrijos, M., Godon, J., Moletta, R., Cheikh, R. B., & Touhami, Y. (2004). Two-phases anaerobic digestion of fruit and vegetable wastes: Bioreactors performance. *Biochemical Engineering Journal*, *21*(2), 193–197. doi:10.1016/j.bej.2004.05.001

Bryant, M. (1979). Microbial Methane Production—Theoretical Aspects 1, 2. *Journal of Animal Science*, *48*(1), 193–201. doi:10.2527/jas1979.481193x

Coker, C. (2006). Environmental remediation by composting. *BioCycle*, *47*(12), 18–23.

Cournoyer, M. (1996). Sanitation and stabilization of slaughterhouse sludges through composting. *Proceedings of the Canadian meat research institute technology symposium*.

Dung, T. N. B., Sen, B., Chen, C. C., Kumar, G., & Lin, C. Y. (2014). Food waste to bioenergy via anaerobic processes. *Energy Procedia*, *61*, 307–312. doi:10.1016/j.egypro.2014.11.1113

Ermgassen, E., Phalan, B., Green, R., & Balmford, A. (2016). Reducing the land use of EU pork production: Where there's will, there's a way. *Food Policy*, *58*, 35–48. doi:10.1016/j.foodpol.2015.11.001 PMID:26949285

Gan, S., Lau, E., & Ng, H. (2009). Remediation of soils contaminated with polycyclic aromatic hydrocarbons (PAHs). *Journal of Hazardous Materials*, *172*(2-3), 532–549. doi:10.1016/j.jhazmat.2009.07.118 PMID:19700241

Giusquiani, P., Pagliai, M., Gigliotti, G., Businelli, D., & Benetti, A. (1995). Urban waste compost: Effects on physical, chemical, and biochemical soil properties. *Journal of Environmental Quality*, *24*(1), 175–182. doi:10.2134/jeq1995.00472425002400010024x

Griffin, M. E., McMahon, K. D., Mackie, R. I., & Raskin, L. (1998). Methanogenic population dynamics during start-up of anaerobic digesters treating municipal solid waste and biosolids. *Biotechnology and Bioengineering*, *57*(3), 342–355. doi:10.1002/(SICI)1097-0290(19980205)57:3<342::AID-BIT11>3.0.CO;2-I PMID:10099211

Grobe, K. (1994). Composter links up with food processor. *BioCycle*, *34*(40), 42–43.

Karakashev, D., Batstone, D. J., & Angelidaki, I. (2005). Influence of environmental conditions on methanogenic compositions in anaerobic biogas reactors. *Applied and Environmental Microbiology*, *71*(1), 331–338. doi:10.1128/AEM.71.1.331-338.2005 PMID:15640206

Khaliq, A., Abbasi, M. K., & Hussain, T. (2006). Effects of integrated use of organic and inorganic nutrient sources with effective microorganisms (EM) on seed cotton yield in Pakistan. *Bioresource Technology*, *97*(8), 967–972. doi:10.1016/j.biortech.2005.05.002 PMID:16023343

Lacey, J. (1989). Pre-and post-harvest ecology of fungi causing spoilage of foods and other stored products. *The Journal of Applied Bacteriology*, *67*, 11s–25s. doi:10.1111/j.1365-2672.1989.tb03766.x PMID:2508232

Lauwers, J. L., Appels, P., Thompson, I., Degr'eve, J., Impe, J. F. V., & Dewil, R. (2013). Mathematical modelling of anaerobic digestion of biomass and waste: Power and limitations. *Progress in Energy and Combustion Science*, *39*(4), 383–402. doi:10.1016/j.pecs.2013.03.003

Mayer, J., Scheid, S., Widmer, F., Fließbach, A., & Oberholzer, H. R. (2010). How effective are 'Effective microorganisms® (EM)'? Results from a field study in temperate climate. *Applied Soil Ecology*, *46*(2), 230–239. doi:10.1016/j.apsoil.2010.08.007

Misra, R. V., Roy, R. N., & Hiraoka, H. (2003). *On-Farm Composting Methods*. Food and Agriculture Organization of the United Nations.

Mittal, K. M. (1997). Biogas Systems, Policies, Progress and Prospects. New Age International.

Morita, M., & Sasaki, K. (2012). Factors influencing the degradation of garbage in methanogenic bioreactors and impacts on biogas formation. *Applied Microbiology and Biotechnology*, *94*(3), 575–582. doi:10.100700253-012-3953-z PMID:22395906

Muttalib, A., Aminah, S., Ismail, S., Norkhadijah, S., & Praveena, S. M. (2016). Application of effective microorganism (EM) in food waste composting: A review. *Asia Pacific Environmental and Occupational Health Journal*, *2*(2), 37–47.

Nagavallemma, K. P., Wani, S. P., Stephane, L., Padmaja, V. V., Vineela, C., & Babu Rao, M. (2006). Vermicomposting: Recycling wastes into valuable organic fertilizer. *International Crops Research Institute for the Semi-Arid Tropics*, *2*, 16.

Namkoong, W., Hwang, E.-Y., Park, J.-S., & Choi, J.-Y. (2002). Bioremediation of diesel-contaminated soil with composting. *Environmental Pollution*, *119*(1), 23–31. doi:10.1016/S0269-7491(01)00328-1 PMID:12125726

Nasir, I. M., Ghazi, T. I. M., & Omar, R. (2012). Production of biogas from solid organic wastes through anaerobic digestion: A review. *Applied Microbiology and Biotechnology*, *95*(2), 321–329. doi:10.100700253-012-4152-7 PMID:22622840

Novinscak, A., Filion, M., Surette, C., & Allain, A. (2008). Application of molecular technologies to monitor the microbial content of biosolids and composted biosolids. *Water Science and Technology*, *57*(4), 471–477. doi:10.2166/wst.2008.019 PMID:18359983

Pandey, A., Tripathi, P. H., Pandey, S. C., Pathak, V. M., & Nailwal, T. K. (2018). Removal of Toxic Pollutants From Soil Using Microbial Biotechnology. In Microbial Biotechnology in Environmental Monitoring and Cleanup (pp. 86-105). Academic Press. doi:10.4018/978-1-5225-3126-5.ch006

Papargyropoulou, E., Lozano, R., Steinberger, J. K., & Wright, N., & bin Ujang, Z. (2014). The food waste hierarchy as a framework for the management of food surplus and food waste. *Journal of Cleaner Production*, *76*, 106–115. doi:10.1016/j.jclepro.2014.04.020

Paritosh, K., Kushwaha, S. K., Yadav, M., Pareek, N., Chawade, A., & Vivekanand, V. (2017). Food waste to energy: An overview of sustainable approaches for food waste management and nutrient recycling. *BioMed Research International, 2017*, 19. doi:10.1155/2017/2370927 PMID:28293629

Pfotzer, G., & Schüler, C. (1997). Effects of different compost amendments on soil biotic and faunal feeding activity in an organic farming system. *Biological Agriculture and Horticulture, 15*(1-4), 177–183. doi:10.1080/01448765.1997.9755192

Riggle, D. (1989). Revival time for composting food industry wastes. *BioCycle, 29*, 35–37.

Romney, A. (1990). *CIP: cleaning in place* (2nd ed.). Society of Dairy Technology.

Saez, L., Perez, J., & Martinez, J. (1992). Low molecular weight phenolics attenuation during simulated treatment of wastewaters from olive oil mill in evaporation ponds. *Water Research, 26*(9), 1261–1266. doi:10.1016/0043-1354(92)90187-9

Samant, M., Pandey, S. C., & Pandey, A. (2018). Impact of Hazardous Waste Material on Environment and Their Management Strategies. In Microbial Biotechnology in Environmental Monitoring and Cleanup (pp. 175-192). Academic Press. doi:10.4018/978-1-5225-3126-5.ch011

Schink, B. (1997). Energetics of syntrophic cooperation in methanogenic degradation. *Microbiology and Molecular Biology Reviews, 61*(2), 262–280. PMID:9184013

Schüler, C., Pikny, J., Nasir, M., & Vogtmann, H. (1993). Effects of composted organic kitchen and garden waste on *Mycosphaerella pinodes* (Berk, et Blox) Vestergr., causal organism of foot rot on peas (Pisum sativum L.). *Biological Agriculture and Horticulture, 9*(4), 353–360. doi:10.1080/01448765.1993.11978505

Sreenivasan, E. (2013). Evaluation of Effective Microorganisms Technology in Industrial Wood Waste Management. *International Journal of Advances in Engineering and Technology, 4*(3), 21–22.

Tandy, S., Healey, J. R., Nason, M. A., Williamson, J. C., & Jones, D. L. (2009). Remediation of metal polluted mine soil with compost: Co-composting versus incorporation. *Environmental Pollution, 157*(2), 690–697. doi:10.1016/j.envpol.2008.08.006 PMID:18819736

Umsakul, K., Dissara, Y., & Srimuang, N. (2010). Chemical, physical and microbiological changes during composting of the water hyacinth. *Pakistan Journal of Biological Sciences, 13*(20), 985–992. doi:10.3923/pjbs.2010.985.992 PMID:21319457

Xu, H. L. (2001). Effects of a microbial inoculant and organic fertilizers on the growth, photosynthesis and yield of sweet corn. *Journal of Crop Production, 3*(1), 183–214. doi:10.1300/J144v03n01_16

Xu, H. L., Wang, R., & Mridha, M. A. U. (2000). Effect of organic fertilizers and a microbial inoculant on leaf photosynthesis and fruit yield and quality of tomato plants. *Journal of Crop Production, 3*(1), 173–182. doi:10.1300/J144v03n01_15

Ziemi'nski, K., & Frąc, M. (2012). Methane fermentation process as anaerobic digestion of biomass: Transformations, stages and microorganisms. *African Journal of Biotechnology, 11*(18), 4127–4139.

Chapter 8
Sustainable Management of Coffee and Cocoa Agro-Waste

Pushpa S. Murthy
Central Food Technological Research Institute, India

Nivas Manohar Desai
Central Food Technological Research Institute, India

Siridevi G. B.
Central Food Technological Research Institute, India

ABSTRACT

Agricultural waste is not only a sustainability problem related to food security but also an economic problem since it has a direct impact on the profitability of entire food supply chain. Sustainable management of agricultural waste is a systematic approach towards reducing waste and its allied impacts over the entire life cycle, starting with the use of natural resources, production, sales, and consumption, and ending either with final disposal or recovery. Management of agro-waste focuses on three main aspects (i.e., recycle, reuse, and reduce [R3]). Building on this familiar concept of "R3" will impact environmental protection and more fully recognize the impacts of the food and agriculture wasted. Thus, in the chapter, the authors highlight the sustainable utilization of waste generated from coffee and cocoa processing for the development of value-added products.

DOI: 10.4018/978-1-5225-7706-5.ch008

INTRODUCTION

"Value Addition Is Prime Need for Sustainable Agro Industry, Environment Conservation and Human Health"

Food and agriculture waste is getting a lot of consideration and has reached alarming levels. Yearly, approximately one-third of produced food for human consumption is getting lost or wasted. To overcome such a problem of food and agriculture waste, several by-products have been studied so as to weigh up their potential to reuse either totally or partially in the food industry or other. Coffee and cocoa industry is one of the classical models of generation of waste during the production process right from the field to the cup. The coffee and cocoa residue has been tried with different purposes to give a new life. Food and Agriculture waste not only imply a major negative issue in balancing food resources but also causes several socio-economical and environmental impacts too. The Public is more interested in increasing food and agricultural production to meet future food crisis, but providing inadequate attention to losses in the global food supply chain.

Approximately one-third (1.3 billion tons/year) of food produced for human consumption is lost or wasted worldwide (FAO, 2011). According to Food and Agriculture Organization of the United Nations (FAO), "food loss" refers to a decrease in mass (dry matter) or nutritional value (quality) of food that was originally intended for human consumption. The term "food wastage" embraces both food loss and food waste (FAO, 2013). In general, food or agricultural loss is either lost or wasted right through their supply chain, from early production down to ultimate household consumption.

Food or agriculture loss can be mitigated by

1. **Food Wastage Reduction:** Beyond improving losses of crops on farms due to poor practices, doing more to better balance production with demand.
2. **Re-usage Within the Human Food Chain:** Finding secondary markets or donating extra food to feed vulnerable members of society. If the food is not fit for human consumption, it should be diverted for livestock feed.
3. **Recycling and Recovery:** Waste recycling, anaerobic digestion, composting, and incineration with energy recovery allow energy and nutrients to be recovered from food waste.

The spent/residue is otherwise defined as by product that is generated during the production or manufacturing process that can be of value addition and not just waste. It can be further used directly other than normal industrial practice. However, it must fulfil all pertinent product, environmental and public health protection requirements for the specific use with no adverse environmental and/or human health impacts. According to the International Coffee Organization, around 84.3 million bags of Arabica and 59.1 million bags of Robusta coffee were produced worldwide. The highest per capita coffee consumption is owed by Europe among the world coffee segments. The EU consumes 2.5 million tons of coffee per year, which equates to 4 kilos of roasted coffee per EU inhabitant per year. As a result, large amounts of spent coffee grounds are generated (Ayala and Fernadez, 2017) Proper management of spent coffee grounds has become a challenging problem as the production of this waste residue has increased rapidly worldwide (Kim et al., 2017). Used coffee grounds are inexpensive and widely available materials that can be utilized to reduce the levels of nitrite, nitrate and ammonia in wastewater (Mariana et al., 2018).

Hence, throughout in this chapter, more specific and current attempts to recuperate food wastes or even high value-added components derived from them are discussed with a special emphasis to the coffee industry.

Food By-Products and Their Recovery

The nutritional and/or functional compounds derived from agricultural and food processing by-products can be recovered either by "upstream" wastes or by "downstream" food wastes.

Figure 1. Waste steps in food life cycle

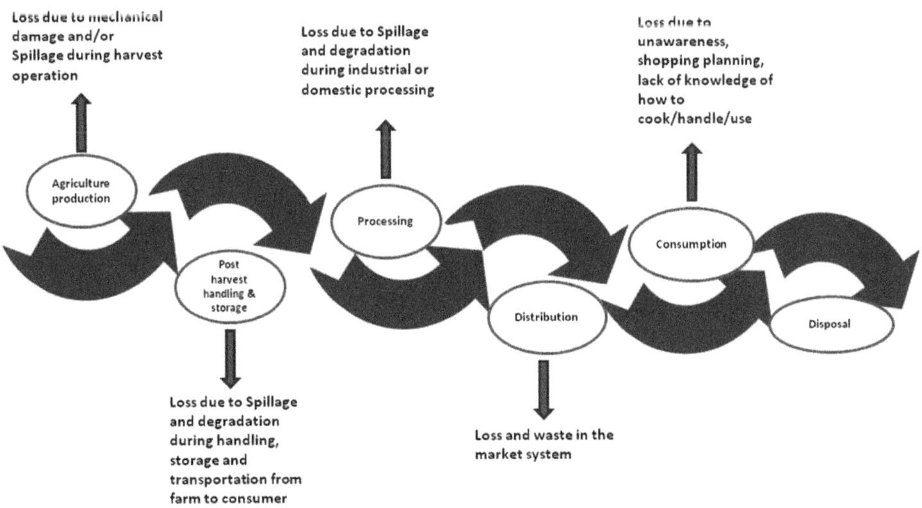

Figure 2. Distribution of food loss and waste

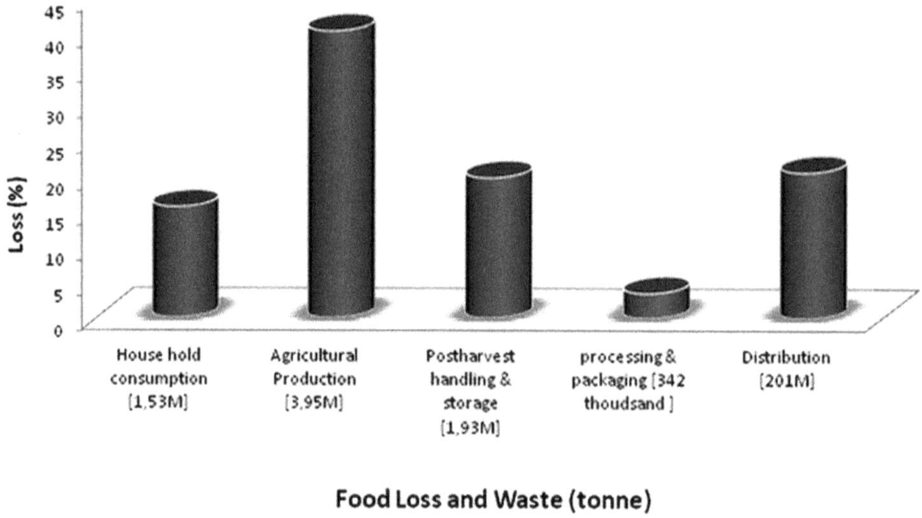

Galanakis, 2012 reports indicate abundant upstream wastes in few locations and vulnerable compared to end product wastes generated in food supply. The "downstream" food wastes are widely dispersed; even they require an additional collection step, which negotiates their valorization as a component recovery source. Food/agriculture wastes originated along the food supply chain and most sources during the industrial level, while lesser created at consumer's stages. Plant processing wastes include the major group and are the most widely studied substrates. The most successfully recovered compounds include the selective extraction of phenols, pectin, and phytosterols, easily utilized in the food and pharmaceutical industries. In general, fruit and vegetable by-products are the most widely and historically investigated area. Once these by-products are formed of soft tissues rich in antioxidants and dietary fibres', the simultaneous extraction in two separate streams is easily performed (Galanakis, 2012). It has also been considered that the largest food wastes volumes occur in the fruit and vegetable group, where coffee is one. Certainly, coffee and cocoa industry is a prime source for huge amounts of wastes all along the entire productive chain, from the field to the cup. Hence, the following chapter aims to elucidate the current status of coffee and cocoa waste volumes and its promising reuses.

Coffee Waste and Its Sustainable Management

Global coffee consumption has increased at an average annual growth rate of 2.4% since 2011 to reach approximately 9 million tons in 2014. Accordingly, the production of spent coffee grounds (SCG), the solid residue remaining after brewing, has also increased steadily (Kim et al., 2017). Thus, the management of SCG has become an increasingly challenging problem.

Coffee Wastes: From Bean to Cup

Coffee a member of family Rubiaceae, widely spread and largely traded commodity. Two coffee varieties gained higher commercial importance – *Coffea arabica*, with greater representativeness, and *Coffea canephora*, usually known as Arabica and Robusta, respectively. Coffee beverage is a complex sequence involving technological processes, begins from the field and ends in the cup. After harvesting, coffee (cherries) undergoes from primary processing, with three different methods i.e. dry, wet and/or semi-dried.

In the dry method, the cherries (beans, mucilage, pulp, and shell) dried either under the sun or in a mechanical dryer, until the dried husk can be easily removed.

In the wet process, the cherries are pulped mechanically and the remnant of mucilage i.e. pectin and sugars are removed by microbial fermentation washed and dried as like dry method.

The semi-dry method is more recent; the cherries are pulped and dried along with mucilage without the fermentation step.

Along with waste a large quantity of contaminated water also produced in several washing steps, which are heavily loaded with carbon and can cause a high impact on the environment.

Actually, two major classes of coffee by-products may be distinguished, one derived from green coffee production, and second one obtained after roasting. Since last decade, the focus is shifted towards the tentative use of these agro-industrial wastes. Each class of coffee waste/residues will be further elaborated in the following subsections along with its prospective novel applications.

Pre-Roasting Coffee By-Products and Their Applications

Dry Processing

Based on the coffee processing method, different residues may be produced. During dry processing, the prime residue generates is coffee cherry husks, includes the coffee cherry outer skin, pulp, and parchment, and contributes about 12% of the berry on a dry-weight basis. From 1 ton of fresh coffee fruits, approximately 0.18 tons of husks are released, producing around 150 to 200 kg of commercial green coffee (Murthy and Naidu, 2012a). Coffee husks are rich in total carbohydrates (58-85%), 8-11% protein, and 0.5-3% lipids (Murthy and Naidu, 2012b). Some other trace amounts of bioactive compounds, like caffeine and chlorogenic acids, are also present in this residue (Franca et al., 2009).

There are several approaches have been made for the re-utilization and applications of coffee cherry husks. The husk can be used as a substrate for biogas and alcohol production, biosorbents in the removal of cationic dyes from aqueous solutions, converted into fuel pellets or extracted for bioactive substances recovery (Jayachandra et al., 2011; Murthy and Naidu, 2012b). Coffee husks are one of the better candidates for direct use as a bed for edible mushroom production and/or composting (da Silva et al., 2012).

Wet and Semi-Dry Processing

The first by-product obtained during wet or semi-dry processing is coffee pulp, contributes 29% dry-weight of the whole cherry (Murthy and Naidu, 2012a). For every 2 tons of commercial green coffee produced, 1 ton of coffee pulp is obtained (Murthy and Naidu, 2012a).

Coffee pulp consists of the outer skin (exocarp) as well as the fleshy portion (Mesocarp). It is basically rich in carbohydrates (32%), proteins (5-13%) and minerals (9%) and also contains significant amounts of tannins, polyphenols, and caffeine (Ulloa Rojas et al., 2003). As like to coffee husks, coffee pulp has also been studied to be reutilized for mushroom production, composting, biosorbent, bioactive compounds extraction purposes (Rodriguez-Duran et al., 2014).

Extraction of the coffee beverage from coffee powder with hot water precipitates residue are called spent coffee grounds(SCG). It can be concluded that tons of coffee waste is generated from cafeterias and domestic production (Fernandes, 2017). SCG is the most abundant coffee by-product (45%) generated 114 not only in the coffee beverage preparation but also during the instant coffee manufacturing. About 2 kg of wet SCG are obtained from each kg of instant coffee produced, with the generation of several million tons worldwide every year (Martinez –Safez, 2017).

Environmental Impact of Coffee Waste

The spent coffee waste contains large amounts of organic compounds such as fatty acids, lignin, cellulose, hemicellulose, and other polysaccharides that justify its valorization, the wet processing of coffee cherries is an alternative, however generates a large amount of coffee processing wastewater (CPW), rich in suspended organic matter, organic and inorganic compounds in solution, with high polluting potential which must be necessarily treated before its release into the environment (Murthy and Naidu 2012b,Chilakala and Ahn, 2017). The International coffee organization estimates that 6,985,680 tons of coffee beans were produced worldwide in 2004. Coffee processing leads to generation of coffee by-products namely parchment husk, pulp, cherry husks contribute to waste disposal and environment pollution (Chilakala and Ahn, 2017). Developing countries are facing a serious problem in properly disposing of the waste produced by the production of coffee (Chilakala and Ahn, 2017). It is estimated that more than two million tons of coffee waste is generated yearly (Pandey et al., 2000). Mexico produced 232,020 tons of Arabica coffee beans and was the 7th leading producer of coffee in 2004. As a result, Mexico faces the challenge of economically treating coffee waste to avoid soil and water pollution while staying competitive in the global coffee marker. In fact all coffee producing countries face this challenge, however generates a large amount of coffee processing wastewater (CPW), rich in suspended organic matter, organic and inorganic compounds in solution, with high polluting potential, the wastewater from coffee industries has high concentration of organic pollutants, and is very harmful for surrounding water bodies, human health and aquatic life if discharged directly into the surface waters (Enden and Calvert, 2002). Direct discharge or untreated effluent to waterbodies have impacted in health issues among the local residents. Besides these characteristics, the coffee processing wastewater has phenolic compounds (Fia et al., 2010), organic, acidic acids and the fermentation of sugars in the mucilage make the wastewater very acid (pH down to 3.8). Under these acid conditions, higher plants and animals will hardly survive and after the first fermentation of sugars in the wastewater took place, the organic substances diluted in the wastewater break down only very slowly by microbiological processes using up oxygen from the water. This process causes problems as the demand for oxygen to break down organic material in the wastewater exceeds the supply, dissolved in the water, thus creating anaerobic conditions.

Green Coffee By-Products

Caffeine

Caffeine is another one chemical constituents generated during decaffeination of green coffee. Caffeine is a water-soluble alkaloid. It plays an important role in the development of immune resistance against bacterial invaders by increasing the concentration of some immune-competent cells and reinforcing the activity of lysozyme (Ramanavičiene et al., 2003). Caffeine intake has been associated with high blood cholesterol, coronary diseases, and cancer, and other studies suggest that its consumption may lower the incidence of suicide and hepatic cirrhosis (Farah and Donangelo 2006). Caffeine metabolites especially 1-methylxanthine and 1-methylurate, have exhibited antioxidant activity *in-vitro, and the in-vivo* iron-reducing capacity of regular coffee is higher than that of decaffeinated coffee (Lee & Clin Chim, 2000). The antibacterial effect of regular coffee against carcinogenic microorganisms is also higher than that of decaffeinated coffee (Antonio et al., 2010).

Green Coffee Oil (GCO)

Products derived from coffee (*Coffea arabica* L., Rubiaceae) have been long used by mankind as beverages, foods, and cosmetics. Most recently, the oil extracted by cold pressing the unroasted beans of coffee was introduced to the cosmetic market with great impact. This so-called green coffee oil, GCO, has been studied for its activity on the skin health (Chiari et al., 2014). This vegetable oil presents a unique composition and previous studies showed an expressive antioxidant activity against lipid peroxidation (Kroyer et al., 1989).

Green Coffee Spent

Green coffee spent is the final residue obtained after processing the green coffee beans considered as the major residue (70%) generated and represent serious environmental problems in the coffee industry. As these residues are rich in polyphenols like trigonelline, caffeine, theobromine, and theophylline, value-added products can be developed from the highly rich polyphenolic substance of spent.

Post-Roasting Coffee By-Products

After primary processing, green coffee is ready for roasting, which will entirely modify the physical and chemical composition, and as a result, the by-products produced. Coffee silver skin is obtained after coffee roasting and spent after brewing in industries/cafeteria.

Coffee Silverskin

One of the coffee industry residue produced during roasting is Coffee silverskin, often known as “chaff”. It consists on the tegument of coffee beans and thus has a very low mass (4.2% w/w) of the green coffee bean, with reduced environmental impact. It is highly rich in soluble dietary fibre (54% of total dietary fibre) and compounds with antioxidant capacity, particularly phenolic compounds (Mussatto et al., 2011).

There is very little information available on the reuse of coffee silver skin. It constitutes a fine source of antioxidants and dietary fibre and may be considered as a new potential functional ingredient (Pourfarzad et al., 2013). In addition, coffee silver skin can also be utilized as a nutrient source for the production of fructo oligosaccharides and β-furctofuranosidase using *Aspergillus japonicus* trough solid-state fermentation. It can also be used as raw material to produce fuel ethanol or as an ingredient in anti-aging cosmetics and functional foods (Bilbao et al., 2014).

Spent Coffee Grounds (SCG)

"Spent coffee grounds" are the residues/waste acquired from the soluble coffee industry/restaurants and home on day to today basis during coffee brewing. High consumption is also a cause of high waste generation in the soluble coffee industries. Industrial spent coffee grounds estimated 6 million tons annual production worldwide. In general about 2 kg of wet spent coffee grounds are obtained to each 1 kg of soluble coffee produced (Mussatto et al., 2011). Given its high organic content (i.e., high calorific value), the potential of using SCG as a feedstock for biofuels is an interesting question in the context of waste-to-energy conversion. Many recent studies have investigated the conversion of SCG into biofuels, such as biodiesel, ethanol, and biogas using physico-chemical or biological processes. An increasing amount of attention is being paid to this potential with the ever-increasing production of SCG (Kim et al., 2017). The ability of spent coffee grounds to remove heavy metal ions from synthetic solutions and from mining waste leachate was investigated by Alaya and Fernadez,(2018). They found that adsorption of Zn^{2+}, Ni^{2+} and Cd^{2+} was significantly affected by pH. The percentage of metal removal gradually increases with increasing dosage of adsorbent. Thus, SCG can be used without pretreatment to treat mining waste leachate.

Sustainable Utilization of Agriculture Waste From Coffee

Production of Mushrooms

Considering the nutritional, organoleptic and therapeutic aspects, improved methods for mushrooms cultivation have been adopted worldwide. Fan et al. (2001), attempted cultivation of mushroom on coffee industry residues. Utilization of coffee waste such as coffee husk and spent ground, individually or in combination for mushroom cultivation using *L. edodes, Pleurotus* spp. and *Flammulina velutipes* have been very well documented (Murthy and Manonmani, 2008). Valorisation of coffee husk and coffee spent as substrates without any pre-treatment for the cultivation of edible fungus is well known.

Organic Acids

Utilization of Coffee husk for the production of citric acid using *A. niger* in a solid-state fermentation system was reported by Shankaranad and Lonsane (1999). An attempt has also been made by Machado et al. (2002) for the production of gibberellins in SmF and SSF by using coffee husk as the carbon source and further detailed studied will help to optimise the process and utilisation significantly.

Organic Manure

Residues generated from agriculture and coffee industry can be utilized for the production of organic manures. The recycling of waste can be achieved by composting and vermicomposting and it is a cost-effective technology. These recycled products can balance and improve soil nutrient, ultimately results in better growth. Coffee husk is one of the suitable medium for composting and vermicomposting (Murthy and Manonmani 2008).

Enzyme Production

Coffee pulp and husk may be used as substrates for bioprocesses. Recent approaches have shown their re-utilization for the production of enzymes, aroma compounds, metabolites. There are some reports on the production of enzymes like pectinase, tannase, and caffeinase from coffee pulp and husk. Battestin and Macedo (2007) reported the tannase production by using coffee husk with *P. variotii* at optimum conditions. Exploitation of coffee waste for the production of enzymes like amylase, protease, xylanase, cellulose, Tannase, Endoglucanase, endoglucanase, β-glucosidase through fungal organisms such as *N. crassa, A. oryzae, Penicillium* sp., *A. niger, Rhizopus, Penicillium etc.,* (Murthy and Naidu, 2010a,b; ; Murthy et al., 2012 Murthy and Naidu, 2011; Navya et al., 2012a; Navya et al., 2012b; Roopali et al,2013a; Roopali et al,2013b; Roopali et al,2013c; Ritika et al., 2015). Coffee pulp is a rich source of organic matter, the pulp serves as an excellent substrate for production of value-added microbial metabolites through solid state fermentation (SSF) system. Microbial enzymes are the secondary metabolites which are of high demand in various industry. In fact, pectinase is one such microbial enzyme, that plays an important role in coffee beans processing. In the wet fermentation method of coffee cherries, the natural pectolytic microflora present on the cherries is allowed to grow and metabolize to facilitate pectinase production. This enzyme helps to hydrolyze the mucilage layer of the coffee bean and consists mainly of pectin. Coffee pectin hydrolysis has prime importance in end coffee quality. For the hydrolysing, the pectic covering on the coffee beans external addition of microbial pectinase is more focused nowadays. As a result, a large quantity of microbial pectinase is being consumed by the coffee processing industry.

The coffee pulp utilisation for pectinase production adds value since commercial pectinase is expensive. Moreover, pectinases also extensively utilized in fruit processing industries for clarification of fruit juices and wines, for the production of pectin-free starch and also in the curing of cocoa and tobacco (Joshi et al. 1991, Murthy et al., 2012). It has been estimated that the pectinase market for various kinds of industrial processes is about 165 million pounds per annum (Fogarty and Kelly 1983).also Anthocyanins from coffee pulp were extracted and has cosmetic value and also can be used as food colorant (Murthy and Naidu, 2012c)

Ethanol/Biodiesel

The production of bio-ethanol from waste has got recent attention. Machado (2009) studied the utilization of spent coffee as raw material for ethanol production. Spent coffee was subjected to acid hydrolysis process and the obtained hydrolysate was used as fermentation medium by *Saccharomyces cerevisiae* for the ethanol production. Similarly, production of a distilled beverage from spent coffee was also reported by Sampaio (2010). Coffee husks present the excellent potential for residue-based ethanol production Gouvea et al. (2009). In recent years, there has been increased interest in the production of biodiesel out of SCG as a sustainable practice for waste reduction. Coffee is the second largest traded commodity worldwide, and the world's coffee production in 2016/2017 is estimated at 9.34 million tons (Liu, 2017). Extraction of coffee oil from coffee spent or low grade coffee *via* transesterification is significant and cost effective for biodiesel production in comparison with other wastes utilisation. It is less expensive, exhibits higher stability (due to its high antioxidant content) and a pleasant smell (Haile 2014). However, the ester based biodiesel has many limitations, such as producing excessive glycerol, low energy content, low oxidative stability and plugging from the tank to the engine. Moreover, application of biodiesel along with unmodified diesel due to molecular oxygen present in the structure of biodiesel is economical.

Bio-Energy (Biogas)

Coffee husk, a prime agro-industrial waste, can be reutilized for biomethanation using thermophilic *Mycotypha* (Jayachandra et al., 2011). The water sapped from coffee cherry extract is one more possible resource for biogas production. The biogas can be used for the electricity generation, in addition, all the lower grade waste heat from cooling and exhaust can still be utilized for coffee drying (Rathanivelu and Graziosi, 2005).

Prebiotics

Prebiotics are dietary fibres having the positive influence on the intestinal microflora. there are several other effects of prebiotics also well known today which includes prevention of diarrhoea or constipation, modulation of the metabolism of the intestinal flora, cancer prevention, positive effects on lipid metabolism, stimulation of mineral adsorption and immunomodulatory properties they are indirect, i.e. mediated by the intestinal microflora, hence less-well proven. Now a day's research in the food industry is focusing on the use of prebiotic as functional food ingredients to manipulate the composition of colonic microflora to improve health (Aryana & McGrew, 2007). The prebiotic products are safe and are effective in supporting human health.

Melanoidins

Melanoidins are nitrogen-containing, brown-coloured, high molecular weight, compounds generated during the final stages of the Maillard reaction. These are mainly associated with specific colour and characteristic appearance of processed foods. However, the recent signings of melanoidins is linked with their health-promoting properties.

Figure 3. Prebiotics: mechanisms of action

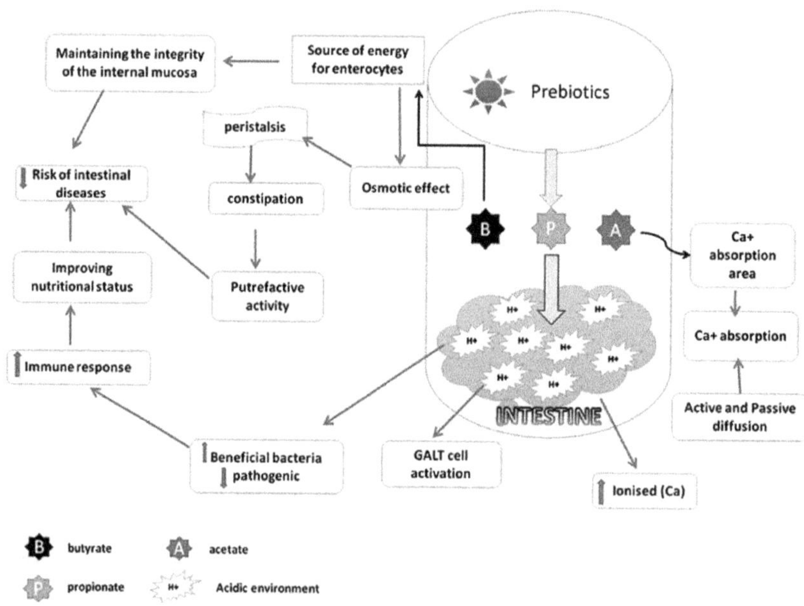

Use of Melanoidins as a Functional Ingredient in New Food Production

Melanoidins are not only responsible for the appearance of processed foods but the importance of these compounds is mainly due to their biological properties. Melanoidins, have been associated with several health beneficial effects including antioxidant, antimicrobial, anti-inflammatory, antihypertensive and prebiotic activity, among others (Echavarrı´a et al., 2012; Morales et al., 2012).

Melanoidins can be produced in model systems but also extracted from processed foods, where are widely distributed as a consequence of thermal treatment applied to food. In addition, these compounds are present in by-products of certain processed foods, such as coffee. In this sense, the exploitation of coffee by-products including coffee silverskin and spent coffee has been proposed as an important source of melanoidins which, in turn, allows revaluing these leftovers and decreasing the environmental problem. The products and by-products containing melanoidins as functional food component has been considered as potential food ingredients for healthier and tasty foods.

Coffee Wine

The most abundantly available agro-industrial waste after coffee cherry processing is Coffee cherry husk (CCH). Adam and Dougan, (1981) reported that for the processing of every tonne of cherries, nearly 0.18 ton of husk is produced. The annual production of CCH is approximately 48.2 thousand tonnes. CCH is a major source of pollutants (Jayachandra et al., 2011) which contributes 23–37% of fermentable sugar along with 3–5% polyphenols on the dry weight basis (Zuluga, 1989). Recently efficient use of CCH is in progress, several attempts have also been made to develop value-added products such as ethanol, SCP, mushroom, enzymes, compost and organic acids using CCH as raw material (Anu Appaiah and Ganesh, 2006). Application of CCH in bioprocess not only provides alternate substrate also helps to overcome pollution problem.

Other Constituents and Uses

Green coffee also contains minor constituents that may be undesirable both for flavour and bioactivity of the brew. Most of these products are microbial by-products that occur due to inappropriate harvesting, weather conditions during primary processing or improper storage. Examples of such incidental compounds are ochratoxin A (OTA) and specific biogenic amines. Other minor undesirable compounds, especially in terms of health concerns, are acrylamide and polycyclic aromatic hydrocarbons

(PAHs) formed by high roasting temperatures. In addition, carbolines Harman and Norharman are formed during coffee roasting; although studies are inconclusive regarding the health effects of these carbolines, they appear to be beneficial.

These industrial spent coffee grounds are a coffee by-product with fine particle size, high moisture (80-85%), high organic load, and acidity (Mussatto et al., 2011a). Due to its chemical attributes, direct disposal into the environment is hazardous. Therefore, plenty of eco- friendly attempts has been proposed recently. Industrial spent coffee may be used as burning fuel in the industrial soluble industry directly, as a source to produce low-cost CO_2 adsorbents, dyes or heavy metals adsorbents, to produce biodiesel and fuel pellets or, other value-added products as H_2 and ethanol, and as a subtract for edible fungus production (Machado et al., 2012; Plaza et al., 2012). Mariana et al., (2018) used the coffee grounds as raw material to be prepared as activated carbon. Furthermore, activated carbon was used to reduce the content of ammonia, nitrite and nitrate in the fertilizer industry wastewater effluent.

Cocoa Waste and It's Sustainable Management

Cocoa Waste and Its By-Products

Cocoa is a globally traded commodity connected in a lengthy and complex value chain – finally ending up in the form of a chocolate product on retail shelves worldwide. Cocoa is having vital importance to the economy of West Africa. It has become the primary agricultural export of the region, and in countries like Cote d'Ivoire, Ghana, Nigeria, and Cameroon it can account for significant portions of national exports and household incomes. It is estimated that throughout West Africa there are more than two million cocoa smallholders and that those touched by cocoa farming may number as many as 10 million.

Cocoa products can be very nutritious with large quantities of sugar and calorie-rich fillers. Chocolate is not high in cholesterol. Cocoa and its components (cocoa solids and cocoa butter) are not recognized as a source of trans fat in the diet. A 50 g milk chocolate bar provides 10 per cent of the UK Estimated Average Requirement of energy, 9 per cent of protein, nearly 22 per cent of calcium, more than 10 per cent of iron and 25 per cent of riboflavin. Chocolate is the richest source of the mineral magnesium, essential for mental health and heart function along with the seeds which are rich in copper, sulfur and Vitamin C thereby contributing to allied health benefits such as

- Promotion of cardiovascular health
- Decreased oxidation of LDL to prevent atherosclerosis or plaque formation
- Reduction in LDL cholesterol

- Elevation in HDL cholesterol
- Suppression of decay-causing bacteria and plaque formation (Water-soluble cocoa extracts)
- Anti-Depressant and euphoric effects (from Tryptophan in chocolate)
- Stimulant effects (theobromine, phenylethylamine in cocoa)
- General improvement in the health and well-being of elderly men (Strandberg et al., 2007).
- Flavanols such as catechin and epicatechin can make up as much as 10 per cent of the dry weight in a cocoa bean. Epicatechin is linked to the reported vascular effects observed after the consumption of flavanol-rich cocoa (Schroeter et al., 2005). This and other polyphenols are said to prevent fat-like substances in the bloodstream from oxidizing and clogging the arteries.

Cocoa Pod Husk (CPH), Cocoa Beans and Shells

The cocoa meal, cocoa bean shells and pod husks all have nutritive value and can be considered as animal feed materials but their use is severely restricted by the theobromine content which is toxic to livestock. Dried fresh CPH can be fed to cattle up to 7 kg per day without toxic effects and up to 2 kg per day to pigs without toxic symptoms. Up to 0.8 kg of cocoa shells (a good source of vitamin D) is acceptable to cows but they are more dangerous to pigs and poultry. In parts of West Africa, CPH is burnt and the ash used as a source of potassium carbonate for the manufacture of soft soap. If the theobromine is removed through cooking for 90 min in water, filtering and drying it will be harmless. Up to 25% of the treated product can be included in rations for pigs without the reduction in weight gain or feed efficiency (Sobamiwa, 1996).

Cocoa Husk (testa)

- Mouthwash and toothpaste - Cocoa bean husk, the outer part of the cocoa beans which is usually discarded during chocolate production, has an antibacterial effect on the mouth and can fight effectively against plaque and other damaging agents.
- Cacao husk pigments - extracted and utilized in Japanese food industry
- Pod gums
- Mulch – for example, Hershey's and Vitasoil cocoa shell mulches

Cocoa Pods

The possibility of using cocoa pods as a cheap source of pectin has been investigated. Pectin is a gel-forming material, which frequently occurs in fruits and has many applications in the food, pharmaceutical, and textile industries. The husks of immature cocoa pods are found to contain 25 to 30% crude pectin (dry weight basis) and mature pod husks 6 to 12% pectin. Cocoa pods husks can be used as a compost or mulch if left to rot in the fields on cocoa estates where they recycle nutrients back into the soil as manure and also serve as a breeding ground for midges. Midges are the chief pollinators of cocoa and increasing the number of midges enhances pollination efficiency and ultimately pod yields (Freire et al., 1996). Cocoa pod husk and cocoa beans shells have relatively high potassium contents and may be used to manufacture fertilizers or composts. Cocoa pod husks may also be burnt and the ash used to manufacture a potassium containing fertilizers.

Pulp/Juice

Pectin is also present in cocoa pulp and juice and in many cocoa producing countries spin-off industries have been created utilizing the pulp and juice of cocoa.

The pulp and juice are also fermented to give a good quality wine and liqueur. Cocoa mucilage has also been used to provide alcohol, vinegar, and other products.

Cocoa Bean

The unfermented cocoa bean has limited food use but can be ground and pressed or passed through an expeller to extract cocoa butter. Cocoa butter has a range of commercial uses in the food, cosmetic and pharmaceutical industries. It is the most valuable product that can be extracted from the cocoa bean and accounts for up to 55% of the mass of the bean. The press cake from unfermented beans can be used as a feedstock but may be too bitter and unpalatable to some animals. The press cake is also used as manure. Waste cocoa beans and cocoa bean shells can be used as a source of theobromine which is then either used directly in medicinal preparations or converted to caffeine. However, these products might have difficulty competing in price with synthetic theobromine and caffeine.

Cocoa bean shells, when used as mulch, contains approximately 2.5% nitrogen, 1% phosphate and 3% potash as well as a natural gum that is activated when watered. This enables the cocoa shell mulch to sluggish soil moisture loss due to evaporation as well as retards weed growth. The cocoa shell texture also deters slugs and snails and helps to prevent plant damage (VitaSoil, 2003).

Cocoa Butter

- Cocoa butter has a range of commercial uses in the food, cosmetic and pharmaceutical industries.
- In the cosmetic industry it is used directly as a skin toner and moisturizer, Indirectly as a base for other cosmetics, it has been used in sunscreens and " anti-aging" creams.
- In the pharmaceutical industry, it is used as a base for many medicinal creams and lotions since it melts at skin temperature. Recent research has shown that cocoa beans contain beneficial antioxidant properties.
- Cocoa flavanols found in wine, stimulate the processing of Nitric Oxide in the body which greatly reduces the incidence of high blood pressure (Warner, 2002).

Cocoa Honey

The mucilaginous layer released due to fermentation is "cocoa honey" which envelops the cocoa grains. This liquid has a sweet-sour flavour and is rich in sugars and bioactive compounds. The residual material in the seeds, rich in carbohydrates, is a source of substrate for the development of micro-organisms responsible for fermentation, important in the formation of the aromas and flavours of the end product (Efraim et al., 2010; Ouattara et al., 2014).

Uses of Cocoa By-Products

- Reduces hair loss and prevents male pattern baldness
- Improve skin health and also reduces ageing signs
- Reduces skin scars and inflammation

Re-Utilization and Applications of Coffee and Cocoa

The waste generated from coffee and cocoa can be utilized for the development of different value-added products and by-products.

CONCLUSION

The present book chapter emphasizes on the valorization of coffee and cocoa by-products and its waste management have been discussed. Though some significant utilization is possible in coffee and cocoa agro waste, further perspective in this area will certainly be more helpful so as to maintain the environment and ecological aspects. In this regards more research has to be initiated to develop efficient methods to control waste by developing value-added quality products. However, considering the waste generated during coffee and cocoa processing, there is a need to appropriate utilization of solid residues for the production of some functional food ingredients.

REFERENCES

Adams, M. R., & Dougan, J. (1981). Waste Products. In R. J. Clarke & R. Macrae (Eds.), *Coffee Technology*. New York: Elsevier Applied Science Publishers Ltd.

Antonio, A. G., Moraes, R. S., Perrone, D., Maia, L. C., Santos, K. R. N., I'orio, N. L. P., & Farah, A. (2010). Species, roasting degree and decaffeination influence the antibacterial activity of coffee against *Streptococcus mutans*. *Food Chemistry*, *118*, 782–788.

Anu Appaiah, K. A., & Ganesh, D. (2006). Coffee cherry husk, a feedstock for methane generation. In *ASIC – Proceedings of 20th ASIC Conference, Association Scientifique* (pp.540–545). Internationale Du café.

Battestin, V., & Macedo, G. A. (2007). Tannase production by *Paecilomyces variotii*. *Bioresource Technology*, *98*(9), 1832–1837.

Bhoite, Navya, & Murthy. (2013a), Statistical optimization, partial purification and characterization of coffee pulp β-glucosidase and its application in ethanol production. *Food Science and Biotechnology, 22*(S), 205-212.

Bilbao, C., Ezequiel, M. E. I., Benavent, M. A., Calleja, M. H., del Moral, M. P., & Ariz, M. V. (2014). *Application of products of coffee silverskin in anti-ageing cosmetics and functional food*. Patent Number: EP2730171 A1.

Chiari, B. G., Trovatti, E., Pecoraro, E., Corrêa, M. A., Cicarelli, R. M. B., Ribeiro, S. J. L., & Isaac, V. L. B. (2014). Synergistic effect of green coffee oil and synthetic sunscreen for health care application. *Industrial Crops and Products*, *52*, 389–393.

Chilakala, R., & Ahn, A. (2017). Environmental Effect of the Coffee Waste and Anti-Microbial Property of Oyster Shell Waste Treatment. *Journal of Energy Engineering*, *26*(2), 39–49.

da Silva, M. C. S., Naozuka, J., da Luz, J. M. R., de Assunção, L. S., Oliveira, P. V., & Vanetti, M. C. D. (2012). Enrichment of *Pleurotus ostreatus* mushrooms with selenium in coffee husks. *Food Chemistry*, *131*(2), 558–563.

Delgado-Andrade, C., Rufia´ n-Henares, J.A., & Morales, F.J. (2005). Assessing the antioxidant activity of melanoidins from coffee brews by different antioxidant methods. *Journal of Agricultural and Food Chemistry, 53*, 7832–7836.

Efraim, P., Pezoa-García, N. H., Jardim, D. C. P., Nishikawa, A., Haddad, R., & Eberlin, M. N. (2010). Influência da fermentação e secagem damêndoas de cacau no teor de compostos fenólicos e na aceitação sensorial. *Food Science and Technology (Campinas), 30*(1), 142–150.

Enden, J.C., & Calvert, K.C. (2002). *Limit environmental damage by basic knowledge of coffee wastewaters*. GTZ-PPP Project-improvement of coffee quality and sustainability of coffee production in Vietnam.

Fan, L., Pandey, A., & Soccol, C. R. (2001). *Flammulina velutipes* on coffee husk and coffee spentground. *Brazilian Archives of Biology and Technology, 44*, 205–212.

FAO. (2011). Global food losses and food waste – Extent, causes and prevention. FAO.

FAO. (2013). Food and Agriculture Organization of the United Nations. Food wastage footprint – Impacts on natural resources, Summary Report. FAO.

Farah, A., & Donangelo, C. M. (2006). Phenolic compounds in coffee. *Brazilian Journal of Plant Physiology, 18*, 23–36.

Fernandes, A. S. (2017). Impacts of discarded coffee waste on human and environmental health. *Ecotoxicology and Environmental Safety, 141*, 30–36.

Fia, F.R.L., Matos, A.T., Borges, A.C., Moreira, D.A., Fia, R., & Eustaquio, V.Junior. (2010). "Remoção, "decompostos fenólicos em reatores anaeróbios de leito fixo com diferentes materiais suporte. *Revista Brasileira de Engenharia Agrícola e Ambiental, 14*(10), 1079–1086.

Fogarty, W. M., & Kelly, C. T. (1983). Pectic enzymes. In W. M. Fogarty (Ed.), *Microbial Enzymes and Biotechnology*. London: Applied Science Publisher.

Franca, A. S., Oliveira, L. S., & Ferreira, M. E. (2009). Kinetics and equilibrium studies of methylene blue adsorption by spent coffee grounds. *Desalination, 249*, 267–272.

Freire, E. S., Mororó, R. C., & Schwan, R. F. (1996). The cacao-pulp agroindustry and the uses of its residues in Bahia: Progress achieved in the last ten years. *Proc. 12th International Cocoa Research Conference*.

Galanakis, C. M. (2012). Recovery of high added-value components from food wastes: Conventional, emerging technologies and commercialized applications. *Trends in Food Science & Technology*, *26*, 68–87.

Gibson, G. R., Probert, H. M., Van Loo, J. A. E., & Roberfroid, M. B. (2004). Dietary modulation of the human colonic microbiota: Updating the concept of prebiotics. *Nutrition Research Reviews*, *17*(2), 257–259.

Gouvea, B. M., Torres, C., Franca, A. S., Oliveira, L. S., & Oliveira, E. S. (2009). Feasibility of ethanol production from coffee husks. *Biotechnology Letters*, *31*, 1315–1319.

Haile, M. (2014). Integrated volarization of spent coffee grounds to biofuels. *Biofuel Research Journal*, *2*, 65–69.

Jayachandra, C., & Venugopal, K.A., & Appaiah, A. (2011). Utilization of phytotoxic agro waste – coffee cherry husk through pretreatment by the ascomycetes fungi *Mycotypha* for biomethanation. *Energy for Sustainable Development*, *15*(1), 104–108.

Joshi, V. K., Chauhan, S. K., & Lal, B. B. (1991). Extraction of juices from peaches, plums and apricots by pectinolytìc treatment. *Journal of Food Science and Technology*, *28*, 64–65.

Kim, J., Kim, H., Baek, G., & Lee, C. (2017). Anaerobic co-digestion of spent coffee grounds with different waste feedstocks for biogas production. *Waste Management (New York, N.Y.)*, *60*, 322–328.

Kroyer, G. T., Kretschmer, L., & Washiittl, J. (1989). Antioxidant properties of tea and coffee extracts. *Proceedings of 5th European Conference of Food Chemistry*, 2, 433–437.

Lee, C., & Clin Chim, A. (2000). Antioxidant ability of caffeine and its metabolites based on the study of oxygen radical absorbing capacity and inhibition of LDL peroxidation. *Clinica Chimica Acta*, *295*, 141–154.

Liu, Y. (2017). Direct transesterification of spent coffee grounds for biodiesel production. *Fuel*, *199*, 157–161.

Machado, C. M. M., Soccol, C. R., Oliveira, B. H., & Pandey, A. (2002). Gibberellic acid production by solid-state fermentation in coffee husk. *Applied Biochemistry and Biotechnology*, *15*, 102–106.

Machado, E.M.S., & Rosa, M., Rodriguez-Jasso, J. A., & Teixeira, S. I. M. (2012). Growth of fungal strains on coffee industry residues with removal of polyphenolic compounds. *Biochemical Engineering Journal*, *60*, 87–90.

Machado, E. S. M. (2009). *Reaproveitamento de resíduos da indústria do café como matériaprimapara a produc¸ ão de etanol* (MSc thesis). Braga, Portugal: Department of Biological Engineering, University of Minho.

Mariana, M., Mahidin, M., Mulana, F., & Aman, F. (2018). Utilization of Activated Carbon Prepared from Aceh Coffee Grounds as Bio-sorbent for Treatment of Fertilizer Industrial Wastewater. *Materials Science and Engineering, 358*, 12-27.

Martinez-Saez, N. (2017). Use of spent coffee grounds as food ingredient in bakery products. *Food Chemistry*, *216*, 114–122.

Mathur, Navya, Basavaraj, & Murthy. (2015). Bioprocess of Robusta cherry coffee with polyphenol oxidase and quality enhancement. *European Journal of Food Research and Technology*, *240*, 319–325.

Monente, C., Bravo, J., Vitas, A.I., & Arbillaga, L., De Pen˜a, M.P., & Cid, C. (2015). Coffee and spent coffee extracts protect against cell mutagens and inhibit growth of food-borne pathogen microorganisms. *Journal of Functional Foods*, *12*, 365–374.

Morales, F. J., Somoza, V., & Fogliano, V. (2012). Physiological relevance of dietary melanoidins. *Amino Acids*, *42*, 1097–1109.

Murthy, P. S., Manjunatha, M. R., Sulochannama, G., & Naidu, M. M. (2012). Extraction, Characterization and Bioactivity of Coffee Anthocyanins. *European Journal of Biological Sciences*, *4*(1), 13–19.

Murthy, P. S., & Manonmani, H. K. (2008). Recycling of spent mushroom substrate to vermicompost. *Journal of Environmental Sciences (China)*, *3*(2), 212–216.

Murthy, P. S., & Naidu, M. M. (2010). Protease production by *Aspergillus oryzae* in solid-state fermentation utilizing coffee by-products. *World Applied Sciences Journal*, *8*(2), 199–205.

Murthy, P. S., & Naidu, M. M. (2011). Improvement of Robusta Coffee Fermentation with Microbial Enzymes. *European Journal of Applied Sciences*, *3*(4), 130–139.

Murthy, P. S., & Naidu, M. M. (2012a). Production and application of Xylanase from *Penicillium* sp. utilizing coffee by-products. *Food and Bioprocess Technology*, *5*(2), 657–664.

Murthy, P. S., & Naidu, M. M. (2012b). Sustainable management of coffee industry byproducts and value addition – A review. *Resources, Conservation and Recycling*, *66*, 45–58.

Murthy, P. S., & Naidu, M. M. (2012c). Recovery of phenolic antioxidants and functional compounds from coffee industry by-products. *Food and Bioprocess Technology*, *5*(3), 897–903.

Murthy, P. S., Naidu, M. M., & Srinivas, P. (2009). Production of amylase under solid–state fermentation utilizing coffee waste. *Journal of Chemical Technology and Biotechnology (Oxford, Oxfordshire)*, *84*, 1246–1249.

Mussatto, S. I., Carneiro, L. M., Silva, J. P. A., Roberto, I. C., & Teixeira, J. A. (2011). A study on chemical constituents and sugars extraction from spent coffee grounds. *Carbohydrate Polymers*, *83*, 368–374.

Navya, P.N., & Bhoite, R. N., & Murthy, P. S. (2012a). Improved β-glucosidase production from *Rhizopus stolonifer* utilizing coffee husk. *International Journal of Current Research*, *4*(8), 123–129.

Navya, P.N., & Bhoite, R.N., & Murthy, P. S. (2012b). Bioconversion of Coffee Husk Cellulose and Statistical Optimization of Process for Production of Exoglucanase by *Rhizopus stolonifer. World Applied Sciences Journal*, *20*(6), 781–789.

Navya, P. N., & Pushpa, S. M. (2013). Production, statistical optimization and application of endoglucanase from *Rhizopus stolonifer* utilizing coffee husk. *Bioprocess and Biosystems Engineering*, *36*, 1115–1123.

Nestel, A. J., & Fernandez, B. (2018). Treatment of mining waste leachate by the adsorption process using spent coffee grounds. *Environmental Technology*, *15*, 1–15.

Ouattara, D. H., Ouattara, H. G., Goualie, B. G., Kouame, L. M., & Niamke, S. L. (2014). Biochemical and functional properties of lactic acid bacteria isolated from Ivorian cocoa fermenting beans. *Journal of Applied Biosciences*, *77*, 6489–6499.

Pandey, A., Soccol, C. R., Nigam, P., Brand, D., Mohan, R., & Roussos, S. (2000). Biotechnological potential of coffee pulp and coffee husk for bioprocesses. *Biochemical Engineering Journal*, *6*, 153–162.

Phimsen, S. (2016). Oil extracted from spent coffee grounds for bio-hydrotreated diesel production. *Energy Conversion and Management*, *126*, 1028–1036.

Pourfarzad, A., Mahdavian-Mehr, H., & Sedaghat, N. (2013). Coffee silverskin as a source of dietary fiber in bread-making: Optimization of chemical treatment using response surface methodology. *Food Science and Technology (Campinas)*, *50*(2), 599–606.

Ramanavičiene, A., Mostovojus, V., Bachmatova, I., & Ramanavičiene, A. (2003). Anti-bacterial effect of caffeine on *Escherichia coli* and *Pseudomonas fluorescens. Acta Medica Lithuanica, 10*, 185–188.

Rathanivelu, R., & Graziosi, G. (2005). Potential alternative use of coffee waste and by-products. ICO Proceedings, ED 1967/05.

Rodriguez-Duran, L. V., Ramirez-Coronel, M. A., Aranda-Delgado, E., Nampoothiri, K. M., Favela-Torres, E., & Aguilar, C. N. (2014). Soluble and bound hydroxycinnamates in coffee pulp (*Coffea arabica*) from seven cultivars at three ripening stages. *Journal of Agricultural and Food Chemistry, 62*(31), 7869–7876.

Roopali, Navya, & Murthy. (2013c). Purification and characterization of a Coffee Pulp Tannase Produced by *Penicillium verrucosum. Journal of Food Science and Engineering, 3*, 323–331.

Sampaio, A. R. M. (2010). *Desenvolvimento, de tecnologias para produc¸ ão de etanol a partir do hidrolisado da borra de café* (MSc thesis). Braga, Portugal: Department of Biological Engineering, University of Minho.

Schroeter, H., Christian, H., Balzer, J., Kleinbongard, P., Keen, C. L., Hollenberg, N. K., ... Schmitz, H. H. (2006). Epicatechin mediates beneficial effects of flavanol-rich cocoa on vascular function in humans. *Proceedings of the National Academy of Sciences of the United States of America, 103*(4), 1024-1029.

Shankaranad, V. S., & Lonsane, B. K. (1999). Coffee husk: An inexpensive substrate for production of citric acid by *Aspergillus niger* in a solid-state fermentation system. *World Journal of Microbiology & Biotechnology, 10*(2), 165–168.

Sobamiwa, O. (1996). Cocoa pod husk utilization in animal feeds: summaries and strategies. *Proc. 12th International Cocoa Research Conference.*

Bhoite, Navya, & Murthy. (2013b). Statistical optimization of Bioprocess parameters for enhanced Gallic acid production from coffee pulp tannins by *Penicillium verrucosum. Preparative Biochemistry & Biotechnology, 43*(4), 350–363.

Strandberg, T. E., Strandberg, A. Y., Pitkala, K., Salomaa, V. V., Tilvis, R. S., & Miettinen, T. A. (2007). Chocolate, well-being and health among elderly men. *European Journal of Clinical Nutrition, 62*(2), 247–253.

Ulloa-Rojas, J. B., Verreth, J. A. J., Amato, S., & Huisman, E. A. (2003). Biological treatments affect the chemical composition of coffee pulp. *Bioresource Technology*, *89*, 267–274.

VitaSoil. (2003). *Cocoa Sell Mulch* Retrieved from http://www.vitasoil.com. cocoa%20main.html

Warner, J. (2002). *Chocolate: The next health drinks?* WebMD Health Inc. Retrieved from http://www.content.health.msn.com/article/1671.52817

Zuluaga, V. J. (1989). Utilizacion integral de los sub products del cafe. In *Proceedings of I Seminario Internacional Sobre Biotechnologia en la Agro industria Cafetalera (SIBAC).* Xalapa, Mexico: ORSTOM.

Chapter 9
Utilization and Management of Food Waste

Shriram M. Naikare
SNDT College of Home Science, India

ABSTRACT

The food industry generates a huge amount of waste annually around the globe from a variety of sources. Approximately one third of all food produced today goes to landfill as waste. The food waste is not only a humanitarian problem, but also a serious economic and environmental pollution problem. The global volume of food wastage has been reported to around 1.3bn tones worth to about $165 bn. In India, about 40% of the food produced is wasted, which is estimated to about Rs. 50,000 crores worth every year. The important types of food wastes generated are agricultural residue, processed food, fruit and vegetable processing, marine food, dairy processing, meat and poultry, hotel and restaurant, etc. The food industrial waste can be converted into byproducts mainly based on the processing of fruits and vegetables and allied food manufacturing, supply and distribution, livestock feed, using it as source of bioactive compounds, useful bioenergy production, artificial fertilizer and decomposed manure, a variety of chemicals, antioxidant, nutraceuticals, etc.

INTRODUCTION

Food supply and waste management are the emerging challenges for the policy makers and companies in the food supply and processing. The global population is expected to grow 9 billion and demand for food upto 77% by 2050. Over the same period, food production will be under threat from climate change, competing

DOI: 10.4018/978-1-5225-7706-5.ch009

land uses, and erosion and diminishing supplies of clean water. The food which we consume has to undergo a series of food processing operations soon after harvesting at the farm level.

The agro-food industry generate huge amount of wastage annually around the globe from a variety of sources. Food is a basic need of human beings, while food waste has been identified a major crucial challenge faced by human community today (Gustavsson et al, 2011).

Over 4.2 million tons of food waste is dispersed to landfill in Australia each year. 2.7 million tons of this is from households and around 1.5 million tons of this is from commercial and industrial sector, (DEWHA, 2009) costing around $ 10.5 billion in waste disposal charges and lost product. The largest single contributor in the commercial and industrial sector is food service activities.(Example- Cafes, restaurants, fast food outlets), which generate 661,000 tons of food waste per year, followed by manufacturing (312,000 tones) and food retail (179,000 tons). Most waste in food manufacturing is unavoidable, and almost 90% is already recovered as animal feed, compost or bio-energy. (Verghese et al 2013)

Presently, around 21,000 people die every day due to hunger related causes (Vandermeersch et al, 2014) and globally one in nine people go to bed each night hungry(http.//www.fao). Nevertheless, approximately one third of all the food produced goes to landfill as waste (Memon,2010). The vast amount of food ending up as waste is not only a humanitarian problem but also serious economic, nutritional and environmental pollution problem (Sakai et al,2011, Autrey et al, 2007).

At global statistics, according to the British Institute of Mechanical Engineers (IME) half of the food produced is wasted worldwide at different stages. The global volume of the food wastage has been reported to around 1.3 billion tons. The total volume of water used each year to produce food that is lost or wasted (250 km3) i.e. equivalent to the annual flow of Russian's Volga river or three times the Lake Geneva. Similarly, 1.4 billion hectares of land 28% of the world's agriculture area is used annually to produce food that is lost or wasted (FAO, 2015). About $ 165 billion worth of food waste enters landfills each year.

In India, according to UN Development program 40% of the food produced is wasted at pre- and post-harvest stages. Ministry of Food Processing Industries, Government of India's resources about Rs. 58,000 crore worth of food is wasted every year. About 25% of fresh water used to produce food is ultimately wasted as millions of people still don't have access to drinking water. About 300 million of barrels of oil are used to produce food that is ultimately wasted. As a result, a large quantity of food is wasted and being thrown away around the world while a child dies every five seconds because of hunger. In terms of food waste- agricultural produce, meat, poultry and milk- India ranks seventh, with the Russian Federation at the top in the list. India's major land is under agriculture, hence there is highest wastage of cereals, pulses, fruits and vegetables. Meat accounts for just four percent of the food wastage but contributes 20% of the economic cost of the wastage. Wastage of fruits and vegetables is 70% of the total produce, but translated into only 40% of the economic losses. Also, rice crop emits methane, a potent global warming gas, because of the decomposition of organic matter in submerged paddy fields. Food loss and waste costs the world about $ 940 billion a year.

However, the utilization and disposal of food waste is difficult due to its inadequate biological stability, potentially pathogenic nature, high water content, potential for rapid autoxidation, microbial decomposition through high level of enzymatic activity. The world population will reach to 9.6 billion by 2050 (FAO, 2015).

WORLD ENVIRONMENTAL PROBLEMS

Population growth contributes to GHG (Green House Gas) emission through its effect on deforestation as land is grabbed for enhancing food production (Lambin and Moyfroidt, 2011). As the world's population grows and becomes more affluent, waste production rises and might double by 2025 (Hoornweg et.al, 2013). According to the US Environmental Protection Agency (EPA), food wastage currently represents the single largest type of waste entering landfills (Nishida, 2014) Wasted food leads to over utilization of water and fossil fuels and to increasing greenhouse gas emission i.e. methane and carbon di oxide arising from degradation of food in landfills (Hall et. al.,2009).

Therefore, the environmental impact of food waste is twofold (Morane, 2016)

1. It is associated with the depletion of natural resources used for its production (example soil depletion) and distribution.
2. It relates to the costs associated with waste disposal. There is a growing awareness needed to minimize the amount of food wasted at the end of the food supply chain- an issue particularly relevant in high-income countries where more than 40% of the food losses occur at retail and consumer level (FAO, 2015).

Globally per capita food waste by consumers amounts to 95-115 kg/ year in Europe and North America compared to 6-11 kg/year in South or South East Asia and Sub-saharian Africa (Gustavsson et. al., 2011). Food waste reduction at the consumption level represents indeed a large target for medium and high income countries, where evidence shows that the main source of the problem is the domestic setting (Monier et. al. 2010; Braun 2012).

Reasons for Food Losses and Food Wastage

Agricultural Production: Destruction from insects, pests, diseases, inappropriate crop cultivation practices, changing agro-climatic conditions, not meeting the quality specifications, low yielding varieties, lack of inputs, poor crop yield due to draught and natural calamities, etc.

1. **Post-Harvest Handling and Storage Practices:** Not meeting the specifications for quality and/or poor or lack of post-harvest handling, packaging, storage facilities may lead to damage due to insect, pest, spillage, germination and degradation (lack of pack houses, packaging materials, pre-cooling facilities, storage and transport facilities (cold chain, cold storages, poor supply chain management, etc.)
2. **Lack of Primary Processing and Packaging Facilities:** Inadequate infrastructure such as godowns, ware houses, cold storages for perishable commodities, referred vans for high value commodities like grapes, strawberry, broccoli, milk and milk products, poultry, meat, fish, etc. these operations create trimmings and other food preparation waste. Inedible portions, wet or dry material, their storage and transport or proper utilization at proper stage. Wet or dry garbage may create severe problems of their proper disposal, failure may create air pollution and health hazards.
3. **Food Processing Industry Sector Waste**
4. **Distribution and Logistics (Wholesale and Retail):** Damage or loss of food in transit/ storage due to packaging failures, shelf life of processed, fresh food commodities, poor road facilities, transit storage (warehouse/ cold storage) at the port or metro cities hub. Packaging failures, product spoilage, fresh produce (perishable), may get damaged during handling, storage and distribution, short shelf life hence low sales.
5. **Food Service Sector:** Food wastage generated in the hotels, restaurants, institutional kitchens, poor management of such wet food wastages, their packaging, boxes, plastics, improper food handling, left over or stale food items.
6. **At Home:** Trimmings, cuttings, peels, stones, seeds, and other food preparation waste, damaged or spoiled food items, preparing too much food, leftover food, improper stored food and food items. The overall food loss and wastage costs the world about $ 940 billion a year. The food losses are reported to be higher in developing countries than the developed nations. However to overcome and handle the food wastage problem is a huge challenge and task all over the globe.

The overall food loss (waste) in USA alone, annually people throw away 30% of the food produced which corresponds to 40 billion liters of water. Whereas in UK, the household waste estimated to be 6.7 million MT purchased. This means that approximately 32% of all food purchased every year is not eaten. Most of this (5.9 million MT or 88%) is currently collected by local authorities. Most of the food waste (4.1 million MT or 61% is avoidable and could have been eaten if had been better managed

- The annual food losses and waste are estimated to be about 30% for cereals, 40-50% for root crops, 30% for fish and 20% for oilseeds and meat
- On globe scale, just 43% of the fruits and vegetables produced are consumed and the remaining 57% are wasted
- Food waste accounts for roughly US $680 billion in industrialized countries and US $ 310 billion in developing countries
- Roughly one-third of the food is lost or wasted that translates into 1.30 billion MT each year worth nearly one trillion US dollars and equivalent of 6-10% of human generated greenhouse gas emission (Bos and Hamelinck, 2014).

Classification of Food Waste (Based on Nature of Waste)

1. Solid Waste (Organic and Inorganic) Sources- domestic waste, factory waste, waste from oil industry, e-waste, agricultural waste, food processing waste, variety of plastic based waste, packaging material (industry and domestic waste) etc. (Mackensine et al). out of the total solid waste generated, 44% is wet (organic)
2. Wet Waste
    a. Kitchen waste (food waste, cooked and uncooked food, egg shells, meat and bones, fish, fruit and vegetable inedible portion etc.
    b. Flower, fruit and vegetable waste
    c. Garden, tree, leaves, branches, straws, trash waste
    d. Sanitary waste (drainage waste)
    e. Food industry waste (raw materials and finished goods)
    f. Food waste (left over, stale, spoiled food)
    g. Wet garbage and industry (sewage) waste
3. Dry Waste
    a. Paper, plastic (all kinds), laminates, foils
    b. Card boards, cartoons, packaging, glass bottles, metal tins and containers, strappings, foils, rags, rubber, houses, pipes, sweepings, ashes, wrappings, discarded clothes, etc.

4. Domestic Hazard Waste
    a. Compact florescent lamps, tubes, glasses
    b. Chemicals, detergents, etc.
5. Non-Hazard Waste
    a. Glass bottles, iron containers/ wares, plastic bottles/ wares and materials

Food Waste From Different Food Groups

- Cereals (grains), pulses, fruits and vegetables, meat, dairy products, marine, sugarcane, winery, plantation by-products, slaughter house, canning industry.
- Wastes are untreated and underutilized; therefore its disposal is widely adopted through burning, dumping or land filling.
- Juice industry produced a large amount of waste as peels, pulp, seeds, fiber.
- Fruit and vegetable processing industry waste.

Reasons for Food Waste Generation: Scenario

As per the FAO report, around one third of the food produced for human consumption is lost or wasted globally, which is equivalent to 1.3 billion ton each year (Gustavsson et al, 2011; WRAP, 2011). In the United States, the figure is likely to be closer to 40% (Hall et al, 2009). The per capita food loss for North America and Oceania combined is estimated to be around 280-300 kg/year, which is equivalent to around6.5 million tons of food waste in Australia (ABS, 2013).

Around 4.2 million tons of food waste is disposed to landfill in Australia each year with almost half of the commercial and industrial waste coming from the food service sector. Source: (DEWHA, 2009).

While considering the food losses/ wastes at different stages of processing, in less developed economies, foods tend to be lost at the agricultural cultivation and post-harvest stages (Kummu et al, 2012) due to the inefficient harvesting, storage, transport and processing. Waste tends to move up the distribution to the retail and consumer levels as the standard of development improves (IME, 2013, Kummu et al, 2012). This is where food is much more likely to be thrown away when it is still edible (Gustavsson, 2011). Verghese et al, 2013 reported that the largest single contributor to food waste in Australia is the food service sector (Food and beverage services) such as hotels, pubs, restaurants, cafes and commercial caterers, which recycles only 2% of the food waste they generate and send approximately 645,000 tons of landfill each year.

The second largest contributor is the food retail sector which also recycles very little (5%) and sends around 170,000 tons to landfill each year. The areas of high loss are the perishable products such as fruits, vegetables, meat, bread and cut flowers. Another 75,000 tons is sent to landfill from wholesale trade sector.

Further they observed that the food manufacturing sector generates a significant amount of food waste but with a recycling rate of 88% sends very little to landfill. A large proportion of this waste is unavoidable, for example skin, bones and other inedible food components. One of the reasons for the high recovery rate for food waste is that manufacturers produce relatively consistent and uncontaminated wastes that can be used for animal feed or as feed stock for composting.

Finally, the team reported that the remaining food waste is generated in the manufacturing and service organizations that are largely outside the food supply. Most of this waste is related to employee consumption, i.e. generated in canteens and kitchens.

Low recovery rates for commercial and industry waste sector can be attributed to inadequate infrastructure for recovery, difficulties in on-site handling, storage and collection and low value of this material compared to other recyclables (US Report, 2012). This waste represents a significant cost to business. In addition to the costs of waste disposal and recycling, the value of the food inputs that are ultimately thrown away or recycled by the commercial and industrial sector in Australia is estimated to be around $ 10.5 billion (E CSRU, 2012).

The edible components of food wasted at each stage of the supply chain in North America and Oceania. For example, wastage rates for fruits and vegetables in the supply chain are 4% in post-harvest handling and storage, 12% in distribution including retail. Overall wastage rates are highest in consumption 35% followed by agricultural production sector 20% (Gustavsson et al, 2011). Perishable products (high moisture) have a short shelf life such as fresh fruits and vegetables, baked goods, meat and seafood have a higher tendency to become waste (Mera, 2011).

Food waste is the food not suitable for human consumption, no longer fit for sale, which is subjected to livestock feed or fertilizer through decomposition. Major food waste generates during distribution or storage processes at warehousing or in-store display.

In food service operations, more food is been consumed away from home in restaurants, cafes or 'take-away' (home delivery) food (IME, 2013).

The Sustainable Restaurants Association (SRA) in UK identified three main sources of food waste and estimated that if an average restaurant reduced its waste by 20%, it could save more than 2,000 pound from avoided food costs and up to 1700 on avoided waste collection costs, 65% from preparation, 30% from customers' plates and 5% spoilage (out of date). (SRA, 2010)

In industrial countries, the large amount of food wasted is generated by households, Australians waste about $ 5.2 billion worth of food every year (Baker et al, 2009). The research on other countries has revealed some interesting insights, that perishable foods such as fruits, vegetables, dairy products and pre-prepared meals are the largest contributors to food waste. (Ventour, 2008, Williams et al, 2012)

Solutions for Reduction of Food Industry Waste

- Effective supply chain management practices to fresh agro produce (fruits and vegetables, dairy products)
- Reduction in food wastage (at processing, storage, distribution)
- Improvement in post-harvest handling practices, transport, storage and distribution of food through appropriate technologies (cold chain, improved packaging etc.)
- Value addition of the by-products generated in the food industry
- Quick and appropriate disposal of food industry wastage, garbage, effluents, sewage, etc.
- Food lost or wasted should be discarded to avoid environmental pollution (each year it accounts for 3.3 billion tons of carbon di oxide emission globally) (FAO, 2015)
- Government and Community must work collaboratively to achieve policy of zero waste or policy "No to food waste".
- The agro-industrial residue have high nutritional potential, therefore it can be utilized for production of a variety of by-products, chemicals (Grawinha et al, 2008) or any suitable disposal.
- Conversion of waste into valuable product through biodegradation/ decomposting.
- Fermentation of the solids/ semi-solid waste.
- Formation of 'Food Banks' and its timely distribution to the needy/ hungry population
- Bio gas (fuel gas) production
- Composting through earthworms/ microbes into manure

Utilization and Management of Food Waste

The food waste can be categorized as solid (organic and inorganic), semi-solid waste, dry waste and liquid (wet waste). The food processing industry generates vast, hazardous either by-product waste or material ready for discard causing harmful effect to human beings and animals, creating severe environmental pollution (solid, liquid, gas pollution)

The present scenario of overall waste management in India indicates that the waste used for biogas production 5%, composting 18% and vermicomposting 32% (Matkar and Singh, 2007)

Classification of waste according to their properties is shown in Figure 1. Worldwide the food waste, garbage processing has become a crucial problem. Garbage processing countries (%)

- Austria – 63%
- Germany – 62%
- Taiwan – 60%
- Singapore – 59%
- South Korea – 49%
- Britain – 39%
- Italy – 36%
- France – 35%

Source: Strategy paper on Solid Waste management (Nitri, Nagpur)

The agro industrial waste produced by food industries is mainly based on the processing of fruits and vegetables. It is estimated that the food industry in Europe generates about 250 million tons/year of byproducts, waste and effluents and 6% of them represented by fruits and vegetables (De Los Fuenters et al, 2004). Waste and byproducts include damaged fruits, leaves, unripe, immature, peels, stones, stalks,

Figure 1.Classification of waste

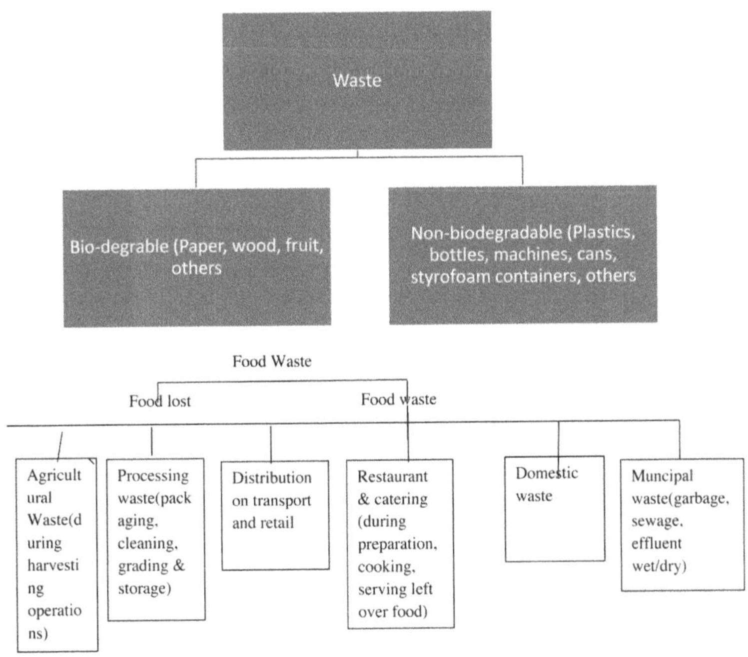

Table 1. Different types of Garbage generated in India

TYPE OF GARBAGE	%
Organic	50%
Plastic	33%
Metal	1%
Paper	4%
Glass	6%
Others	6%

Source: World Bank Report on 'What a Waste', 2012

etc. A huge amount of waste are also generated during food processing of juices, canned foods, sauces, liquors, dehydration as solid and liquid residue that are usually disposed into landfills or used as compost or animal feed.

All the wastes pose increasing problems of disposal and potentially severe pollution problems. In Italy, total production of tomato accounts to nearly 9,000,000 tons/year and about 1.8% to 2.3% of them (i.e., 162,000 to 207000 tons) is discarded as waste (unripe, damaged, peels and seeds). Agro Industrial wastes represents a cheap, chemical, feed stock for extraction of useful chemicals, byproducts, since they are rich in high value components like lipids, fibers, natural pigment, carotenoids and oxidants, nutraceuticals, phenolic compounds (Velentia et al, 2014)

and reservoir of complex carbohydrates, proteins which can be utilized for the production of commercially important metabolites. Agro-Industrial wastes are useful for manufacturing of bio-fuels, enzymes, vitamins, antioxidants, animal feed, antibiotics and other chemical through solid state fermentation (SSF). A variety of micro-organisms are used for the production of these valuable products through SSF processes.

Types of Agri-Industrial Waste

It includes leaves, stalk, seed, pods, stem, peels, stubbles, plant, branches, husks, seed coat, stones, cores, etc. while the process residue include husks, immature seeds, roots, molasses, etc. the industrial waste contains peels, fruits, seeds, stones, pod shells, coconut shells and fibers, soya bean pod shells, cake, etc.

Sugar cane, rice straw, corn stalks, saw dust, sugar beet waste, pomace, barley straw, cotton stalk, oat straw, soya stalks, sunflower heads, stalks, wheat straw, food industry waste – potato, sweet potato, mango, orange peels, pineapple peels, coffee skins (hulls).

Table 2. Food and agricultural commodity waste world production (Million)

Sr.no	Commodity	Production yield	Waste
1	Apple	61.9	4.9
2	Banana	71.3	8.4
3	Cassava	202.6	26.8
4	Coconut	54.7	2.5
5	Citrus fruit	108.8	8.6
6	Coffee	7.8	0.1
7	Corn	221.4	26.1
8	Grape	66.6	2.2
9	Olive	16.0	0.2
10	Onion	55.1	4.7
11	Pineapple	15.3	1.6
12	Potato	327.6	22.4
13	Soybean	204.20	4.2
14	Sugar beet	249.2	0.5
15	Sugar cane	1324.0	117.5
16	Tea	3.3	0.07
17	Tomato	120.4	9.7
18	Wheat	627.1	20.0
19	Total cereals	2264.0	78.2
20	Total fruits	503.3	42.4
21	Total oil crop	132.7	10.4
22	Total vegetables	865.8	70.2

Source: FAO, 2004

Global Food Waste by Commodity

1. Fish and sea food : 2%
2. Oil seed and pulses : 3%
3. Meat : 4%
4. Milk : 8%
5. Cereals : 20%
6. Fruits and vegetables : 45%

Table 3. Major processed food and types of waste generated

Sr. no	Food crop	Food product	Waste
1	Rice, wheat, corn	Grain, flour, bread, biscuits, roti, cake, starch, flakes, bakery products	Straw, stem, leaves, husk, comb, hulls, fibers, brans, germ, gluten, todder
2	Fruits and vegetables	Juice, pulp, preserved products, vegetable oil, potato products, fruits, roots, tubers, bulbs, sugar dehydrated, pickles, fermented products	Rotten fruits, vegetables and their parts, pomace, skin, seeds, stones, fibers.
3	Fish and sea food	Canned, salted fish, smoked fish, processed form, dehydrated, frozen	Scales, fins, shells, bones, guts, fish oil, skeleton.
4	Meat and poultry	Processed meat(beef, pork, poultry, eggs and their products)	Blood, hairs, head, skin, horn, bones, carcass, fat, feet, guts, wide intestinal parts.
5	Dairy products	Milk, butter, cheese, milk powder, cream, ghee, paneer, ice cream,	Whey, processed water, solids, waste material, effluents etc.
6	Beverages	Cocoa, coffee, tea, fruits, alcohol(wine), molasses, grain based alcohol	Shells, seed coat, molasses, sewage water.
7	Oils	Oil, hydrogenated fat, fatty acids	Cake, solid impurities, water effluents, rancid spoiled seeds oil
8	Sugar	Sugar. jaggery, confectionary,	Solid wastage, sugar industry effluents, waste.

Source: FAO ,2015

EPA Has Given Food Recovery Hierarchy

1. **Source of Reduction:** Reduce the huge volume of food generated
2. **Feed Hungry People:** - Donate extra food to food banks, kitchens, shelters
3. **Feed Animals:** Divert food surplus to animal feed.
4. **Industrial Uses:** Extraction oils, chemicals, valuable nutrients and industrial aids from waste, conversion into fuels, drying and improving storage through powders, use in pharmaceuticals, allied uses.
5. **Composting:** Use for bio fermentation, biogas, fertilizers, composting through bacteria, earthworms, solid waste fermentation
6. **Landfill:** Incineration

The percent wastage reported along the food supply chain at different stages is

1. Pre harvest – 25% (food cost at pre harvest)
2. Post-harvest stage – 20- 40% (post-harvest losses)

3. Processing losses – 30% (grain and others – cleaning, grading, packaging)
4. Transportation – 20% (during transport spoilage, storage, poor packaging, over loading, lack of cold storages and chain transport for perishables F & V)
5. Retailing - 10% (handling, food cost, number of handlers, middle man are increased, short distance and local movement),
6. Consumption – 40% (produced, prepared food wasted during eating, kitchen storage, quality loss due to excessive purchasing).

Consequences of Food Loss

1. Wastage of valuable bulk
2. Loss of bulk nutrients
3. Loss of functional nutraceuticals nutrients (natural ingredients/ nutrients)
4. Severe problem of their disposal, transport, movement
5. Being wet/ perishable likely to undergo fermentation on quickly; need to provide additional attention.
6. Emission of toxic gases (CO2, CO, methane), microbes, when wasted food is kept open as such or buried in landfills.
7. Air, water, atmosphere get severely polluted due to improper disposal.
8. Loss of energy, manpower, water, land, etc. for growing of the food being lost.
9. Heavy financial loss to the community/ government on disposal.
10. Loss of soil fertility (that soil remain as waste land)
11. Emission of greenhouse gases (methane, CO2, CO, SO2, H2S, etc.)
12. Food industry causes health hazards and air pollution to human beings.

Measures for Reducing Food Losses

Harvesting the agricultural commodities at optimum maturity stage by adopting suitable harvesting aids.

- Proper handling and threshing, grading, drying and bagging of the produce at proper storage conditions.
- For packaging reusable plastic crates (woven bags with plastic liners) be introduced as primary or secondary packaging in supply chain operation to improve efficiencies and extend shelf life, particularly for fresh perishable produce, (Chonhenchob and Singh, 2003) and it can produce the environmental benefits of reusable, more robust structure for food supplies in the food supply chain. (Lee and Xu, 2004, Singh et.al, 2006.)

- Supply of surplus and unsalable processed food, commodities to food rescue organization from farm / food industry / food storages to food recovery organization. (Varghese et.al, 2013.)
- Pre-processing and packaging of food produce can reduce food waste in supply chain and in the home by extending the shelf life.
- Application of the improved packaging technology to fresh processed food products extended their shelf life significantly through multi-layer barrier packaging, modified atmosphere packaging (MAP), edible coatings, ethylene scavengers, oxygen and carbon dioxide scavengers, moisture absorbers, aseptic packaging, tetra pack packaging, retortable pouch packaging, vacuum packaging, N2 gas flushing packaging, grape guard pad in package etc. (Varghese et.al, 2013)
- Adoption of cold chain supply and cold storage will help to reduce the losses of perishable food commodities significantly.

Utilization of Fruits and Vegetables Waste

- Fresh wet waste utilization is worldwide concern to dictating the improvement of alternative cleaner and renewable bioenergy resources (Okonko et.al, 2009) these waste causes serious disposal problem.(Rodrigoue 2008).
- Over the last decade the annual production of fruits and vegetable have been increased by 70%. However, the economy is suffering the loss of about $ 750 billion (i.e. Rs. 47 lakh crore), as 1/3rd of the food produced goes wasted.
- The juice industries produce huge amount of waste as peels, coffee pulp waste, pomace, fruit seeds, stones, etc. All over the world the fiber sources are found to the tune of 147.2 million metric tons in 1990. (Belewu and Babalola, 2009)
- As per the composition of these agro industry waste / residue are concerned, they constitute high nutritional profile and hence being used for agro industry by products. (Grawinhna et.al, 2008)
- Various studies reported that different kinds of wastes such as pomegranate peels, lemon peels and green walnut husks can be used as natural antimicrobials (Adame et.al.)
- Same food industry by products / wastes contain high amount of proteins (soya cake ground cake), sugars (molasses) and minerals (rice bran, wheat bran). Due to high nutritional composition these residues not described as wastes but considered as raw material/ by product for other product formation and development. (Nguyen et.al, 2010).
- Fruit and vegetable processing industry has accounted 25% losses and wastes in the form of organic waste such as peel, stem, core, stones, seeds and

pomace generated either from fruit discarded into the sorting operation or pomace from juice extraction.

- Waste is the potential source of functional dietary fiber for food applications some of the waste goes to animal feed for e.g. 10,000 tons of apple pomace out of total production 1 million tons is being utilized for by product processing. This by products can be utilized as a valuable source of natural food additives of high nutritional value (Husain et.al, 2015)
- The statistical figures of fruit and vegetable waste produced reported by NHB (2014-15) and Djillas (2009) were apple peel, pomace, seed- 25%, mango peels stones -45%, banana peel- 35%, citrus peel, rag, seeds – 50%, pineapple-skin, core-33%, grapes- stem, skin, seeds – 20%, guava-peel, core, seeds 10%, tomato – peel, core, seeds – 20% potato- peel- 15% and peas- shells – 40%.

By-products resulting from processing of papaya, pineapple and mango represent approx. 10 – 16% of fruit weight. In case of citrus fruits, amount residues accounts for about 50% of the original fruit weight, seeds constitute considerable proportion of grape ranging from 38 -52% on dry mater bases. (Kaur et.al, 2017).

NUTRITIONAL COMPOSITION OF FRUIT AND VEGETABLE POMACE

The nutritional value of fruits and vegetable are reported to be rich in dietary fiber, vitamins A and C, minerals (Ca, Fe, Zn, K, Cu, Ph, Mg and Mn they are good source of phytochemicals, antioxidants, L alpha tocopherol, carotenoids, beta carotene, lycopene, cryptoxanthin, zeaxanthin and lutein). (Gopalan et.al, 2016). The nutritional value apple pomace assessed by Sudha et.al. (2007) and revealed that it contains 51.1% dietary fibers, 7.31 -8.53% of fruit protein, 3.85 – 4.7% total ash, high amounts poly phenols (7000 mg / kg), flavon-3-ols (1850-2550 mg/ kg) hydroxycinnmates and hydroxylchalcones and pectin (10-15%).apple peels were found to contain up to 33,00 mg / 100gm of phenolic compounds. Majority of total fibers was located in the peel of apple (0.91%).

Guava pomace was reported to contain high amount of total dietary fiber (63.949 / 100gm), reduced calorie content (182kcal /100gm), iron (13.8 mg /100gm), zinc (3.31mg/ 100gm) and considerable amount of ascorbic acid (Vitamin. C 87.44 mg /100gm), total carotenoids (1.25 mg/100gm), an insoluble dietary fiber (63.55mg/100gm).

In case of citrus fruits, citrus pulp obtain after juice extraction contains 41-42% dietary fiber, 6% crude protein, 6.3% ash, minerals like Calcium (7.7 gm./ kg), phosphorus(1.6 gm./ Kg) (Silva et.al, 1992). It was also found that total phenolic compounds in peels of oranges and lemon were 15% higher than that of pulp of these fruits (Gopalan et al, 2014).

Pineapple pomace has good nutritive value, rich in dietary fibers, contains calcium, phosphorus and iron. About 25% of fresh fruit is lost as pomace. Pomace contains about 1.8% ash, 21.5mg / 100gm ascorbic acid and 0.41% crude fiber (Husain et.al, 2015).

Pomegranate peels contain 249.4mg/gm. of phenolic compounds as compare to only 24.4 mg/gm of phenolic compounds found in the pulp of pomegranate.

Banana peels constituting about 40% of total weight of fresh banana as a major waste. It is rich source of starch (3%), crude protein (6-9%) total dietary fiber (43.2-49.7%) and crude fat (3.8-11.0%). Banana peels is a good source of micronutrients (K, P Ca, Mg) PUFA (linolenic acid and alpha linolenic acid) and essential amino acids (leucin, valine, phenylalanine, threonine). Moreover significant amount of lignin (6-12%) pectin (10-11%), cellulose (7.6-9.6%).

Hemi cellulose (6.4-9.4%) and galacturonic acid is found in banana peel as dietary fiber. Moreover, Shyamala and Jamuna (2011) stated that peel had good antioxidant components and activity where the free radical scavenging activity of tannic acid (90-62%) and polyphenols (200-850mg equivalent to tannic acid /100gm) were found.

Grapes (Vitis vinifera) constitute seeds 38-52%. The seed oil is rich in unsaturated fatty acids (particularly linoleic acid) and phenolic compounds. (80% grapes used for wine world wide). During tomato processing, about 3-7% of the raw material is lost as waste. Tomato pomace generally consists of the crushed are dried skin and seeds of the fruit. Appropriately, the seeds account for 10% of fruit and 60% of the total waste. The seeds are reported to be good source of protein (35%) and fat (25%). Tomato seed oil is found to be rich in unsaturated fatty acids such as linolenic acid that has largely attracted the interest of researchers (Eller et al, 2010). As compared to seeds and pulp, the tomato peel contains higher levels of total flavonoids, total phenolic compounds, lycopene and ascorbic acid exhibiting higher antioxidant activity.

Carrot pomace, generated during processing, contains 14.75% soluble fiber, 30% insoluble fiber, 6.50 proteins, 5.12% ash, 5456 μg total carotenes and 607 μg β-carotene.

Chemically, the agricultural wastes contain 31-60% cellulose, 11-38% pentosane and 12-28% lignin. This product has been reported to be used in the alcohol production (Aappaiah, 2017). Fruits are very rich in carbohydrate and sugar content which can be a very good source of alcohol production.

Grape and wine making industry generate a number of waste and by products. These material include wine pruning, grape stalks, grape pomace, grape seeds, yeasts, tartrate, carbon dioxide and waste matter, every by- product will become fertilizers, animal feed or fuel. (Nerantzis and Tetaridis, 2006). The grape seed extracts have gained ground as nutritional supplement in view of its antioxidant activity (Arvanitoyannis et al, 2006).

Enzyme Production

Grape pomace, main polluting waste from the wine industry, is a good natural medium for solid state fermentation that is used for production of hydrolytic enzymes such as cellulases, xylanses, and pectinases using Aspergillus awamori (Botella et al, 2007). Proteolytic enzymes such as bromelain is recovered from pineapple pomace and papain from papaya latex. Moreover, orange peel and orange finished pulp, sugar beet pulping and peas waste are good substrates for polygalacturonase production. Apple pomace, a waste from the apple processing industry is also used as a substrate for pectinase production by aspergillus spp. in solid state fermentation.

Pectin Production

Pectin a heteropolysaccharide having properties like capacity to make gels, emulsify and stabilize. The major waste during processing is peel (citrus) which is widely used for the producing pectin powder; other sources of pectin are mango peels, residue of sunflower and guava (Kaur et al, 2017). Lal et al, (1988) have given the detail information about utilizing waste of fruits and vegetables. Apple peel, pomace for pectin, guava peels for preparation of guava cheese, water melon rind for pickle making, jackfruit for pectin, pineapple for vinegar production, limes for citric acid, seeds and for oil, orange, lime peel can be used for extraction of essential oil, Citrus oil/ orange oil. Banana pseudostem, leaves for preparation of paper pulp and banana fiber (for clothes) and ecofriendly containers, green papaya for latex and tutti fruiti preparation, other waste and garbage can be used in feed or decomposition of compost manuring.

1. Salad dressing- orange peels and orange waste pulp.
2. Yoghurt- added with grape pomace extract for enrichment of bioactive compound (Tseng and Zhao, 2013).
3. Grape seed oil rich in polyphenols, antioxidants and vitamins used in cooking oil.
4. Mango seed kernel oil/ fat to be used as cocoa butter equivalent.
5. Pulpy waste in ethanol production by fermentation.
6. Tomato pomace can be used in extruded products.
7. Others for recovery of fiber, vitamins, β-carotene.

8. Natural colouring pigments - beetroot (red), leaves (green), paprika red (chilli powder), turmeric (yellow), carrot (orange red), kesar (pink-yellow), radish (anthocyanin).
9. Brewery and wine industry waste- the brewery industry waste are the spent grain, the trub, and the residual yeast. Brewer spent grain (BSG) is the main by-product of brewing industry representing approximately 85% of the total by product generated. It is rich in cellulose and non-cellulosic polysaccharides (Aliyu and Bala, 2011).

Marine Industry Waste

Sea food by-products could serve as important value added nutraceuticals and functional food ingredients (Gormley, 2013). By-products from sea food processing may account for up to 80% of the harvest depending on the species. These include w-3 PFA from the livers of white lean fish waste flesh parts of fatty fish, blubber of marine animals, hydrolysates from fish guts, cleaning, peptides and products from crustaceans such as chitosan, chitosan oligomers and glucosamines. Hence, by-products from sea foods could serve as important value added nutraceuticals and functional food ingredients (Gormley, 2013). Gelatin, a thickening polysaccharide, is obtained from sea animal carcasses. Like wise the moss, agar are also obtained from sea weeds, cod liver oil from cod fish liver (Kadam and Prabha Sankar, 2010).

Meat Industry Waste

According to the European Commission (EC) the animal by-products may be defined as whole bodies or parts of creatures, products of animal origin or other products obtained from animals as carcasses, skin, bones, meat, trimmings, blood, fatty tissues, horns, feet, hoofs or intestinal organs. Meat by-products are reported to be rich in lipids, polysaccharides, proteins, and the bioactive peptides which are known to have antimicrobial, antioxidative, antithrombic, anti-hypertensive properties (Lafarga and Teagase, 2014).

Grain Processing Industry Waste

Rice bran,10 percent of the weight of rice grain, in rice milling yields the by-products20% husk, 8% bran and 2% germ. Rice bran is rich in antioxidant (polyphenols, Vit. E (alfa tocopherol) and carotenoids).Rice bran is presently used for extraction of edible rice bran oil after refining. Rice bran used in other products are bread, biscuits, pasta, noodles a n ice-creams having more functional and textural properties (Gul et al, 2015). Rice husk is a major protective covering of paddy grain which accounts to about 14% - 28% of the grain, estimated to 80 million tones (average 20% of paddy) must disposed of annually worldwide. The major application of the husk are in the production of husk ash, silicon (husk contains about 90% silica), fuel, briquettes, poultry litter, traditionally used in cattle feed, composite press boards, furfural, silicon tetrachloride, activated carbon, cement concrete, husk as fuel, electricity generation, etc. (Pillaiyar, 1988; Juliano, 1985). B- glucan extracted from grain flour which progress lipid metabolism, reduce the glycemic index and lower plasma cholesterol, lignan concentrate from flax seed which act as anticancer, antioxidant, antibacterial, antiviral and anti-inflammatory agent and phenolic compounds extracted from cereal bran which provide antioxidants resistance against free radical damage, cancer and cardiovascular diseases. Flax seed super rich in lignans can be added to different cereal based formulations like bread, muffins and other bakery products (Bainao, 2014). Maize germ obtained during grain milling is used for extraction of maize oil and it is further used for edible purposes after refining as 'Mazola oil' (Helkar. 2016)

Dairy Processing Industry Waste

Whey is a liquid by-product of dairy industry obtained during the preparation of chhana, paneer, cheese and contain casein. World whey production is estimated to about 180 to 190 X 106 tons per year with an annual increment of 1 to 2% and only 50% and only 50% of whey is utilized or processed (Roman et al, 2012). Whey contains 45 to 50% total milk solids, 70% milk sugar (lactose and galactose), 20% milk proteins (casein) and 70 – 90% milk minerals and almost all the water soluble vitamins originally present in the milk (Horton 1995). Whey disposal becomes a serious environmental pollutant being loaded with high amount of organic matter. Whey posses preventive and curative elements responsible for treatment of ailments such as arthritis, anemia and liver complaints (Cruz et al, 2009).Fruit and dairy waste based on products are attaining considerable attention due to delicious taste and market for such food products has incredible potential (Ismail et al,2011). Whey based fruit beverages are more suitable for health as compared to other drinks because of probiotic effect (Kumar 2005). Production of nourishing pleasant whey based on fruit RTS (ready- to- serve) beverages is one of the most promising trend in utilization of dairy waste whey. Whey powders are rich source of protein of high biological value (Ramos et al, 2016).

Ur is one of the by-products of milk industry having high probiotic functional, medicinal and nutritional properties (Homayouni, et al, 2012).

Summary

In general, the food industry waste could be utilized by various ways such as—

- Reduction in pre- and post- harvest losses,
- Efficient storage, packaging and distribution of food products,
- Supply chain management, cold chain management, cold storages,
- Proper modern packaging technologies be adopted widely,
- Efficient and proper collection and disposal of waste after proper segregation.
- Ultrafiltration and recovery of food waste,
- Recovery of fruit and vegetable waste (semi solid),
- Recovery of protein and other fermentable value added chemicals,
- Extraction of fat and other ingredients for manufacturing valuable items,
- Utilization of waste in animal feed after proper processing,
- Utilization in fuel, electricity energy, bio- gas generation projects,
- Decomposition of waste through microorganisms, earth worms, (composting), bio- gas production for domestic use.
- Recycling of the food waste, sewage water, effluents after proper treatment, or can be send into landfills (very little share, but care must be taken to avoid environmental pollution).
- Reduction, reuse and recycling of food waste must be mandatory enforced by the governments of the countries to their society so as to improve and protect their people's life.
- **Benefits of the recycling of waste:**
    - It reduces the amount/ expenditure required for disposal,
    - It saves natural resources,
    - It reduces the amount of energy needed for manufacturing New Products,
    - It reduces pollution and destruction,
    - It provides employment opportunities,
    - It helps to National Economy
    - It helps to maintain Zero Food Wastage

REFERENCES

Adame, Samino, & Sanchez, , & Gonzalez. (2012). In vitro estimation of the antibacterial activity and antioxidant capacity of aqueous extracts from grape seeds. *Food Control*, *24*, 136 – 141.

Aliyu, S., & Bela, M. (2011). Brewer's spent grain: A review of its potential and applications. *African Journal of Biotechnology*, *10*, 324–331.

Appaiah Anu, K. A. (2017). Fruit and vegetable wastes: An alternative feed stock for alcohol production. *Indian Food Industry*, *36*(6), 30–41.

Arvanitoyannis, S., Ladas, D., & Mavromattes, A. (2006). Potential Uses and application of treated wine waste: A Review. *International Journal of Food Science & Technology*, *41*(5), 475–487. doi:10.1111/j.1365-2621.2005.01111.x

Autrel, E., Betrhier, F., Lutze Zajac, A., & Nicolas, F. (2007). Inceneration of municipal and material recovery performances. *Journal of Hazardous Materials*, *139*, 569. doi:10.1016/j.jhazmat.2006.02.065 PMID:16707217

Baker, D., Fear, J., & Denniss, R. (2009). *What a waste: An analysis of house hold expenditure on food, in policy Brief no. 6*. Canberra: The Australia Institute.

Belewu, M. A., & Babalola, F. T. (2009). Nutrient enrichment of some waste agriculture residue after solid state fermentation: Application to animal nutrition. *Animal Feed Science and Technology*, *14*(4), 122.

Bos, A., & Hamelinck, C. (2014). *Green House Impact of marginal fossil fuel use*. Project no. BIEN 14973.

Butella, C., Diaz, A., Ory, I. W., Webb, C., & Blandino, A. (2007). Xylanase and pectinase production by Aspergillus awamori on grape pomace in solid state fermentation. *Process Biochem.*, *42*, 98-101.

Chandrasekaran. (2012). *Valorization of food processing by-products*. CRC Press.

Chonhenchob, V., & Singh, S. P. (2003). A comparison of corrugated boxes and reusable plastic containers for mango distribution. *Packaging Technology & Science*, *16*(6), 231–237. doi:10.1002/pts.630

Chonhenchob, V., & Singh, S. P. (2005). Packaging performance comparison for distribution and export of papaya fruit. *Packaging Technology & Science*, *18*(3), 125–131. doi:10.1002/pts.681

Cruz, A. G., Ana, A. S. J., Marchione, M. M., Teixeira, A. M., & Schmaltz, F. I. (2009). Milk drink using whey butter, cheese and acerola juice as a potential source of Vit C. *Food and Bioprocess Technology*, *28*, 368–373. doi:10.100711947-008-0059-9

De Las Fuentes, L. B., Sanders, B., Lorenzo, A., & Aber, S. (2004). Awareness Agrofood wastes minimization and reduction network. Total Food Proceedings, 233 – 244.

DEWHA, National Waste Policy. (2009). Less waste more resources. Department of Environment, Heritage and the Arts, Editor, Common Wealth of Australia.

Djilas, S., Canadanovic-Brunet, J., & Cetkovic, G. (2009). By products of fruit processing as source of phytochemicals. *Chem Ind.Chem. Eng., 15*(4), 191–202. doi:10.2298/CICEQ0904191D

Eller, F. J., Mosser, J. K., Kenar, J. A., & Taylor, S. L. (2010). Extraction and analysis of tomato seed oil. *Journal of the American Oil Chemists' Society, 87*(7), 755–762. doi:10.100711746-010-1563-4

FAO. (2011). *Global food losses and waste, extent, causes and prevention.* Rome: FAO.

FAO. (2015). *The state of food insecurity in the world.* Rome: FAO.

Ganders, D. (2012). *Wasted: How America is losing up to 40 percent of its food from farm to fork to landfill.* New York: Natural Resources Defense Council.

Gopalan, C., Rama Sastri, B. V., & Balasubramanian, S. C. (2014). *Nutritive value of Indian Foods* (p. 161). Hyderabad: NIN.

Gormley, R. (2013). *Fish as a functional food: some issues and outcomes, Sea Health.* UCD.

Graminha, Goncalves, Pirota, Balsalobre, Silva, & Gomes. (2008). Enzyme production by solid feed. *Sci. Technol, 144*, 1–22.

Gul, K., Yousuf, B., Singh, A. K., Sing, P., & Wane, A. (2015). Rice Bran, Nutritional value and its emerging potential for development of functional food: A review. *J of Bioactive Carbohydrates and Dietary Fibers, 6*(1), 24–30. doi:10.1016/j.bcdf.2015.06.002

Gustavsson, J., Cederberge, Sonesson, U., Van O Hedrick, R., & Me beck, A. (2011). *Global Food issues and food waste: Extent causes and prevention.* Rome: FAO, UNO.

Hall, K. D., Quo, J., Dore, M., & Chow, C. (2009). The progressive increase of food waste in America and its environmental impact. National Institute of Diabetes and Digestive and Kidney Diseases.

Helkar, P. B., Sahoo, A. K., & Patil, N. J. (2016). Review: Food industry by-products used as a functional food ingredients. *J. of Waste Resource, 6*(3), 1-6.

Homayouni, A., Alizadeh, M., Alikhan, H., & Zijah, V. (2012). *Functional dairy probiotic development trends, concepts and products.* Intech. doi:10.5772/48797

Hoornweg, D., Bhada-Ta-ta, P., & Kennedy, C. (2013). Environment: waste production must peak this century. *Nature, 502*, 615 – 617.

Horton, B. S. (1995). Whey processing and utilization. *IDF Bulletin, 308*, 2–6.

Hossain, M. F., Akhtar, S., & Anwar, M. (2015). Nutritional value and medicinal benefits of pineapple. *Int. J. Nutri. Food Sci., 4*(1), 84–88. doi:10.11648/j.ijnfs.20150401.22

Hussein, A. M. S., Kamil, M. M., Hegazy, N. A., Mahmoud, K. F., & Ibrahim, M. A. (2015). Utilization of some fruits and vegetables by-products to produce high dietary fibre jam. *Food Sci. Quality and Mgmt, 37*, 39–45.

Institution of Mechanical Engineers (IME). (2013). *Global food waste not want not*. Author.

Ismali,, A. E., & Abdellader,, M. O., & Ali, A. (2011). Microbial and chemical evolution of whey based mango beverage, Advances. *Journal of Food Science and Technology, 38*, 250–253.

Juliano, B. O. (1985). *Rice Chemistry and Technology*. American Association of Cereal Chemistry.

Kadam, S., & Prabha Sankar, P. (2010). Marine foods as functional ingredients in bakery and pasta products. *Food Research International, 3*(8), 1975–1980. doi:10.1016/j.foodres.2010.06.007

Kaur, R., Kapoor, S., & Sharma, S. (2017). Utilization potential of fruit and vegetable pomace. *Indian Food Industry, 36*(2), 24 – 30.

Kumar, R. S. (2005). Whey beverage: A review. *Beverage and Food World*, 58-60.

Kummu, M., de Moel, H., Porkka, M., Siebert, S., Varis, O., & Ward, P. (2012). Lost food, wasted resources: Global food supply chain losses and their impact on freshwater, cropland and fertilizer use. *The Science of the Total Environment, 43*, 477–489. doi:10.1016/j.scitotenv.2012.08.092 PMID:23032564

Lafarga, T., & Teagase, M. H. (2014). Bioactive peptides from meat muscle and by products: Generation functionality and application as functional ingredients. *Meat Science, 98*(2), 227–239. doi:10.1016/j.meatsci.2014.05.036 PMID:24971811

Lal, G., Siddappa, G. S., & Tandon, G. L. (1988). *Preservation of Fruits and Vegetables*. New Delhi: ICAR Pub.

Lambing, E. F., & May Froidt, P. (2011). Global land use change, deglobalization and booming land scarcity. *Proceedings of the National Academy of Sciences of the United States of America, 108*(9), 3465–3472. doi:10.1073/pnas.1100480108 PMID:21321211

Lee, S. G., & Xu, X. (2004). A simplified life cycle assessment of reusable and single -use bulk transit packaging. *Packaging Technology & Science, 17*(2), 67–83. doi:10.1002/pts.643

Mena, C. B., Adenso-Diaz, B., & Yurt, O. (2011). The causes of food waste in the supplier – retail interface: Evidence from the UK and Spain. *Resources, Conservation and Recycling, 55*(6), 648–654. doi:10.1016/j.resconrec.2010.09.006

Menon, M. A. (2010). Integrated solid waste management based on the '3R' approach. *J. Motor Cycles Waste, 12*(1), 30–40. doi:10.100710163-009-0274-0

Monier, V., Shailendra, M., Escalon, V. O., Connor, C., Gibbon, T., Anderson, G., . . . Reisinger, H. (2010). Preparatory study on food waste across EU 27. European Commission (DGENV) Directorate Industry, Final Report.

Moroney, P. (2016). Recycling, recovering and preventing food waste competing solutions for food systems substantially in the United States and France. Academic Press.

Nerantzis, E. T., & Tetaridis, P. (2006). *Integrated enology: utilization of winery by-products into high value added products.* Academic Press.

Nguyen, T. A. D., Kim, K. R., Han, S. J., Cho, H., Kim, J. W., Park, S. M., & Sim, S. J. (2010). Pretreatment of rice straw with ammonia and ionic liquid for lingo cellulose conversion to fermentable sugars. *Bioresource Technology, 101*(19), 7432–7438. doi:10.1016/j.biortech.2010.04.053 PMID:20466540

Nishida, J. (2014). Reducing food waste and promoting two recovery, globally, EPA c connect. *The Official Blog of the EPA Leadership.* Retrieved from https:/blog.epa.gov/blog/2014/reducingfood-waste-and-promotingfood-recovery-gglobally

Pillaiyar, P. (1988). *Rice post production Manual.* New Delhi: Willey Eastern Ltd.

Ramos, O. L., Rodrigues, R. M., Texeira, J. A., & Vicentre, A. A. (2016). Whey and whey powders: Production and uses. Food Science Encyclopedia of Food and Health, 498-505.

Rodríguez Couto, S. (2008). Exploitation of biological waste for the production of value added products under solid state fermentation conditions. *Biotechnology Journal, 3*(7), 859–870. doi:10.1002/biot.200800031 PMID:18543242

Roman, A., Vetal, G., Illzes, A., Kovacs, Z., & Czermak, P. (2012). Modeling of dia filteration process for determination of acid whey: an empirical approach. *J. of food Process Engg, 35*(5), 708-714.

Sakai, S. J., Yoshida, H., Hirai, Y., & Ansari, M. (2011). International comparative study of '3R's and waste management policy developments. *Master Cycles Waste, 13*(2), 86–102. doi:10.100710163-011-0009-x

Shyamala, B. N., & Jamuna, P. (2010). Nutritional content and antinutritional properties of pulp waste from *Daucus carota* and *Beta vulgaris*. *Mal, 16*(3), 397-408.

Silva, A. G., Wanderley, R. C., Pedroso, A. F., & Ashbell, G. (1997). Ruminal digestion kinetics of citrus peel. *Animal Feed Science and Technology, 68*(3-4), 247–257. doi:10.1016/S0377-8401(97)00056-4

Singh, Chonhenchob, & Singh. (2006). Life cycle inventory and analysis of reusable plastic containers and display-ready corrugated containers used for packaging fresh fruits and vegetables. *Packaging Technology and Science, 19*, 279-293.

Stuart, T. (2009). *Waste: Uncovering the global food scandal*. New York: Penguin.

Sudha, M. L., Baskaran, V., & Leelavathi, K. (2007). Apple pomace as source of dietary fibre and Polyphenols and its effect on the rheological characteristics and cake m making. *Food Chemistry, 104*(2), 689–692. doi:10.1016/j.foodchem.2006.12.016

Sustainable Restaurant Association. (2010). Too good to waste; Restaurant food waste survey Report 2010, UK. Author.

Taurisano, Gianluca, Nicolas, & Di Donato. (2014). Reuse of April waste, recovery of valuable compounds by Eco-friendly techniques. *International J. of performation to Engg., 10*(4), 419-425.

Tseng, A., & Zhao, Y. (2003). wine grape pomace as antioxidant dietary fibre for enhancing nutritional value and improving stability of yoghurt and salad dressing. *Food Chemistry, 138*(1), 356–365. doi:10.1016/j.foodchem.2012.09.148 PMID:23265499

Vandermeerch, T., Alvarenga, R.A.F., Ragart, P., & Dewalf, J. (2014). Environmental sustainability assessment of food waste valorization, options, resource conserve. *Recy, 87*, 57.

Varghese, K., Lewis, H., Lockroy, S., & Williams, H. (2013). *The role of packaging in minimizing food waste in the supply chain of the future, Final report, Centre for design RMIT University*. CHEP.

Ventour, L. (2008). Food waste report v2, in the food we waste wrap and exodus market research. Weston – Super-Mare.

Williams, H., Wikstrom, F., Otterbring, T., Lofgren, M., & Gustafsson, A. (2012). Reasons for household food waste with special attention to packaging, *J. of Cleaner Production, 24*, 148.

WRAP. (2011). Fruit and Vegetable resource maps, Final Report 201., Waste and Resources Action Program (WRAP).

Chapter 10
Various Approaches for Food Waste Processing and Its Management

Anupam Pandey
Kumaun University Nainital, India

Ankita Harishchandra Tripathi
Kumaun University Nainital, India

Priyanka Harishchandra Tripathi
National Institute of Pathology (ICMR), India

Satish Chandra Pandey
Kumaun University Nainital, India

Tushar Joshi
Kumaun University Nainital, India

Ashutosh Paliwal
Kumaun University Nainital, India

Veena Pande
Kumaun University Nainital, India

ABSTRACT

Food wastage is a huge crisis arising in today's world. An extensive amount of waste generation has become a serious concern of our society in the past years that affects developing and developed countries equally, and according to the Food and Agriculture Organization (FAO), as much as one-third of the food intentionally grown for human consumption is never consumed and is therefore wasted, with significant environmental, social, and economic ramifications. By wasting food, we also waste the time and energy that we have used to produce the food and as well our natural resources and the limited available agricultural land will be used up which could be handled in a much better and sustainable way. Additionally, waste has a strong financial impact and affects the environment including the overall greenhouse gas emission. In an increasingly resource-constrained world, it is imperative to reduce the high environmental, social, and economic impacts associated with this type of waste.

DOI: 10.4018/978-1-5225-7706-5.ch010

INTRODUCTION

Every year approximately 1.3 billion tons of food which equals one third of total food production worldwide is lost or wasted (Gustavsson *et al.*, 2011). Food waste is predominantly challenging in industrialized countries that have a major contribution to household food waste. As food production is resource intensive, food loses and wastes indirectly cost the environment and the major effect of this can be seen in the environmental burden in the form of, water and air pollution, deforestation, soil erosion as well as greenhouse gas emissions that occur during the processes of food production, stowage, conveyance, and waste-management (Mourad, 2016). Owing to these rising environmental burdens, social and economic concerns towards food waste is progressively accredited as a crucial issue between governments, academics, NGOs, businesses, and the general public (Beretta *et al.*, 2013; Edjabou *et al.*, 2016). Humans are totally depended on plants and animals for their nutritional assistances. The Global Food Report, by the Institute of Mechanical Engineers, has claimed that there could be a whopping three billion to be fed with food by the end of this century. In that period, one can expect great changes in the areas of wealth, calorific intake and dietary preferences of people in developing countries across the globe. Hence, it lies in our hands to focus in producing food in safer quantities by availing the best technologies.

Food waste is generally defined as the loss of materials planned for human ingestion that are afterwardeither discharged, which thereby get contaminated, degrade and are subsequently lost. As per the Food and Agriculture Organisation (FAO) of United Nations, food is "Any modification in the accessibility, edibility, wholesomeness or quality of eatable material that averts it from being eaten by people". This definition was stated for the period of post-harvest of food ending, when the point is of proprietorship of the final consumer (FAO, 1981). Another definition of FL provided by Gustavsson *et al.,* (2011) included description of food supply chain (FSC) production stage along with postharvest and processing stage. According to Parfitt *et al.,* (2010), "Food waste (FW) is the loss of food taking place either at the market stage or at final consumption and utilization stages and is generated due to the negligent behaviour on the part of retailer as well as consumer. European Project FUSIONS defines food waste as''any form of food, edible or inedible, aloof from (diverted or lost from) the food supply chain that is to be either disposed or improved (includes anaerobic digestion, incineration, composted crops, co-generation, bio-energy production, sewer disposal, landfill or discarded into the sea)'' (Östergren *et al.,* 2014). For proper metabolic functioning and cellular activities, cell needs energy and this energy comes from food. All human beings depend on food for both energy constraint and survival.

Research study carried by Smil, 2010 explained that when the losses and food wastage along the food supply chain was taken into account along with the transformation of food production into animal feed, it was reported that 43% out of the total food cultivated worldwide is directly consumed by humans. According to the Unites States Department of Agriculture USDA (2007), in the United States a total of 30% of the food intended for human feeding is wasted every year, mostly in the houses, restaurants and food service establishments. According to Eurostat data (2006) the quantity of food wasted annually in Europe is 89 million tons, equivalent to 180 kg per capita, but this figure is not inclusive of the losses that occur all through the food production and harvesting stages. Looking only at waste in the houses, and using various national data sources, it was found that the amount wasted per person per year is: 110 kg in Great Britain, 109 kg in the United States, 108 kg in Italy, 99 kg in France, 82 kg in Germany and 72 kg in Sweden.

According to (BIOIS, 2010), in food supply chain, the largest food-waste fraction is contributed by private households, therefore prevention of food waste over the final stages of supply chain is of extreme importance to prevent further changes in climate (Parfitt *et al.,* 2010). In emerging republics, the great losses are suffered at the initial level of the food supply chain, mainly due to limitation of techniques used for cultivation, harvesting, and preservation, or due to a lack of adequate transportation and storage infrastructures. In commercial countries, the largest share of waste in the food supply chain occurs in the final phase (household consumption, restaurants and food service establishments). However, even in these countries, the losses recorded at the agricultural level are not insignificant for instance, in Italy in 2009, 17.7 million tons of agricultural foodstuffs was left in the fields, representing 3.25% of total production (Segrè and Gaiani, 2011). In agriculture, research shows that the food losses are attributable to climatic and environmental dynamics, and also to disease and parasites. But there are discrepancies at this stage when we compare between emerging and advanced countries which can be credited to the accessible technology, agricultural skills and the techniques used for preparation of agricultural field, infrastructures, sowing, cultivation, harvesting, processing, and storage. In developed countries, and sometimes also in emerging countries, regulatory and economic factors play a part. However, there is undoubtedly still a long way to go in understanding the causes of the initial stages losses in the food supply chains. The primary identified reasons of waste produced during the initial processing stages of the agricultural product and semi-finished goods are inefficiencies and technical malfunctions in the production processes - commonly known as "production waste". There are many causes for food waste production throughout the distribution and sales of food (both wholesale as well as retail) which includes improper ordering and false estimation of consumer requirements.

According to a survey conducted in October 2011 by Coldiretti-SWG, Italians have reduced food waste by 57% because of the economic crisis. To combat waste and thus save more food, as many as three out of four Italians spend more carefully than before crisis. Among the measures taken to reduce food waste are to shop more wisely, reduce the quantity of food purchased (31%), increase use of leftover products in meals (24%) and pay more attention to expiration dates (18%).

Food Waste Management

Presently, due to ever increasing population, food systems have become very inefficient: it is projected that one half of all the food produced would get lost before reaching human mouth. In 2015 United Nations has established "The Sustainable Development Goal 12" that 'Safeguard maintainable production and consumption patterns' including a target specific for food waste reduction: at retail and consumer levels, halve per capita reduction in global food waste till 2030 (The Agenda For Sustainable Development 2030 (2015)). During prehistoric time, when techniques and knowledge were less, traditional methodology was used by people for waste management. The traditional method included three steps: 1. Reduce, 2. Reuse, 3. recycle.

- **Reduce:** Take minimum food which can be finished by an individual.
- **Reuse:** Food that has been cooked can be used more than once in a day.
- **Recycle:** Even after consumption, leftover food can further be used for production of some commercial products such as biofuel, ethanol etc.

Sustainable management of food waste is a importantexploration area that has speedily grown over recent years. Methodology of food waste management includes classification of food waste. Food waste can be categorized according to the type of food: drinks, meat, cereals, fish, fruits, etc. this classification is beneficial in order to quantify the quantity of food wasted on the basis of mass, economic cost, and energy content (Flores *et al.,* 1999).

Lin *et al.* classified food waste as organic crop residue (including fruits and vegetables), animal by-products, domestic waste packaging, mixed food waste and catering waste. Edjabou *et al.* included two new factors: vegetable/ animal-derived food waste and avoidable-processed/ avoidable-unprocessed food waste.

In the UK, WRAP identified supply chain stages, where food waste is generated (e.g. manufacturer, retailer) and had assess the edibility of the waste. Accordingly, food waste can be avoidable (food parts that are edible) and unavoidable (food parts that are inedible such as fruits skin, bones etc.). Furthermore, food waste can be divided at household level as cooked/uncooked, packaged/unpackaged or opened/

unopened packaging or leftovers and untouched food which usually is thrown and wasted (Matsuda *et al.*, 2012).

According to a published research in "The Food We Waste" (WRAP, 2007), the main reasons for food wastes in the home include:

- Buying too much: Due to the special offers such as "buy one get one free"
- Buying more perishable food
- Poor storage and management: Not eating food in date order, preparing too much.
- Sensitivity to food hygiene: Not taking a chance with food close to its "best before" date

Attitudes and behaviour towards food waste are driven by a number of factors at consumer level including:

- Retail practices that encourage people to buy more than they need.
- Lack of economic incentives to avoid food waste.
- Lack of advance planning, skills and knowledge regarding food storage and preparation.

Figure 1. Food waste management

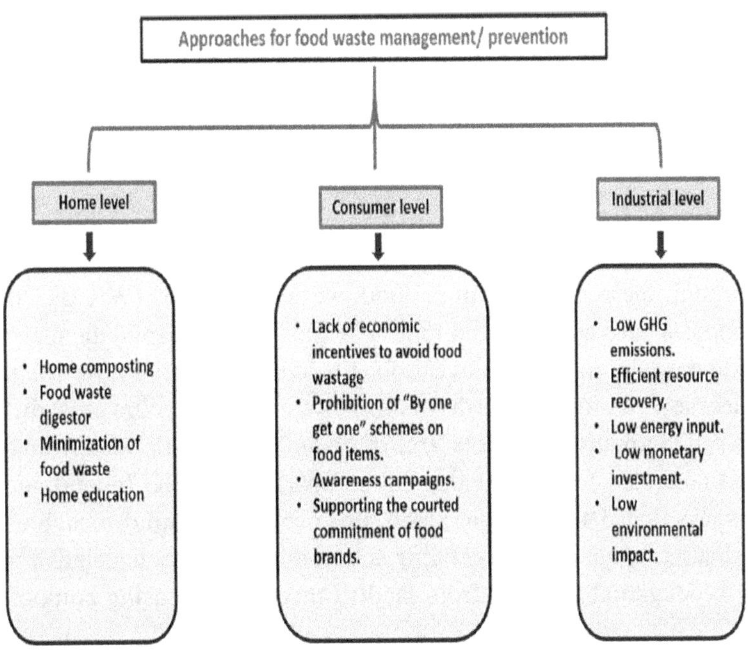

Ideas to sanitize people for management of food waste, supermarkets-initiated programs to manage food waste at consumer stage. The programs included initiatives on moving away from Buy One Get One Free promotions (these promotions are not common in other European countries) to "Buy one get one free". The buyer still benefits from the free items of food, but they can be collected at a later date when they are more likely to be consumed as the consumer will purchase smaller quantity at one time. Plans by the former government entitled "war on waste" suggested scrapping best before dates, limiting sell by labels and creating new food packaging sizes in an attempt to save $300. With "best before" dates and other labelling – The Food standards agency (FSA) found that only one-third of people correctly interpreted these terms and more than one quarter thought that food, past its best before date could be unsafe and should be thrown away (FSA, 2006). Other supermarket initiatives include "Love your leftovers" and "Great taste less waste" involving recipes cards, vouchers and website information. Many of these ideas are being introduced to reduce packaging and food waste. This is a promise signed up to by more than 40 retailers and brand owners to reduce waste and provide choices to consumers regarding the products they buy. This relates to the fact that a substantial amount of food waste is literally waste food which has not been touched by the purchaser (Hogg et al., 2007).

The second phase of the obligation was launched in March 2010 with a more focused aim to achieve a better sustainable use of resources over their entire lifecycle. The main targets of phase second are to reduce supply chain product and packaging waste by 5% and to reduce UK household food and drink waste by 4% by 2010. A report by the British Retail Consortium revealed that retailers in the UK have managed to halve waste sent to landfill from 48% to 23% since 2005. This has been achieved through re-use, using energy recovery technology and reducing consumer food waste (Barton, 2010).

Collections of Food Wastes

In UK, through the refuse stream or food waste collections (WRAP, 2009b) the local authorities have collected 5.8 tonnes of wastefood. Despite the fact, the trash stream is disposed of in landfill, in Europe the local authorities have been asked to limit the degradable waste in thrash stream. Now, though there are great differences, in Austria and Germany,near about 75% of organic waste is divided from the waste stream and collected for composting, in comparison to UK, Ireland and Greece where it is less than 10% (ACR+, 2009). For many local authorities, presenting a waste food assemblage offers a genuine solution for meeting legislative marks for collecting biodegradable waste from landfill thereby, increasing composting and recycling rates. It has been found that greater than 100kg per inhabitant of organic collection is made every year (ACR+, 2009).

In 2009, 137 local authorities in UK provided a food waste collection, with 47% offering food only collections (Brook Lyndhurst, 2009). In France, green waste collections are used widely by native authorities to capture biodegradable waste. As kerbside recycling collections now capture a broad range of dry recyclables, a large proportion of the remaining refuse stream consists of biodegradable material. In Bournemouth, food waste makes up to 38% of the refuse stream, while green waste makes up 11% (Resource Futures, 2009). Implementing a separate food waste collection would have a significant impact on the quantity and quality of material remaining in the refuse stream and ultimately its cost of disposal.

Drivers for Collections of Food Waste

A number of strategies forsubstantialassistances to accumulating food waste separately from the waste stream. Numerous are directly related to the alteration of biodegradable waste obtained from landfill by:

- Improving recycling rates
- Reducing waste disposal costs as landfill cost increases
- Reducing the impactsassociated with landfill (toxicity in leachate, gas emissions from landfill) on environment.
- Reducing greenhouse gas emissions by eliminating putrescent content from landfill sites.

Food waste collections can also help local authorities who presently practice alternatives to landfill for the treatment of the waste stream, such as Mechanical Biological Treatment (MBT) Energy from Waste (EW) by:

- Improving recycle rates
- By producing improved quality liquid fertilizers and composts, so that they can be used for improving soil quality.
- Generating large amount of power and heat through anaerobic digestion (AD).
- Reducing annoyance created by rats, flies and vermin attracted to food in refuse stream.

The benefits will vary in different areas, depending on the local authority's current performance and collection systems already in place.

Figure 2. Food related practices and routines

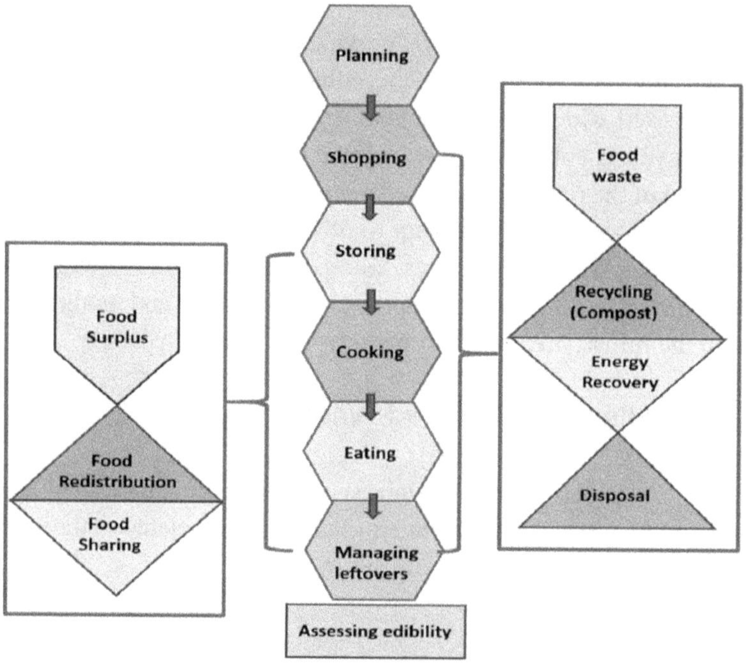

Policy Initiatives

Economic Instruments

According to Driesen, 2006; FUSIONS, 2016, economic enticements, goal is to lessen food waste through various market signals and costs. It can be in the form of taxes, subsidies, fees etc. Financial instruments are considered as an influential means to change consumption forms towards additional sustainable food practices (Reisch *et al.*, 2013). It is expected that if the actual cost of natural reserve use is replicated in charges, consumers are more expected to become active in prevention of food waste (UNEP, 2014). The volume- or weight-based fee system "Pay-As-You-Throw" (PAYT) is a mutual method that has been prompted in various countries, such as the, Sweden, Japan, Canada, United States, Taiwan, Thailand, Vietnam Korea, and China (UNEP, 2014). In these countries, implicating households for personally produced waste has been found to be an operative scheme for reduction of food waste (Chalak *et al.*, 2016; Dahlen and Lagerkvist, 2010; EEA, 2009).

Regulations

For reduction of food waste, regulatory approaches with an aim of waste reduction targets has been implemented, which includes mandatory management plans, aim to induce waste reduction, laws and standards, restrictions or covenants, penalties for those who do not obey the regulatory provisions. So far, regulations have been accepted in many countries, such as Italy, Belgium, France, and the Netherlands. The National Pact against Food Waste in France, for example, summaries eleven measures to accomplish a food waste reduction of 50% by 2025 (Mourad, 2016). One possible regulatory instrument is the evaluation and abolition of needless food-safety values that lead to high rates of food waste. In contrast to fiscal and economic incentives, well-defined principles appear to be a more operative tool to battle household food waste generation (Chalak *et al.*, 2016).

Information and Education Campaigns

Information campaigns present one of the utmost common tools used for food waste anticipation and reduction (Priefer *et al.*, 2016). Information and education movements, information stages and face-to-face door-stepping operations have been employed all over Europe to raise awareness and improve consumer's knowledge about prevention of food waste (Schanes *et al.*, 2018).

Industrial Application

Increasing efforts are currently being focused on defining effective and stable means of obtaining biofuel and bio-products from waste food materials. These opportunities could pay for benefits from an environmental point of view due to the reduction of methane gas emissions from landfills and the conservation of natural resources such as coal and fossil fuels, from a social point of view since lack of food verses fuel competition, and from an economical point of view thanks to costs saving linked to surplus food production and specific investments in establishing non-food crops dedicated to biofuel or bio-plastic production. Bio-refineries are the concept underlying industrial waste food consumption. Similarly, to the transformation by oil refineries of petroleum into fuels and ingredients for use in a wide variety of consumer products, bio-refineries convert organic waste and biomasses (corn, sugar cane and other plant-based materials) into a range of ingredients for bio-based fuels or products. Waste food produced from agriculture and food processing is abundant and concentrated in specific locations. These materials could be less susceptible to deterioration if compared to food waste produced at household level at the end of the FSC (Galanakis, 2012). These characteristics highlight the potential to develop

Figure 3. Approaches towards food waste management

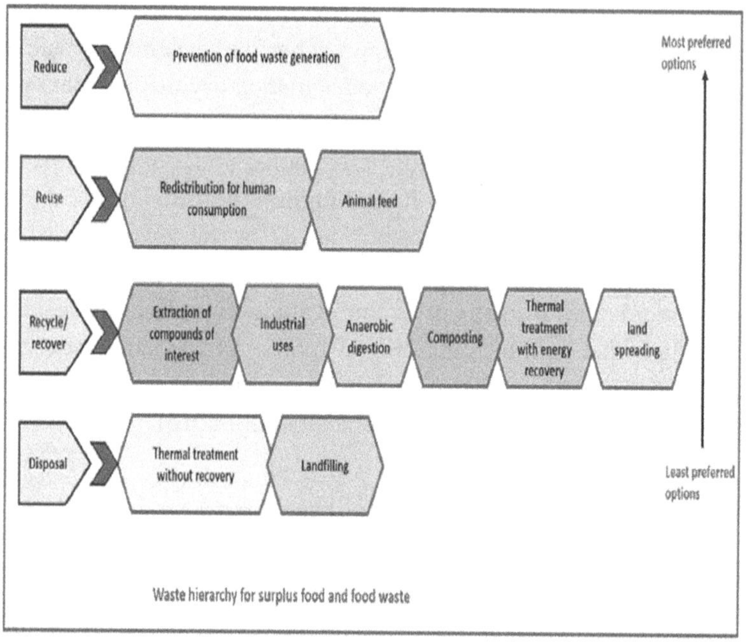

industrial utilisation processes based on symbiosis where the wastes from one sector are inputs for other sectors. Accessibility of food waste and location of potential users define the feasibility of industrial symbiosis (Mirabella *et al.*, 2014). Therefore, precise efforts will be required from the agricultural and the industrial sectors to define sustainable and inventive processes for residues use and conversion, and from governments to stimulate and support this new vision with specific legislations. The potential profitability of chemicals and biofuels produced from food waste will stimulate investments on bio-refinery chains rather than treatments of food waste in traditional waste management processes. Valorisation routes of food wastes in bio-refinery chains include both extraction of high-value components already present in the substrates to be used for nutrition or pharmaceutical applications and conversion into chemicals, materials or biofuels by the use of chemical or biological processes. Type, origin, seasonal generation and territorial distribution of food waste will affect transport logistic for its utilisation and its compatibility with the transformation process. High and concentrated volumes of food waste will be generally required to sustain large production capacities and meet economy of scale. Cost-effectiveness of conversion processes will then be ensured by security of supply at regional scale, low heterogeneity of substrates and large variety of extractable chemicals, biopolymers and biofuels. For these reasons, large fluxes of agro-industrial wastes seem to be more suitable for bio-refinery chains where stability of supply and substrate homogeneity are required for extraction or production of specific commodities while source segregated organic waste from household or restaurants would be more indicated for treatment processes where composition variability, origins and contaminations do not represent limits for the selected process (Pfaltzgraff *et al.,* 2013).

Developing Composting Technology

Sustainable food waste management system also includes development of composting technologies. Compost is generally a dark black or brown colour organic mixture formed from degradation and breakdown of organic components and is rich in humus. In agriculture, Composts are extensively used as manure/fertilizer and for soil amendments. Compost are prepared by mixing organic wastes such as leaves, food waste, garden waste with bulking agents such as wood chips. This forms an ideal environment for fungi and anaerobic bacteria to undergo the process of chemical decomposition. Maturation and curing process stabilizes the compost.

According to US EPA, composting process serve may benefits. These include:

- Decrease and exclusion of the requirement for chemical fertilizers
- Crop yields increase
- Assistance of habitat revitalization, wetland restoration, and reforestation, by modifying polluted, compressed, and marginal soils
- Capturing and destruction of 99.6 percent of industrial volatile organic chemicals (VOCs) in contaminated air;
- More cost-effective soil, water, and air remediation compared to conventional technologies;
- Extension of municipal landfill life by diverting organic materials from landfills.

Biofuel and Bioenergy Production

Food waste is characterised by a variable chemical composition depending on its source of production. Waste food materials may therefore comprise a mixture of carbohydrates, lipids and proteins, or, if generated from specific agro-industrial sectors, may be rich in one of these constituents. Different biofuels are therefore produced from food waste using bioprocesses or thermo-chemical processes, depending on their chemical composition. The use of food waste for energy production was recently reviewed by Pham *et al.,* (2015) and by Kiran *et al.,* (2014). Waste food can be converted into biofuels or energy by means of the following processes:

- Trans-esterification of oils and fats to produce biodiesel;
- Fermentation of carbohydrates to produce bioethanol or bio-butanol;
- Anaerobic digestion to produce biogas (methane rich gas);
- Dark fermentation to produce hydrogen;
- Pyrolysis and gasification;
- Hydrothermal carbonisation
- Incineration;

Not all the listed processes are currently developed at industrial level in full running mode. For example, waste food is widely studied as a substrate for the biological production of hydrogen by dark fermentation, although no full-scale applications have been realised to date (Alibardi *et al.,* 2014; De Gioannis *et al.,* 2013). Incineration is a mature technology in which waste food material directly applied to flame for decay. Incineration is applied to diminish waste volumes and produce electrical energy and heat; however, the high moisture contents of food waste limit its application together with the concerns of local communities on air emissions (Pham *et al.,* 2015). Anaerobic digestion, on the contrary, is a technology facing growing interests and large applications (Clarke and Alibardi, 2010; Levis *et al.,* 2010). The high biodegradability and moisture content of waste food are ideal characteristics for production of bio-gas and digestion residues can be used as soil conditioner or amendment or as nutrient source (e.g. ammonia or struvite). Biodiesel can be defined as short-chain alcohols and long-chain fatty acids of alkyl esters (methyl/ethyl esters) of derived from natural biological lipid sources such as vegetable oils or animal fats, which had their viscosity reduced by means of a process known as trans-esterification and are suited to use in conventional diesel engines and distributed through existing fuel infrastructure". Any fatty acid source may be used to prepare biodiesel (Refaat, 2012). Thus, any animal or plant lipid should represent a ready substrate for the production of biodiesel. However, the use of edible vegetable oils and animal fats for biodiesel production has traditionally been of high concern due to their competing with food materials. The use of non-edible vegetable oils in biodiesel production is likewise questionable, as the production of crops for fuel implies an inappropriate use of land, water, and energy resources vital for the production of food for human consumption; the use of waste oil may therefore represent a more realistic and effective element for use in the production of biodiesel (Gasparatos *et al.,* 2011; Refaat, 2012). The new technologies developed in recent years have enabled the production of biodiesel from recycled frying oils, resulting in a final quality comparable to that obtained with virgin vegetable oil biodiesel. Canakci (2007) claimed the annual production of oils, greases and animal fats from restaurants in the United States could replace more than 5 million litres of diesel fuel if collected and converted to biodiesel. Waste cooking oil requires a series of pre-treatment steps to eliminate solid impurities and reduce water contents and free fatty acids. The pre-treatment process may include washing, centrifugation, flash evaporation, and acid esterification. Final ester yield could be up to 80% (Yaakob *et al.,* 2013). These results are expected to encourage the public and private sectors to improve the collection and recycling of used cooking oil to produce biodiesel. Various microorganisms like Protists, Molds and bacteria rapidly grow on waste food from which bio-diesel can be prepared. (Ghanavati *et al.*, 2015; Kiran *et al.*, 2014). First-generation bioethanol can be derived from renewable sources of virgin feedstock; typically starch and sugar crops such as corn, wheat, or sugarcane. Indeed, most of the feed stocks used in first generation biofuel production are crops. For this reason, biofuel expansion may compete with food production both directly and indirectly (Gasparatos *et al.,* 2011). One potential advantage of cellulosic ethanol technologies is that they avoid direct competition for crops used in the food supply chain, as the materials used are non-edible; this option however should be limited to cases in which an overt sustainable surplus of crops occurs or where crop wastes and wood wastes are available as feedstock. (Pirozzi *et al.,* 2012; Refaat, 2012). Cellulosic ethanol has a number of potential benefits over corn grain ethanol, but although the cost of biomass is low, releasing fermentable sugars from these materials remains challenging. Bioethanol can be produced from FW and agricultural waste, the latter being cost-effective, renewable and abundant substrates (Kiran *et al.,* 2014).

Clostridium acetobutylicum bacteria is one of the prominent bacteria involved in the production of butanol by fermenting waste food. This bacterium poses a number of inimitable properties, including the ability to produce high yield of acetone and butanol by utilizing various starchy substances in comparison to Fernbach's original culture method of butanol and acetone production (Stoeberl *et al.,* 2011). Butanol poses greater advantages over ethanol as a fuel or merger/blending component. They require a lower vapour pressure, which enhances combustion competency, greater energy density. They even reduce viscosity of vegetable oils when mixed together in any ratio.

Thermal processes such as Pyrolysis and gasification have been used as an alternative methods in management of waste food (Pham *et al.,* 2015). The process of pyrolysis of waste food, involves heating at temperatures ranging from 400 to 800 °C, which converts the waste food materials from the solid into gas or liquid products, which can be used as raw materials or fuels proposed to consequent biochemical processes. The solid carbon thus produced is then additionally refined by adding activated charcoal. Products of pyrolysis are usually solid, liquid and gaseous, and their proportion depends upon the reaction parameters and the process of pyrolysis. Inflammable mixture of gas is obtained from the process of gasification through partial oxidation of wastefood. About 800-900°C temperature is provided during the process. The gas formed can be used directly or can be used as a fuel for engines and turbines operated through gas or even can be used as a feedstock in the production of chemicals (e.g. methanol) (Pham *et al.,* 2015).

Applicability and feasibility of these processes are strongly dependent on waste characteristics such as elemental composition, heating values, ash, moisture and volatile solids content, the presence of contaminants, bulk density. These characteristics are crucial for process performances and limit the applicability of gasification and pyrolysis. The majority of gasification technologies for example use pre-treated waste as feed stocks and no gasification/pyrolysis processes have been developed using raw food waste (Pham *et al.,* 2015). Only few researches were published on gasification or pyrolysis of food waste. Liu *et al.,* (2014) investigated the effectiveness of catalytic pyrolysis of food waste by using microwave power for heating. Opatokun *et al.,* (2015) evaluated the pyrolysis of both dry raw waste food and digested waste food after biological anaerobic treatment and concluded that both substrates demonstrated potential for fast degradation due to high volatile matter content. Energy content was for both cases mainly spread into biochar and bio-oil fractions while gases provided significantly lower energy. The use of agro-industrial residues for the extraction of high-value chemicals was recently reviewed by Mirabella *et al.,* (2014).

Biopolymers production is a panorama facing growing interest, as it is pertinent both to organic waste generated at household level and also including agro-industrial generated residues. The constituent monomers are available either through the process of fermentation of carbohydrates obtained from feedstocks by microbes, or are often genetically modified, or are obtained from chemical processing of plant oils (Fuessl *et al.,* 2012). The major focus of food waste management research is production of metabolites that can be used as biodegradable and renewable constituents of petrochemical products. Metabolites used for preparation of biopolymers include lactate which is used for the production of a plastic constituent Polylactate, Succinate used as a major precursor of detergent, plastic production, and even in pharmaceuticals. Polyhydroxyalkanotes, chiefly polyhydroxybutyrate, which are ordinary storage polymer found in many bacterial species having properties comparable to polypropylene and polyethylene (Li *et al.,* 2015). As for the biofuel production from virgin feed stocks, significant discussion surrounds the production of bio-plastics from natural materials, hovering the question as to whether they produce a harmful effect on human food supply. In this situation, the chance of using food waste in the manufacture of bio-plastics appears an extremely achievable choice. The production of pure L-lactic acid (optically active) from waste food has captivated significant attention due to its capacity to treat organic wastes with concurrent regaining of valuable by-products (Li *et al.,* 2015). A novel approach was described for operative production of optically active pure acid, L-lactic acid from waste food at moderate temperature, amending main enzyme activity by supplementation of sewage sludge and irregular basic fermentation. A production of optically pure L-lactic acid was achieved from food waste at ambient temperature with a yield of 0.52 g/g COD (Li *et al.*, 2015). Dairy industries produce large amounts of whey from processing milk for numerous industrial products. Whey is basically a by-product obtained during the process of cheese production, and its removal is presently a chief pollution problem for the dairy industry (Abdel-Rahman *et al.*, 2013). Whey is a potent and suitable raw material for lactic acid production, consisting in lactose, mineral salts, fats, proteins, water-soluble vitamins and other essential nutrients for microbial growth (Panesar *et al.*, 2007). At present, amongst the various types of starch-based biodegradable plastics such as polylactic acid (PLA) and polyvinyl acetate (PVA), the group of polyhydroxyalkanoates (PHAs) is one of the most promising. Polyhydroxyalkanoates (PHAs) are linear polyesters of hydroxyacids (hydroxyalkanoate monomers) synthesised by a wide variety of bacteria through bacterial fermentation (Reis *et al.*, 2011). The strength and toughness of PHAs are good, and they are completely resistant to moisture and feature a very low oxygen permeability. Accordingly, PHA is suitable for use in the production of bottles and water-resistant film (Van Wegen *et al.*, 1998). The simplest type of PHA is polyhydroxybutyrate (PHB). PHAs accumulate in bacteria cytoplasm as a high molecular weight polymer forming intracellular granules of 0.2– 0.7 mm in diameter. Typically, PHAs accumulate to a significant proportion of the cell dry weight when bacteria are grown in a media that is limited in a nutrient essential for growth (typically nitrogen or phosphorus), but with an abundant supply of carbon (for example glucose). Under these conditions, bacteria convert the extracellular carbon into an intracellular storage form, namely PHA. When the limiting nutrient is resupplied, intracellular PHA is degraded and the resulting carbon is used for growth (Reis *et al.*, 2011).

The main limitation in using bacterial PHAs as a source of biodegradable polymers is their production cost. In particular the average cost is by far the most significant contributor to overall PHB price, approximately two and a quarter times greater than the capital cost of equipment (Van Wegen *et al.*, 1998). Using agro-industrial food waste as substrate instead of virgin feedstock of refined sugar such as glucose, sucrose and corn steep liquor could represent a turning point. Sugarcane and beet molasses, cheese whey effluents, plant oils, swine waste liquor, vegetable and fruit wastes, effluents of palm oil mill, olive oil mill, paper mill, pull mill and hydrolysates of starch (e.g., corn and tapioca), cellulose and hemicellulose are all excellent alternatives characterised by a high organic fraction (Reis *et al.*, 2011).

CONCLUSION

Food wastage has become a major problem in today's world. In the last few years, food wastage has become a serious issue that has affected both "developed and developing countries" equally. According to Food and Agriculture Organization (FAO), about one-third or one-fifth of all the food produced remains unconsumed. Moreover, food waste has a robust financial influence and effects the environment with the total greenhouse gas emission. A number of resolutions may be applied in the suitable organization of waste food and prioritised in a parallel way to waste food management. The greatest desirable solutions are characterized by circumvention and donation of palatable portions to social services. Food waste is also employed in industrial processes for the production of biofuels or biopolymers. Further steps predict the repossession of nutrients and fixation of carbon by composting. Therefore, it is expected that there will be an increasing number of initiatives, campaigns and legislative developments in order to reach the aforementioned objectives.

REFERENCES

Abdel-Rahman, M. A., Tashiro, Y., & Sonomoto, K. (2013). Recent advances in lactic acid production by microbial fermentation processes. *Biotechnology Advances*, *31*(6), 877–902. doi:10.1016/j.biotechadv.2013.04.002 PMID:23624242

ACR+. (2009). Municipal Waste in Europe – Towards a European Recycling Society. *Victoires Editions.*

Alibardi, L., Muntoni, A., & Polettini, A. (2014). Hydrogen and waste: Illusions, challenges and perspectives. *Waste Management (New York, N.Y.)*, *34*(12), 2425–2426. doi:10.1016/j.wasman.2014.09.001 PMID:25442106

Bajón Fernández, Y., Soares, A., Villa, R., Vale, P., & Cartmell, E. (2014). Carbon capture and biogas enhancement by carbon dioxide enrichment of anaerobic digesters treating sewage sludge or food waste. *Bioresource Technology*, *159*, 1–7. doi:10.1016/j.biortech.2014.02.010 PMID:24632434

Barton, S. (2010, September 17). *Retailers halve waste to landfill since 2005.* Retrieved from https://www.letsrecycle.com/news/latest-news/retailers-halve-waste-to-landfill-since-2005/

Beretta, C., Stoessel, F., Baier, U., & Hellweg, S. (2013). Quantifying food losses and the potential for reduction in Switzerland. *Waste Management (New York, N.Y.)*, *33*(3), 764–773. doi:10.1016/j.wasman.2012.11.007 PMID:23270687

Brook Lyndhurst Blog. (2009). *Enhancing Participation in Kitchen Waste Collections*. Retrieved from http://www.brooklyndhurst.co.uk/enhancing-participation-in-kitchen-waste-collections-_119

Burkhardt, M., Koschack, T., & Busch, G. (2015). Biocatalytic methanation of hydrogen and carbon dioxide in an anaerobic three-phase system. *Bioresource Technology, 178*, 330–333. doi:10.1016/j.biortech.2014.08.023 PMID:25193088

Canakci, M. (2007). The potential of restaurant waste lipids as biodiesel feedstocks. *Bioresource Technology, 98*(1), 183–190. doi:10.1016/j.biortech.2005.11.022 PMID:16412631

Chalak, A., Abou-Daher, C., Chaaban, J., & Abiad, M. G. (2016). The global economic and regulatory determinants of household food waste generation: A cross-country analysis. *Waste Management (New York, N.Y.), 48*, 418–422. doi:10.1016/j.wasman.2015.11.040 PMID:26680687

Clarke, W. P., & Alibardi, L. (2010). Anaerobic digestion for the treatment of solid organic waste: What's hot and what's not. *Waste Management (New York, N.Y.), 30*(10), 1761–1762. doi:10.1016/j.wasman.2010.06.019 PMID:20638829

Dahlén, L., & Lagerkvist, A. (2010). Evaluation of recycling programmes in household waste collection systems. *Waste Management & Research, 28*(7), 577–586. doi:10.1177/0734242X09341193 PMID:19748961

De Gioannis, G., Muntoni, A., Polettini, A., & Pomi, R. (2013). A review of dark fermentative hydrogen production from biodegradable municipal waste fractions. *Waste Management (New York, N.Y.), 33*(6), 1345–1361. doi:10.1016/j.wasman.2013.02.019 PMID:23558084

Edjabou, M. E., Petersen, C., Scheutz, C., & Astrup, T. F. (2016). Food waste from Danish households: Generation and composition. *Waste Management (New York, N.Y.), 52*, 256–268. doi:10.1016/j.wasman.2016.03.032 PMID:27026492

Exodus Market Research. (2006). *A quantitative assessment of the nature, scale and origin of post consumer food waste arising in Great Britain 2006*. Author. (unpublished)

FAO. (1981). Food loss prevention in perishable crops. FAO Agricultural Services Bulletin, 43, 72.

Favaro, L., Alibardi, L., Lavagnolo, M. C., Casella, S., & Basaglia, M. (2013). Effects of inoculum and indigenous microflora on hydrogen production from the organic fraction of municipal solid waste. *International Journal of Hydrogen Energy*, *38*(27), 11774–11779. doi:10.1016/j.ijhydene.2013.06.137

Flores, R. A., Shanklin, C. W., Loza-Garay, M., & Wie, S. H. (1999). Quantification and characterization of food processing wastes/ residues. *Compost Science & Utilization*, *7*(1), 63–71. doi:10.1080/1065657X.1999.10701954

Food Standards Agency. (2006). *Consumer Attitudes to Food Safety 2005*. Author.

Friends of the Earth. (2006). *Briefing: food waste collections*. Retrieved from http://www.foe.co.uk/resource/briefings/food_waste.pdf

Fuessl, A., Yamamoto, M., & Schneller, A. (2012). *Opportunities in bi-based building blocks for polycondensates and vinyl polymers. In Polymer Science: A Comprehensive Reference* (Vol. 5, pp. 49–70). Elsevier.

FUSIONS. (2016). *EU FUSIONS website*. Retrieved from www.eu-fusions.org

Galanakis, C. M. (2012). Recovery of high added-value components from food wastes: Conventional, emerging technologies and commercialized applications. *Trends in Food Science & Technology*, *26*(2), 68–87. doi:10.1016/j.tifs.2012.03.003

Garcia-Garcia, G., Woolley, E., Rahimifard, S., Colwill, J., White, R., & Needham, L. (2017). A Methodology for Sustainable Management of Food Waste. *Waste and Biomass Valorization*, *8*(6), 2209–2227. doi:10.100712649-016-9720-0

Gasparatos, A., Stromberg, P., & Takeuchi, K. (2011). Biofuels, ecosystem services and human wellbeing: Putting biofuels in the ecosystem services narrative. *Agriculture, Ecosystems & Environment*, *142*(3-4), 111–128. doi:10.1016/j.agee.2011.04.020

Ghanavati, H., Nahvi, I., & Karimi, K. (2015). Organic fraction of municipal solid waste as a suitable feedstock for the production of lipid by oleaginous yeast Cryptococcus aerius. *Waste Management (New York, N.Y.)*, *38*, 141–148. doi:10.1016/j.wasman.2014.12.007 PMID:25595390

Gustavsson, J., Cederberg, C., Sonesson, U., Van Otterdijk, R., & Meybeck, A. (2011). *Global food losses and food waste: extent, causes and prevention*. Rome: FAO.

Hogg, D., Barth, J., Schleiss, K., & Favoino, E. (2007). *Dealing with food waste in the UK*. Eunomia Research and Consulting Limited. WRAP. Retrieved from http://www.wrap.org.uk/sites/files/wrap/Dealing_with_Food_Waste_-_Final_2_March_07.pdf

Kiran, E. U., Trzcinski, A. P., Ng, W. J., & Liu, Y. (2014). Bioconversion of food waste to energy: A review. *Fuel, 134*, 389–399. doi:10.1016/j.fuel.2014.05.074

Levis, J. W., Barlaz, M. A., Themelis, N. J., & Ulloa, P. (2010). Assessment of the state of food waste treatment in the United States and Canada. *Waste Management (New York, N.Y.), 30*(8-9), 1486–1494. doi:10.1016/j.wasman.2010.01.031 PMID:20171867

Li, X., Chen, Y., Zhao, S., Chen, H., Zheng, X., Luo, J., & Liu, Y. (2015). Efficient production of optically pure l-lactic acid from food waste at ambient temperature by regulating key enzyme activity. *Water Research, 70*, 148–157. doi:10.1016/j.watres.2014.11.049 PMID:25528545

Liu, H., Ma, X., Li, L., Hu, Z., Guo, P., & Jiang, Y. (2014). The catalytic pyrolysis of food waste by microwave heating. *Bioresource Technology, 166*, 45–50. doi:10.1016/j.biortech.2014.05.020 PMID:24905041

Matsuda, T., Yano, J., Hirai, Y., & Sakai, S. (2012). Life-cycle greenhouse gas inventory analysis of household waste management and food waste reduction activities in Kyoto, Japan. *The International Journal of Life Cycle Assessment, 17*(6), 743–752. doi:10.100711367-012-0400-4

Mirabella, N., Castellani, V., & Sala, S. (2014). Current options for the valorization of food manufacturing waste: A review. *Journal of Cleaner Production, 65*, 28–41. doi:10.1016/j.jclepro.2013.10.051

Mourad, M. (2016). Recycling, recovering and preventing "food waste": Competing solutions for food systems sustainability in the United States and France. *Journal of Cleaner Production, 126*, 461–477. doi:10.1016/j.jclepro.2016.03.084

Nanqi, R., Wanqian, G., Bingfeng, L., Guangli, C., & Jie, D. (2011). Biological hydrogen production by dark fermentation: Challenges and prospects towards scaled-up production. *Current Opinion in Biotechnology, 22*(3), 365–370. doi:10.1016/j.copbio.2011.04.022 PMID:21612910

Opatokun, S. A., Strezov, V., & Kan, T. (2015). Product based evaluation of pyrolysis of food waste and its digestate. *Energy, 92*, 349–354. doi:10.1016/j.energy.2015.02.098

Östergren, K., Gustavsson, J., Bos-Brouwers, H., Timmermans, T., Hansen, O.-J., Møller, H., . . . Redlingshöfer, B. (2014). FUSIONS Definitional Framework for Food Waste. Full Report.

Panesar, P. S., Kennedy, J. F., Gandhi, D. N., & Bunko, K. (2007). Bioutilisation of whey for lactic acid production. *Food Chemistry*, *105*(1), 1–14. doi:10.1016/j.foodchem.2007.03.035

Parfitt, J., Barthel, M., & Macnaughton, S. (2010). Food waste within food supply chains: Quantification and potential for change to 2050. *Phil. Trans. R. Soc.*, *365*(1554), 3065–3081. doi:10.1098/rstb.2010.0126 PMID:20713403

Pfaltzgraff, L. A., De bruyn, M., Cooper, E. C., Budarin, V., & Clark, J. H. (2013). Food waste biomass: A resource for high-value chemicals. *Green Chemistry*, *15*(2), 307–3014. doi:10.1039/c2gc36978h

Pham, T. P., Kaushik, R., Parshetti, G. K., Mahmood, R., & Balasubramanian, R. (2015). Food waste-to-energy conversion technologies: Current status and future directions. *Waste Management (New York, N.Y.)*, *38*, 399–408. doi:10.1016/j.wasman.2014.12.004 PMID:25555663

Pirozzi, D., Ausiello, A., & Yousuf, A. (2012). Exploitation of Lignocellulosic Materials for the Production of II Generation Biodiesel. In *Proceeding Venice 2012, Fourth International Symposium on Energy from Biomass and Waste*. Cini Foundation.

Priefer, C., Jörissen, J., & Bräutigam, K.-R. (2016). Food waste prevention in Europe – A cause-driven approach to identify the most relevant leverage points for action. *Resources, Conservation and Recycling*, *109*, 155–165. doi:10.1016/j.resconrec.2016.03.004

Refaat, A. A. (2012). Biofuels from Waste Materials. *J. Compr. Renew. Energy*, *5*, 217–261. doi:10.1016/B978-0-08-087872-0.00518-7

Reis, M., Albuquerque, M., Villano, M., Majone, M. (2011). Mixed culture processes for Resource Futures. *Bournemouth Borough Council Waste Composition Analysis Phase 2 and Sciences,* (1), 17-26.

Reisch, L., Eberle, U., & Sylvia Lorek, S. (2013). Sustainable food consumption: an overview of contemporary issues and policies. *Sustainability: Science, Practice and Policy*, *9*(2), 7–25. doi:10.1080/15487733.2013.11908111

Resource Futures. (2009). *Bournemouth Borough Council Waste Composition Analysis Phase 2 and Comparative.* RF Project no.510.

Schanes, K., Dobernig, K., & Gozet, B. (2018). Food Waste Matters—A Systematic Review of Household Food Waste Practices and Their Policy Implications. *Journal of Cleaner Production*, *182*, 978–991. doi:10.1016/j.jclepro.2018.02.030

Segrè, A., & Gaiani, S. (2011). *Transforming Food Waste into Resource*. Cambridge, UK: Royal Society of Chemistry.

Smil, V. (2010). Improving efficiency and reducing waste in our food system. *Environmental Sciences, 1*(1), 17–26. doi:10.1076/evms.1.1.17.23766

Stoeberl, M., Werkmeistera, R., Faulstichb, M., & Russa, W. (2011). Biobutanol from food wastes – fermentative production, use as biofuel and the influence on the emissions. *Procedia Food Waste, 1*, 1868–1974.

UN General Assembly. (2015). *Resolution adopted by the General Assembly on 25 September 2015: Transforming our world: the 2030 Agenda for Sustainable Development A/RES/70/1*. Retrieved from www.un.org/ga/search/view_doc.asp?symbol=A/RES/70/1&Lang=E

UNEP. (2014). *Prevention and reduction of food and drink waste in businesses and households - Guidance for governments, local authorities, businesses and other organisations, Version 1.0*. UNEP.

Van Wegen, R. J., Ling, Y., & Middelberg, A. P. J. (1998). Industrial production of polyhydroxyalkanoates using escherichia coli: An economic analysis. *Trans. IChem, E76*(3), 417–426. doi:10.1205/026387698524848

WRAP. (2007). *Understanding Food Waste*. Research Summary. WRAP, Banbury.

WRAP. (2009a). *Household Food and Drink Waste in the UK*. WRAP, Banbury.

Yaakob, Z., Mohammad, M., Alherbawi, M., Alam, Z., & Sopian, K. (2013). Overview of the production of biodiesel from waste cooking oil. *Renewable & Sustainable Energy Reviews, 18*, 184–193. doi:10.1016/j.rser.2012.10.016

Zhang, C., Su, H., Baeyens, J., & Tan, T. (2015). Reviewing the anaerobic digestion of food waste for biogas production. *Renewable & Sustainable Energy Reviews, 38*, 383–392. doi:10.1016/j.rser.2014.05.038

Chapter 11
Understanding the Composition of Food Waste:
An "-Omics" Approach to Food Waste Management

Matthew Chidozie Ogwu
https://orcid.org/0000-0001-6054-1667
Seoul National University, South Korea

ABSTRACT

Due to existing application gap, the diverse microorganisms and their processes in food waste have not been maximally explored or harnessed. This chapter addresses the possible application of "-omics" technologies to understand food waste microbial composition and metabolic processes, to stimulate future research in mining food waste for important microorganisms and bioactive compounds. The chapter highlights potential "-omics" procedures for food waste assessments. These innovative, culture-independent, high-throughput technologies have already revolutionized diverse fields of research and human endeavors. This chapter also introduces the concept of food wastomics to detect, identify, and measure the different molecules and microorganisms that are present and expressed in food waste such as DNA, RNA, proteins, lipids, and metabolites. This knowledge will create a greater understanding of how discarded food is degraded and what useful products or organisms may be harvested from it. Finally, the chapter recommends the integration of food wastomics into foodomics.

DOI: 10.4018/978-1-5225-7706-5.ch011

INTRODUCTION

Food is any substance, whether liquid, concentrated, solid, semi-solid, frozen, dried or dehydrated excluding alcoholic beverages, dietary supplements, soft drinks or tobacco that are prepared or sold for ingestion or chewing by humans and other animals for their taste or nutritional value (ODT, 2004). Besides taste and nutrition, food is also a socio-economic vehicle for distinction and establishing links as well as other symbolic, psychological, physiological, moral, environmental and cultural expressions (Rozin, 2005; Stajcic, 2013). Different methods have evolved for food production, packaging, consumption and storage depending on the nature of the food. Food is an essential part of animal and human lives as it provides the energy required to fulfil daily obligations in good health and for survival.

There is a global target to make food available, easily accessible, and inexpensive despite current food insecurities. Food security is when in the absence of hunger, the population have physical and economic access to sufficient, safe and nutritious food for an active and healthy life at all times (FAO, 2001). According to Shi (2014), food security is a multidimensional challenge embedding social, economic and environmental complexities that require an examination of gaps within agricultural and socio-economic systems. Global food security goals seek to significantly increase agricultural productivity and food supply, minimize the secondary use of food crop as biofuels, reduce food loss and waste, ensure efficient distribution of food, eliminate hunger and malnutrition, produce nutritious food and make food production systems more resilient. Among other factors, food loss and waste (FLW) have a negative influence on food security and global development goal.

Food loss is the decrease in the quality and quantity of nutritional, economic and food safety value of food whereas food waste occurs when the entire or a part of safe and nutritious food is discarded or subjected to alternative (non-food) use (FAO, 2014). In both cases, food suitable for human consumption is not eaten and can occur at any point within the food systems (i.e. production, packaging, distribution, consumption and storage). Food loss may be accidental due to inadequate technology, lack of knowledge and skills, poor logistics and malfunctioning markets, whereas food waste is considered the outcome of an intended or unintended behaviour such as the deliberate removal of food by choice or negligence (FAO, 2017). Leach and Swannell (2017) opined that the FAO's definition is vague because of the emphasis on loss and waste resulting from food that is fit for consumption, citing chicken feet, which are eaten in some part of the world and not in others. Moreover, the definition has ambiguous connotations when considered with the International Food Waste and Loss Standard for evaluating FLW as outlined in FLWP (2016) wherein a detailed sustainable approach for measuring food waste and loss is outlined. Nonetheless, the main drawback of this multi-organization collaboration is insufficient information at various levels. HLPE (2014) rightly considered FLW as an integral part of the food system and the consequences of associated functions and techniques.

The problems of FLW is considered among challenges hampering global food security and sustainable economic development as well as contributing to biodiversity decline, reduced quality of the environment, soil, ecosystem, and earth's natural resource base. Hence, FLW remains a topical issue in global food systems due to the numerous initiatives proposed for their sustainable management (FAO, 2014). It is ironical that a food insecure world does not efficiently manage its food waste. In addition, the intensification of food production and estimated population explosion is positioned to compound this challenge unless a sustainable approach is incorporated into the food system to checkmate wastage. FLW also translate into economic losses for farmers and other stakeholders within the food value chain, and higher prices for consumers, both of which contribute to food insecurity by making food less accessible for vulnerable groups within the populace (FAO, 2017). According to Meyers et al. (2012), increasing food production by dedicating more land to the underlying process could be achieved without imposing an undue environmental cost. However, poor investment decisions in food production infrastructures and technology will cause an increase in food waste and an overall reduction in efficient food waste management. So far, this has produced new dimensions and increased old challenges in land use systems, pollution, and public health, which are a common concern without identifying specific country or regional level challenges or drawing any country-specific conclusions. Due to these growing environmental but also social and economic concerns, food waste is increasingly acknowledged as an urgent issue among governments, businesses, non-governmental organizations (NGOs), academics, and the general public (Schanes et al., 2018). Therefore, addressing food wastage is paramount in the efforts to combat hunger, raise income and improve food security, quality and safety, and bring about sustainable economic and environmental development (FAO, 2011).

Food wastage is a burden that requires urgent and innovative standard operating practices and appropriate measures to rescue human society (Paritosh et al., 2017). Food waste is receiving attention from local, national and global policymakers, as well as international organisations, NGOs and academics from various disciplinary fields due to increasing concerns about food security and environmental impacts (Schanes et al., 2018). However, this attention is devoid of innovative approach that is capable of transforming food waste to economic, scientific and industrial wealth by incorporating "-omics" techniques into food waste management.

With a focus on understanding food waste composition, this chapter sets out to address issues related to the management and reduction of food waste. Food waste accounts for a sizable proportion (approximately one-third) of agricultural output, and reducing them would lessen the need for production increases as it would only require an increase in the supply of available food to strengthen global food security (HLPE, 2014; FAO, 2017). Hence, the unsustainable management of food waste is a major factor militating against global food security and development goals. Within food production systems, "-omics" strategies and techniques are already contributing to the production process, storage, cleaning and overall sanitation of the food manufacturing process, tracing, identification, detection and quantification of microorganisms in food as well as the careful monitoring of microbial contamination in the final products (Josic et al., 2016). Although the potential application of "-omics" in elucidating the biological composition and activity of food waste is possible, it is yet to be explored, at least to its maximum capacity. Thus, the objective of this chapter is to highlight how "-omics" strategies and techniques can contribute to the sustainable management of food waste. This chapter is important for a number of reasons, including the introduction of a new method for managing food waste, and mining or exploring food waste for new and useful products using "-omics" technologies. It will also contribute to reducing environmental contamination resulting from food waste, maximize the socio-economic benefits from enhanced efficiencies in food production systems, and contribute to the knowledge of food waste microdiversity (i.e. diversity of microorganisms). Most foods are wholly or partly derived from biological products and previous studies, including Dorman and Deans (2008); Riya et al. (2013); Tarazona-Dıaz and Aguayo (2013); Barbulova et al. (2015); Guil-Guerrero et al. (2016); Varzakas et al. (2016) have reported that wastes from biological products contain valuable products including bioactive compounds. More so, food waste is considered as the biodegradable output from various sources including food processing industries, households, and hospitality sector (Paritosh et al., 2017) that is a haven for microdiversity. Most compositional analysis of food waste (Edjabou, et al., 2015; Kadir et al., 2017) reflects the compositional fractions of materials in the wasted food. The organic and inorganic compounds that are present in food waste make it a viable growth substrate for microorganisms.

CONTEMPORARY ISSUES ASSOCIATED WITH FOOD WASTE

OLIO (2017) rightly pointed out that food waste is reflective of the massive inefficiencies in food production and marketing systems, the kind that does not persist in other industries and sectors. The outcome is a potential environmental hazard from piles of decomposing wasted invaluable food and fruits (such as in Figure 1), comprising of avoidable (i.e. fit for consumption but discarded food) and unavoidable (i.e. food scraps and crumbs not fit for consumption that is discarded) food waste.

Figure 1. Food waste resulting from unconsumed fruits
Source: OLIO, 2017

Managing such piles of decomposing waste is becoming increasingly challenging and costly. The study by FAO (2013) indicated that the wastage of cereals, meat, fruits and vegetables pose a significant environmental problem in terms of land occupation and carbon footprint especially in parts of Asia, Europe and Latin America. Currently, two-thirds of food waste goes to landfills, with the rest diverted to other useful purposes such as composting, energy and fuel generation, animal feed, and redistribution as food. Most importantly, land continues to be naturally limiting and a limited resource. Thus, it remains pertinent to reduce the amount of food waste that goes to landfill by converting these waste foods, their products and by-products into useful products. However, at the moment, the global food system contributes to environmental pollution, deforestation, land use change and biodiversity loss (Garnett, 2014). According to Paritosh et al. (2017), the use of cost and eco-friendly approach for food waste conversion process and technology will help reduce the associated burden. For instance, anaerobic digestion approach for food waste management, energy, and nutrient production can contribute to solving the world's ever-increasing energy requirements. Food waste decomposition requires a myriad of living and non-living organism mediated processes but the survival of microorganisms in food waste depends on the numerous secretions released into the extracellular environment that is required for their survival and can be characterized using "-omics" technology (Josic et al., 2016).

Even though food waste occurs in all stages of the food production and supply chain, individual domestic food waste is the greatest (Monier et al., 2010; Braun, 2012). Hence, domestic food waste remains a key actor despite scattered evidence of why the food waste occurs and how to minimize it (Schanes et al., 2018). The main drawback is concerning the measurement of domestic food waste, which is often difficult and complicated due to individual household considerations. At the other points, the difficulty is related to the length of the food supply chains that are often long and involve many players, including small farmers, transporters, processors, retailers and households (FAO, 2018). Parfitt et al. (2010) suggested that due to the enormous amounts of food waste from the household level, the prevention of food waste at the final stages of the supply chain is of utmost importance to prevent further environmental effects such as resource depletion and greenhouse gas emissions. More so, Schanes et al. (2018) opined that there is still a relative paucity of field research on the subject of consumer-generated food waste within the context of private households. Domestic waste from developed countries like the USA, UK and others is far higher than what is obtainable in less developed countries like the majority of countries in Africa and Asia. About 50% of food waste occurs domestically in developed countries (OLIO, 2017). FAO (2011) estimated that per capita food waste by consumers' amounts to 95 - 115 Kg/year in Europe and North America compared to 6 - 11 Kg/year in South/Southeast Asia and Sub-Saharan Africa. Chakona et al. (2017) reported that in South Africa, domestically, there is more food waste from prepared food compared to unprepared food and drinks. The habit of preparing excess quantities of food than can be consumed or stored for long periods is a major contributor to food waste in sub-Saharan Africa (Ramukhwatho et al., 2014). Generally, the other causes of food waste include lack of proper planning, purchasing too much food or food materials, industrial errors, changes in food safety policies, infrastructural challenges, wrong estimation on the part of hoteliers, restaurants and other players in the foodservice industry, improper consumer behaviour and habits, cultural practices, etc. Again, the issue of managing the resulting waste is compounded by gross inability to efficiently measure food waste (FLWP, 2016; Leach and Swannell, 2017).

Besides the immediate impact on food security, food waste has a broad 1). Socio-economic and 2). Environmental Consequences.

1. Socio-Economic Consequences of Food Loss and Waste

Economic losses from food waste are estimated to be more than US$940 billion per year at an estimated 1.3 billion tonnes per year (FAO, 2011; Leach and Swannell, 2017). In addition, the global social costs of food wastage amount to about US$900 billion per annum (FAO, 2018). Hence, the socio-economic cost of managing food waste is considered exorbitant to taxpayers. A significant portion of the annual budget of medium- and high-income countries is dedicated to addressing food waste-related problems. By eliminating avoidable food and drink waste, an average family in the United Kingdom would save British£470 a year (Quested et al., 2013). This, in turn, will have a positive impact on food prices and a reduction in public expenditures to manage food waste (WRAP, 2013; Parry et al., 2015). Moreover, economic gains may be made through the recovery of valuable components of food wastes (Otles et al., 2015). Furthermore, the hidden economic cost of food waste includes lost revenues, energy cost, lost labour, lost materials, liabilities and risks among others (Mason et al., 2011).

2. Environmental Consequences of Food Loss and Waste

Food production and delivery require key inputs such as land, human energy and capital, chemicals and other materials, hence a loss or waste of the finished products translate into considerable environmental impacts (Tonini et al., 2018). Food waste is responsible for 8% of global greenhouse gas (GHG) emissions and this may potentially increase global temperatures by ~2 ^{o}C in the near future (FAO, 2011; Bajželj et al., 2014). Global warming results from the GHGs like methane, nitrates, sulphates, CO, ammonia, ozone and other processes like acidification traceable to food waste (von Schneidemesser et al., 2015). Environmental impacts of food waste also result in the loss or reduction of environmental resources due to imbalance biotic and biogeochemical cycles (Tonini et al., 2018). Direct environmental consequences of food waste include foul odours, landfills, loss of aesthetic, increase carbon footprint, eutrophication, health hazards due to ecotoxicity whereas indirect environmental consequences include land use changes, reduction in biological and non-biological (water and soil) resources, and (industrial) food production processes. They are considered indirect because they most often do not result from the final food product but from the processes of their production. According to Nishida (2014), food waste currently represents the single largest type of waste entering landfills. More so, biodiversity losses occur because of intensifying agricultural productivity, which will have a subsequent effect on the environment through slash and burn and shifting agriculture, deforestation, monocropping and invasion of wildlife sanctuaries. An important consideration is that the environmental impacts of food waste can quickly become economical through fines, management, reclamation, restoration and other regulatory actions (Mason et al., 2011).

SOME CONVENTIONAL PROCESSES OF MANAGING FOOD WASTE

Preventing waste generation is the first priority of managing food waste. Beyond that point, prevailing methods include incineration, anaerobic (digestion or) fermentation, composting, landfill, and the use of food residues for agricultural productions, such as animal feed or fertilizer, as the main strategies for waste minimization and valorisation (Otles et al., 2015). Food waste may be solid, liquid, semi-solid and gaseous compounds containing organic and inorganic compounds with a variable pH depending on the contents and decomposition state. Most food wastes ends up in landfills or incineration facilities. In landfills, food waste may be sorted and removed from the overall waste stream. These landfilling takes up space and represents a failure to utilize the resources in food waste most efficiently. In anaerobic digestion, the efficiency of gas extraction and recovery is significantly decreased compared to controlled conditions in dedicated fermenters (Tonini et al., 2018). These processes contribute to energy saving as well as the production of useful chemicals as they gain further integration into the food waste recycling process.

Most food wastes are highly biodegradable and contain water, large quantities of residues with a high biochemical oxygen demand and chemical oxygen demand contents, which combines to make managing food waste difficult in many aspects (Otles et al., 2015). Microbial activities in food waste have the potential to increase over time due to succession within microbial communities. These processes have physical, chemical, thermal, and biological considerations with specific shortfalls. More so, fat rich food waste may be considered for valorisation and recovery to be used as feed, whereas solid food waste is taken to landfills and may be processed further through degradation, fermentation, etc. Anaerobic digestion and thermochemical treatments (e.g. combustion, gasification, and pyrolysis) are used to convert food waste rich in organic compounds to biofuel and aerobic degradation (composting and vermicomposting) to produce humus and nutrients. Recently, anaerobic digestion of food waste has gained popularity because it turns waste into renewable energy. Household waste sorting and recycling is also garnering support in developing countries. However, these food waste management practices require infrastructures, legislative support and technical know-how to be effective. Treatment of food waste contribute to reducing the volumes of the waste, as well as generating new materials and for energy recovery. In addition, Strotmann et al. (2017) suggested a novel participatory method at the industrial level wherein the methodologies to reduce food waste at each level led to resource efficiency. In the past decades, new management methods and treatments that focus on recovery and utilization of valuable bioactive constituents of food wastes have generated more interest and have been used to produce pectins, flavonoids, polyphenols, fibres, etc. (Otles et al., 215). Yet little to no consideration has been given to "-omics" application in waste management that can utilize these constituents to characterize and mine the waste. More so, the taxonomic composition of microorganism in food waste has not attracted sufficient research attention.

UNDERSTANDING FOOD WASTE COMPOSITION USING AN OMICS APPROACH

Microorganisms are responsible for the processing of food waste, thereby giving off a variety of potentially useful chemicals. There is a growing recognition that food waste problems can be solved through the utilisation of the waste as a resource, by the application of green and sustainable scientific methodologies and technologies (Luque and Clark, 2013; Thi et al., 2016). The use of high throughput, next-generation sequencing techniques is a viable tool. Generally, "-omics" is an umbrella term for diverse high throughput sequencing processes, including (meta) genomics and (meta) transcriptomics, as well as metabolic and (meta) proteomics (Walsh et al., 2017). Other subfields include epigenomics, lipidomics, interactomics, metabolomics, foodomics, diseasomics, wastomics, etc. "-Omics" approaches and technologies have revolutionalize biomarker discovery processes essential for the study of biological molecules. Advances in "-omics" strategies and techniques are associated with parallel breakthroughs in genome sequencing, Bioinformatics, and analytic tools including high-throughput technologies that have made it possible to get useful insights into microbial communities, compounds and processes in different environments (Roemer and Boone; 2013; Tang, 2015; DOS Santos et al., 2016).

Amplicon sequencing (single organism genome sequencing) and whole-metagenome shotgun (i.e. whole environmental community) sequencing are covered by genomics and metagenomics respectively. The sequencing depth of both approaches is different. Nonetheless, from these processes, the taxonomic and functional (i.e. metabolic profile) constituent of a target community can be elucidated. In amplicon sequencing (for example 16s rRNA and internal transcribed spacer (ITS)), regions in the marker genes are PCR amplified from DNA extracted from a mixed microbial community, sequenced and aligned against a reference database to determine the taxonomic composition of a sample (Walsh et al., 2017).

Metagenomics is the application of modern genomic techniques to study communities of microorganisms directly in their natural environments, bypassing the need for isolation and the laboratory cultivation of individual species by cloning and analysing their genomes (Chen and Pachter, 2005). Hence, until recently, general knowledge of microbial life was based on organisms raised in a pure culture but a significant percentage of microorganisms cannot be cultured in the laboratory (Sabree et al., 2009). The study of microorganisms that pervade every part of this planet has encountered many challenges through time such as the discovery of unknown organisms and the understanding of how they interact with their environment and the evolution of sequencing technology is likely to be an important contributor to solve different problems (Escobar-Zepeda et al., 2015). According to Handelsman et al. (1998); Thomas et al. (2012) metagenomics is one of the fastest growing scientific disciplines and it is becoming a central tool for improving life worldwide through the fields of microbial ecology, evolution, and diversity. The reason for promoting the use of metagenomics to understand the taxonomic composition and functional interaction of wasted food is that only a small proportion of organisms have been grown in culture and many species do not live in isolation; rather they form complex communities whose culture media is mostly yet to be deciphered. Thus, more often clonal cultures fail to represent the natural environment of a given organism; therefore, many proteins and protein functions remain undiscovered. Diverse microbial communities of bacteria, archaea, viruses and single-celled eukaryotes have crucial roles in processing food waste.

Through transcriptomics, cDNA sequencing can be used to predict with greater certainty the differentially expressed gene transcripts within a given sample (Tarazona et al., 2011; Franzosa et al., 2015) like food waste. Metabolomics and metaproteomics can be used for the identification and quantification of microbial metabolites and microbial proteins (Aldridge and Rhee 2014; Hettich et al., 2012). Proteomics evolved from using gels to separate proteins in two-dimensional electrophoresis (2-DE), wherein proteins with independent isoelectric points may be separated by their molecular weights (Dos Santos et al., 2016). The systematic evaluation of proteins that are expressed by one particular cell, tissue, biological fluid, or organism is called proteomics while the analysis of gene expression (transcription) to understand a biological process is described as a transcriptomics (Dos Santos et al., 2016). The study of gene function, evolution, genome mapping through characterization and quantification of genes that direct the production of proteins is also within the realm of "-omics". Metabolomics is the study of all metabolites in a biological organelle, cell, tissue, organ, system and organisms and community (Vailati-Riboni et al., 2017).

Through this more efficient metagenome, genomic, protein and cDNA sequencing technology, the complete decoding of various microbial genomes within food waste can be elucidated with a view to making this information available and accessible on global databases. Thereafter, they can be exploited in future studies and for targeted food waste mining process. From the success reported in other sequence-based studies that adopted this approach, it can be hypothesized that their application in food waste management (as food wastomics) will facilitate the discovery of new potentially useful organisms, genes, and metabolic process. Moreover, it will make it possible to process large amounts of generated data for both identification and quantification, as well as for corresponding modelling to understand the combined effects of hitherto hidden functional processes and microorganisms with public health implications on the development of complex diseases (Vlaanderen et al., 2010; Andjelković et al., 2017).

Foodomics is the study of chemical compounds in food using high throughput sequencing techniques. Foodomics involves the application of "-omics" technologies to the different aspects of the food production systems (García-Canas et al., 2012). There are established and highly sensitive methods for elucidating the components of food waste including immunochemical analyses; hence, "-omics" plays a complementary role (Josic et al., 2016). These strategies in this approach facilitate the connection of food components (i.e. nutrients), diet, individual health and diseases to optimized population health and well-being (Capozzi and Bordoni, 2013). The field is a fast developing area in food science that is well positioned to address pre-screening of consumable food, food traceability, sustainability, quality assurance, adulteration identification, safety and integrity (Josic and Giacometti, 2013; Andjelković et al., 2017; Braconi et al., 2018). Thus far, it has provided an opportunity to advance the understanding of nutrition and its effect on health and common diet-related diseases (Roche et al., 2015) in support of the capabilities of "-omics" driven advances to change the face of biological research, especially through applications in new subfields, improvement in sequencing methods and reduction of operational cost. "-Omics" has led to ground-breaking insights about the microbial constituent of different environments including air, food, soil, and water (Frias-Lopez et al., 2008; Fierer et al., 2012; Ercolini 2013; Bokulich et al., 2016; Walsh et al., 2017). Recently, Walsh et al. (2017) examined the application of (high throughput sequencing) "-omics" technologies in general food microbiology. The authors highlighted the possibility of adapting the approach to study food-related microbial isolates and mixed microbial communities in foods from a genomics, metagenomics, and metatranscriptomics sequencing perspective.

Since its discovery in the 1970's, sequencing technology have witnessed improvements, especially in cost, efficiency and speed. The current sequencing technology is popularly referred to as next-generation sequencing. The most common ones are by Illumina, Inc. (Genome Analyzer, HiSeq series, MiSeq, NextSeq 500) and the Ion Torrent Personal Genome Machine, GS FLX by 454 Life Sciences/Roche diagnostics, and SOLiD by ABI (Reuter et al. 2015; Ambardar et al., 2016). The different sequencing platforms have unique chemistry and operational principles for template preparation (source nucleic acid extraction, library preparation and template amplification), and sequencing (by synthesis or by hybridization and ligation) with specific pros and cons, hence the choice depends on specific research aims and objectives (Loman et al., 2012; Ambardar et al., 2016; Walsh et al., 2017).

So throughput read length, error rate, coverage, cost and runtime varies as shown in Table 1.

Wastomics is the application of "-omics" strategies and techniques towards understanding the composition of different biological waste. Specifically, food wastomics is the use of "-omics" technology to understand food waste microdiversity, functional processes with a view to mining it for useful bioactive compounds and organisms. This approach investigates the microbes that inhabit the different categories of food waste using sequencing technology to understand their taxonomic composition and functional interactions. This culture-independent approach has many unexplored potentials. In this food wastomics approach, an integration of diverse "-omics" techniques can find application as presented in Table 2.

Food Waste Genomics or Genotyping

Culture-independent amplicon sequencing by targeting 16S, 18S or ITS regions of organismal DNA within food waste will be useful in identifying specific taxonomic groups, their functional composition and activities in the food waste. Functional annotation for 16S (bacteria) and ITS (fungal) communities can be predicted using the Phylogenetic Investigation of Communities by Reconstruction of Unobserved States (PICRUSt) and FUNGuild respectively (Langille et al. 2013; Nguyen et al., 2016). Regardless of the potential of obtaining valuable insights from this culture-independent approach, their application in compositional analysis is limited due to current short sequence reads generated by current sequencers that provide limited resolutions (Walsh et al., 2017). On the other hand, food waste metagenomics is capable of producing a representative outlook of the taxonomic and functional composition of food waste communities (Figure 2). This aspect of food wastomics can give insights into the roles of specific genes in addition to diverse taxonomic

Table 1. Sequencing platforms and the average throughput length

Sequencing platform	Average read length (in base pairs)	Accuracy	Amount of runs/ reads	Average time
Sanger Chain termination	~ 750 bp	99.9%	NA	0.3-3 hours
PGM Ion Torrent	~ 400 bp	98.0%	~ 80 million	2 hours
Roche 454 pyrosequencing	~ 700 bp	99.9%	1 million	24 hours
Illumina synthesis	50 – 300 bp	98.0%	~ 3 billion	1-10 days
SOLiD ligation	50 + 50 bp	99.9%	1.2 – 1.4 billion	1 – 2 weeks

Table 2. Potential application of major "-omics" strategies in food wastomics

Approach	Target molecule	Focus	Potential influence and outcomes
Genotyping (i.e. amplicon sequencing of 16S rRNA and ITS regions)	DNA	Assessing genomic diversity and variability	Inter and intraspecific taxonomic and functional variations
Metagenomics	DNA	Whole community taxonomic and functional analysis	Novel organisms, processes and bioactive compound
Metabolomics	Small molecules (metabolomes/ metabolites)	Elucidating composition, diversity and richness of metabolomes	Characterization of novel metabolomes (metabolites)
Proteomics	Amino acid, peptides and protein molecules	Expressed protein composition and quantification	Highlight the difference between protein translation and degradation in food waste
Transcriptomics	RNA (total RNA, rRNA or mRNA)	Elucidating composition, diversity and richness of transcriptomes (expressed genes)	Gene expression profiling, identification and characterization. Uncover novel RNA species.

assemblages. Recent technological advancements have led to an increase in assays capable of rapidly assessing genomic variability in different samples that have established the possibility of studying the combined effects of the inherent variations (Vlaanderen et al. 2010).

Food Waste Transcriptomics

The activities of microorganisms within food waste reflect the amounts of active mRNA transcripts. A deeper understanding of the set of gene expressions that are differentially expressed can be done by mRNA identification, profiling and characterization of the food waste. Through this approach, food wastomics is capable of reliably associating the different mRNA transcripts to taxa and functions within the food waste and controllably predict resulting changes or harvest useful components at the right time. Measuring gene expression in food waste will provide information about how genes are regulated and the underlying biology such as RNA biogenesis and metabolism (Hrdlickova et al., 2017; Lowe et al., 2017) in food waste.

Figure 2. Metagenome workflow diagram
Source: Adapted from Thomas et al. (2012)

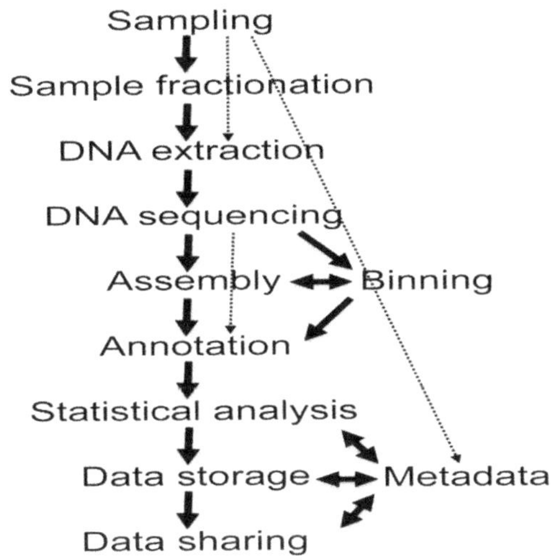

Food Waste Proteomics

A combination of mRNA precursors, posttranslational modification and environmental interactions defines the difficulties associated with the prediction of protein abundance based on expression analysis (transcriptomics) alone (Vlaanderen et al., 2010). Hence, the use of two-dimensional gel electrophoresis, one-dimensional and multidimensional chromatography, combined with high-resolution mass spectrometry will be handy to decipher the protein composition of food waste. The distinction of proteomics from transcriptomics is that it is a more a complex and dynamic approach because proteins are functional molecules that represent actual situations (Hudler et al., 2014). The characterization of proteomes in food waste has been made possible by advances in the different proteomics approaches, which hitherto was hampered by inadequate detection and analytical technologies. Thereby highlighting the function of environmental stimuli within the food waste due to stress levels at a specific point, which may be used to facilitate the identification of new biomarkers and molecular targets (Liebler, 2002; Hudler et al., 2014).

Food Waste Metabolomics

The focus of this food wastomic approach is to characterize food waste metabolomes – small metabolic molecules like lipids and vitamins that are by-products of metabolism in a high throughput manner. These chemical fingerprints underline ongoing biochemical processes, interactions and responses within food waste, which is variably highlighted at different times in relation to activity profiles. An understanding of food waste metabolic molecule composition and their processes is possible through untargeted metabolomics studies. Moreover, the mining of novel metabolite and natural products in food waste can be relevant in other aspects of biology including the development of biomarkers and enzymes. In addition, metabolite profiling also provides a tool to identify secondary metabolites from natural sources (Graziani et al., 2018) as presented by food waste.

Integrated Food Wastomics

This is the complementary combination of different "-omics" approach to provide a greater understanding of food waste composition and processes and may be the future of sustainable food waste management. Compared to single food wastomics study, an integration across multiple layers will help balance the shortfalls of each wastomics approach and enhance individual advantages (Figure 3).

A similar approach is applied to classify organisms (Raupauch et al., 2016), in biomining (Jerez, 2008), disease management through a systems biology approach as reported in Hasin et al. (2017); Langley and Wong (2017) and enrichment of available data (Scelfo et al., 2018). More so, this approach will provide exciting findings

Figure 3. Typical flowchart of an integrated "-omics" approach
Source: Adapted from Dhanapal and Govindaraj (2015).

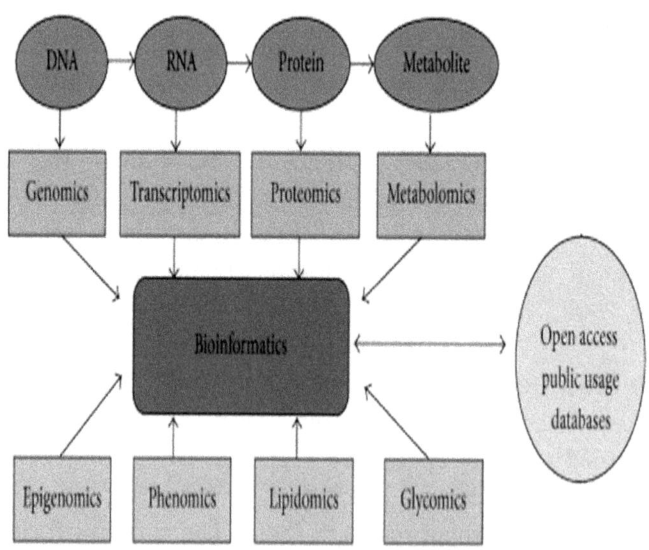

that will make predictions on how to keep the food waste microbial consortium healthy and more efficient with a view to exploring them for novel applications and understandings. This integrated approach has also found application in food safety research (Fratamico, 2008), however, this can be extended to include safety analysis of food waste. The success of this integrated approach relies on the effective and efficient collaboration of scientists and social scientists with diverse backgrounds that are currently key players in the food system chain. More so, as outlined in Whitworth (2013), the combination of genomics and epidemiology is useful in the early detection of emerging microbiological hazards from the accumulation of food waste through detection of virulence factors, resistant genes and genetic diversity.

The science of "-omics" is rapidly expanding to absolve and incorporate diverse everyday applications and needs. Yet, it remains surprising that "-omics" approaches have not been adopted in mining food waste despite the vast potential. Furthermore, it is enjoying vast application in evaluating the extent of ecological effects in diverse environmental stress conditions including contamination from ionizing irradiation. The current limited use in food science is captured by foodomics albeit the techniques are perfectly suited to analyse food waste, since it is able to simply assess variations in the DNA or protein sequence of many microbial genes through a single experiment. Thus, besides understanding food waste composition, the "-omics" approach also presents a new dimension for recycling food waste to reduce the environmental and socio-economic burden through energy and raw material recovery. Furthermore, the nutrient-rich food waste presents exciting "-omics" research opportunities that can contribute to sustainable environmental goals as well as prevent loss of scarce resources.

CONCLUSION

The world has not compensated for food-related changes over several generations (Rozin, 2005). Food waste management needs to be listed as a top priority by local, regional, national and international governments. The improvements in "-omic" strategies and techniques will lead to an increased understanding of key interactions within food waste and enable investigations into waste secretomes (Josic et al., 2016). More effort should be dedicated to generating improved data regarding food waste quantification and breakdown of specific food products within the waste (Tonini et al., 2018). The burden of FLW can be mitigated by sustainable waste management practices through the use of "-omics" technology to characterize the waste and mine for reusable products or organisms. Thereby, any scientific and technological improvements in "-omics" technology will also have positive impacts on waste management.

To control domestic food waste quantity, Schanes et al. (2018) opined that mapping the determinants of food waste generation might help design sustainable food waste prevention strategies. However, after this waste have been generated, "-omics" strategies and techniques will be handy to handle the environmental and economic consequences. Hence, an omics consideration should be added to the integrated approach suggested by Chakona et al. (2017) to address food management with a view to decreasing food waste related hazards. In the same vein, the application of "-omics" strategies to manage food waste has the potential to provide new insights into general knowledge and understanding of food waste decomposition processes and create new perspectives for waste management in future.

The diverse biological waste products may possess the solution to many global problems. Whereas "-omics" technologies offer very exciting opportunities to mine food waste for useful organisms and bioactive compounds. Therefore, integrated studies incorporating the different "-omics" approaches are recommended to reduce the environmental nuisance from food waste. In addition, foodomics may be expanded to include the management of food waste by "-omics". Nonetheless, the field of "-omics" will benefit from validation and standardization especially with regards to consistency of approaches for the generation, processing and interpretation of "-omics" data (Sauer et al., 2017). If not addressed, the lack of standard practice may affect the sustainable application of "-omics" in food waste management, which may be biochemically complicated and at times toxic. More so, improvements in sequencing technology will allow greater resolution (Walsh et al., 2017). The mining of food waste for useful bioactive compounds using "-omics" is a worthy cause because the best of treasures are hidden in waste.

REFERENCES

Aldridge, B. B., & Rhee, K. Y. (2014). Microbial metabolomics: Innovation, application, insight. *Current Opinion in Biotechnology*, *19*, 90–96.

Ambardar, S., Gupta, R., Trakroo, D., Lal, R., & Vakhlu, J. (2016). High Throughput Sequencing: An Overview of Sequencing Chemistry. *Indian Journal of Microbiology*, *56*(4), 394–404. doi:10.100712088-016-0606-4

Andjelković, U., Šrajer Gajdošik, M., Gašo-Sokač, D., Martinović, T., & Josić, D. (2017). Foodomics and Food Safety: Where We Are. *Food Technology and Biotechnology*, *55*(3), 290–307. doi:10.17113/ftb.55.03.17.5044

Bajželj, B., Richards, K. S., Allwood, J. M., Smith, P., Dennis, J. S., Curmi, E., & Gilligan, C. A. (2014). Importance of food-demand management for climate mitigation. *Nature Climate Change*, *4*(10), 924–929. doi:10.1038/nclimate2353

Barbulova, A., Colucci, G., & Apone, F. (2015). New trends in cosmetics: By-products of plant origin and their potential use as cosmetic active ingredients. *Cosmetics*, *2*(2), 82–92. doi:10.3390/cosmetics2020082

Bokulich, N. A., Lewis, Z. T., Boundy-Mills, K., & Mills, D. A. (2016). A new perspective on microbial landscapes within food production. *Current Opinion in Biotechnology*, *37*, 182–189. doi:10.1016/j.copbio.2015.12.008

Braconi, D., Bernardini, G., Millucci, L., & Santucci, A. (2018). Foodomics for human health: Current status and perspectives. *Expert Review of Proteomics*, *15*(2), 153–164. doi:10.1080/14789450.2018.1421072

Braun, S. (2012). *Food Waste Report on the situation and recent activities in Germany, Stuttgart University*. Retrieved from http://ec.europa.eu/dgs/health_foodsafety/dgs_consultations/docs/ag/summary_ahac_05102012_3_susanne_braun_en.pdf

Capozzi, F., & Bordoni, A. (2013). Foodomics: A new comprehensive approach to food and nutrition. *Genes & Nutrition*, *8*(1), 1–4. doi:10.100712263-012-0310-x

Chakona, G., & Shackleton, C. M. (2017). Local setting influences the quantity of household food waste in mid-sized South African towns. *PLoS One*, *12*(12), e0189407. doi:10.1371/journal.pone.0189407

Chen, K., & Pachter, L. (2005). Bioinformatics for Whole-Genome Shotgun Sequencing of Microbial Communities. *PLoS Computational Biology*, *1*(2), e24. doi:10.1371/journal.pcbi.0010024

Dhanapal, A.P., & Govindaraj, M. (2015). Unlimited Thirst for Genome Sequencing, Data Interpretation, and Database Usage in Genomic Era: The Road towards Fast-Track Crop Plant Improvement. *Genetics Research International*. doi:10.1155/2015/684321

Dorman, H. J. D., & Deans, S. G. (2008). Antimicrobial agents from plants: Antibacterial activity of plant volatile oils. *Journal of Applied Microbiology*, *88*(2), 308–316. doi:10.1046/j.1365-2672.2000.00969.x

Dos Santos, B. S., da Silva, L. C. N., da Silva, T. D., Rodrigues, J. F. S., Grisotto, M. A. G., Correia, M. T. dos S., ... Paiva, P. M. G. (2016). Application of Omics Technologies for Evaluation of Antibacterial Mechanisms of Action of Plant-Derived Products. *Frontiers in Microbiology*, *7*, 1466. doi:10.3389/fmicb.2016.01466

Edjabou, V. M. E., Pivnenko, K., Petersen, C., Scheutz, C., & Astrup, T. F. (2015). *Compositional data analysis of household food waste in Denmark.* Abstract from 6th International Workshop on Compositional Data Analysis, Spain. Retrieved from http://orbit.dtu.dk/files/113896186/Vincent_Edajbou.pdf

Ercolini, D. (2013). High-throughput sequencing and metagenomics: Moving forward in the culture-independent analysis of food microbial ecology. *Applied and Environmental Microbiology, 79*(10), 3148–3155. doi:10.1128/AEM.00256-13

Escobar-Zepeda, A., Vera-Ponce de Leon, A., & Sanchez-Flores, A. (2015). The road to metagenomics: From microbiology to DNA sequencing technologies and bioinformatics. *Frontiers in Genetics, 6*, 348. doi:10.3389/fgene.2015.00348

FAO. (2001). *The state of Food Insecurity in the World 2001.* FAO.

FAO. (2011). *Global food losses and food waste - extent, causes and prevention.* FAO. Retrieved from http://www.fao.org/docrep/014/mb060e/mb060e.pdf

FAO. (2013). *Food wastage footprint: impacts on natural resources. Rome.* Retrieved from http://www.fao.org/docrep/018/i3347e/i3347e.pdf

FAO. (2014). *Definitional framework of food loss. Save Food: Global Initiative on Food Loss and Waste Reduction.* Food and Agriculture Organization of the United Nations. Retrieved from http://www.fao.org/fileadmin/user_upload/save-food/PDF/FLW_Definition_and_Scope_2014.pdf

FAO. (2017). *The future of food and agriculture - trends and challenges.* FAO. Retrieved from http://www.fao.org/3/a-i6583e.pdf

FAO. (2018). *Sustainable pathways-foot wastage footprint.* Retrieved from http://www.fao.org/nr/sustainability/food-loss-and-waste/en/

Fierer, N., Leff, J. W., Adams, B. J., Nielsen, U. N., Bates, S. T., Lauber, C. L., ... Caporaso, J. G. (2012). Cross-biome metagenomic analyses of soil microbial communities and their functional attributes. *Proceedings of the National Academy of Sciences of the United States of America, 109*(52), 21390–21395. doi:10.1073/pnas.1215210110

FLWP. (2016). *Food loss and waste accounting and reporting standard.* Retrieved from http://flwprotocol.org/

Franzosa, E. A., Hsu, T., Sirota-Madi, A., Shafquat, A., Abu-Ali, G., Morgan, X. C., & Huttenhower, C. (2015). Sequencing and beyond: Integrating molecular 'omics' for microbial community profiling. *Nature Reviews. Microbiology, 13*(6), 360–372. doi:10.1038/nrmicro3451

Fratamico, P. M. (2008). The application of "Omics" technology for food safety research. *Foodborne Pathogens and Disease*, *5*(4), 369–370. doi:10.1089/fpd.2008.9994

Frias-Lopez, J., Shi, Y., Tyson, G. W., Coleman, M. L., Schuster, S. C., Chiholm, S. W., & Delong, D. F. (2008). Microbial community gene expression in ocean surface waters. *Proceedings of the National Academy of Sciences of the United States of America*, *105*(10), 3805–3810. doi:10.1073/pnas.0708897105

García-Canas, V., Simó, C., Herrero, M., Ibáñez, E., & Cifuentes, A. (2012). Present and future challenges in food analysis: Foodomics. *Analytical Chemistry*, *84*(23), 10150–10159. doi:10.1021/ac301680q

Garnett, T. (2014). *What is a sustainable healthy diet? A discussion paper*. Food Climate Research Network. Retrieved from https://www.fcrn.org.uk/sites/default/files/fcrn_what_is_a_sustainable_healthy_diet_final.pdf

Graziani, V., Scognamiglio, M., Belli, V., Esposito, A., D'Abrosca, B., Chambery, A., ... Fiorentino, A. (2018). Metabolomic approach for a rapid identification of natural products with cytotoxic activity against human colorectal cancer cells. *Scientific Reports*, *5309*. doi:10.103841598-018-23704-9

Guil-Guerrero, J. L., Ramos, L., Moreno, C., Zúñiga-Paredes, J. C., Carlosama-Yepez, M., & Ruales, P. (2016). Plant Foods By-Products as Sources of Health-Promoting Agents for Animal Production: A Review Focusing on the Tropics. *Agronomy Journal*, *108*(5), 1759–1774. doi:10.2134/agronj2015.0555

Handelsman, J., Rondon, M. R., Brady, S. F., Clardy, J., & Goodman, R. M. (1998). Molecular biological access to the chemistry of unknown soil microbes: A new frontier for natural products. *Chemistry & Biology*, *5*(10), R245–R249. doi:10.1016/S1074-5521(98)90108-9

Hasin, Y., Seldin, M., & Lusis, A. (2017). Multi-omics approaches to disease. *Genome Biology*, *18*(1), 83. doi:10.118613059-017-1215-1

Hettich, R. L., Sharma, R., Chourey, K., & Giannone, R. J. (2012). Microbial metaproteomics: Identifying the repertoire of proteins that microorganisms use to compete and cooperate in complex environmental communities. *Current Opinion in Biotechnology*, *15*, 373–380.

HLPE. (2014). *Food losses and waste in the context of sustainable food systems*. A report by the High-Level Panel of Experts on Food Security and Nutrition of the Committee on World Food Security. Retrieved from: http://www.fao.org/fileadmin/user_upload/hlpe/hlpe_documents/HLPE_Reports/HLPE-Report-8_EN.pdf

Hrdlickova, R., Toloue, M., & Tian, B. (2017). RNA-Seq methods for transcriptome analysis. *Wiley Interdisciplinary Reviews. RNA*, *8*(1), e1364. doi:10.1002/wrna.1364

Hudler, P., Kocevar, N., & Komel, R. (2014). Proteomic Approaches in Biomarker Discovery: New Perspectives in Cancer Diagnostics. *The Scientific World Journal*. . doi:10.1155/2014/260348

Jerez, C. A. (2008). The use of genomics, proteomics and other OMICS technologies for the global understanding of biomining microorganisms. *Hydrometallurgy*, *94*(1-4), 162–169. doi:10.1016/j.hydromet.2008.05.032

Josic, D., & Giacometti, J. (2013). Foodomics-Use of Integrated Omics in Nutrition, Food Technology and Biotechnology. *Journal of Data Mining in Genomics & Proteomics*, *4*(02), e106. doi:10.4172/2153-0602.1000e106

Josic, D., Sokac, D. G., Gajdosik, M. S., & Clifton, J. (2016). Microbial omics for food safety. *Journal of Hygienic Engineering and Design*, *67*, 116–129.

Kadir, A. A., Azhari, N. W., & Jamaludin, S. N. (2017). Study on Composition and Generation of Food Waste in Makanan Ringan Mas Industry. *MATEC Web of Conferences*, *103*. DOI: 10.1051/matecconf/20171030

Langille, M. G. I., Zaneveld, J., Caporaso, J. G., McDonald, D., Knights, D., Reyes, J. A., ... Huttenhower, C. (2013). Predictive functional profiling of microbial communities using 16S rRNA marker gene sequences. *Nature Biotechnology*, *31*(9), 814–821. doi:10.1038/nbt.2676

Langley, R. J., & Wong, H. R. (2017). Early diagnosis of sepsis: Is an integrated omic approach the way forward. *Molecular Diagnosis & Therapy*, *21*(5), 525–537. doi:10.100740291-017-0282-z

Leach, B., & Swannell, R. (2017). Food loss and waste: A key issue for our generation. *Environmental Scientist*, *26*(1), 47–51.

Liebler, D. C. (2002). Proteomic approaches to characterize protein modifications: New tools to study the effects of environmental exposures. *Environmental Health Perspectives*, *110*(Suppl 1), 3–9. doi:10.1289/ehp.02110s113

Loman, N. J., & Pallen, M. J. (2015). Twenty years of bacterial genome sequencing. *Nature Reviews. Microbiology*, *13*(12), 787–794. doi:10.1038/nrmicro3565

Lowe, R., Shirley, N., Bleackley, M., Dolan, S., & Shafee, T. (2017). Transcriptomics technologies. *PLoS Computational Biology*, *13*(5), e1005457. doi:10.1371/journal.pcbi.1005457

Luque, R., & Clark, J. H. (2013). Valorisation of food residues: Waste to wealth using green chemical technologies. *Sustainable Chemical Processes*, *1*(1), 10. doi:10.1186/2043-7129-1-10

Mason, L., Boyle, T., Fyfe, J., Smith, T., & Cordell, D. (2011). *National Food Waste Data Assessment: Final Report*. Prepared for the Department of Sustainability, Environment, Water, Population and Communities, by the Institute for Sustainable Futures, University of Technology, Sydney. Retrieved from https://www.environment.gov.au/system/files/resources/128a21f0-5f82-4a7d-b49c-ed0d2f6630c7/files/food-waste.pdf

Meyers, W. H., Ziolkowska, J. R., Tothova, M., & Goychuk, K. (2012). *Issues affecting the future of agriculture and food security for Europe and Central Asia*. FAO Regional Office for Europe and Central Asia. Policy Studies on Rural Transition No. 2012-3. Retrieved from http://www.fao.org/3/a-aq343e.pdf

Monier, V., Shailendra, M., Escalon, V., O'Connor, C., Gibon, T., Anderson, G., . . . Reisinger, H. (2010). Preparatory Study on Food Waste across EU 27. European Commission (DG ENV) Directorate C-Industry. Final Report.

Nguyen, N. H., Song, Z., Bates, S. T., Branco, S., Tedersoo, L., Menke, J., ... Kennedy, P. G. (2016). FUNGuild: An open annotation tool for parsing fungal community datasets by ecological guild. *Fungal Ecology*, *20*, 241–248. doi:10.1016/j.funeco.2015.06.006

Nishida, J. (2014) Reducing Food Waste and Promoting Food Recovery Globally. *EPA Connect. The official blog of the EPA Leadership*. Retrieved from https://blog.epa.gov/blog/2014/10/reducing-food-waste-and-promoting-food-recovery globally/

ODT. (2004). *Food: the new definition*. Retrieved from http://tzhcpas.com/wp-content/uploads/2011/01/streamlined_sales_tax_food_BR.pdf

OLIO. (2017). *The problem of food waste*. Retrieved from: https://olioex.com/food-waste/the-problem-of-food-waste/

Otles, S., Despoudi, S., Bucatariu, C., & Kartal, C. (2015). Food waste management, valorization, and sustainability in the food industry. In C. M. Galanakis (Ed.), *Food Waste Recovery Processing Technologies and Industrial Techniques* (pp. 3–18). London: Academic Press, Elsevier. doi:10.1016/B978-0-12-800351-0.00001-8

Parfitt, J., Barthel, M., & Macnaughton, S. (2010). Food waste within food supply chains: Quantification and potential for change to 2050. *Philosophical Transactions of the Royal Society. Biol. Sci.*, *365*(1554), 3065–3081. doi:10.1098/rstb.2010.0126

Paritosh, K., Kushwaha, S. K., Yadav, M., Pareek, N., Chawade, A., & Vivekanand, V. (2017). Food waste to energy: an overview of sustainable approaches for food waste management and nutrient recycling. BioMed Research International.

Parry, A., Bleazard, P., & Okawa, K. (2015). *Preventing Food Waste: Case Studies of Japan and the United Kingdom. OECD Food, Agriculture and Fisheries Papers, No. 76*. Paris: OECD Publishing. doi:10.1787/5js4w29cf0f7-

Quested, T. E., Ingle, R., & Parry, A. D. (2013). *Household Food and Drink Waste in the UK 2012*. Retrieved from http://www.wrap.org.uk/sites/files/wrap/hhfdw-2012-main.pdf.pdf

Ramukhwatho, F. R., du Plessis, R., & Oelofse, S. (2014). Household Food Wastage in a Developing Country: A Case Study of Mamelodi Township in South Africa. *Proceedings of the 20th WasteCon Conference*. Retrieved from https://researchspace.csir.co.za/dspace/bitstream/handle/10204/7757/Oelofse_2014.pdf?sequence=1&isAllowed=y

Raupauch, M. J., Amann, R., Wheeler, Q. D., & Roos, C. (2016). The application of "-omics" technologies for the classification and identification of animals. *Organisms, Diversity & Evolution*, *16*(1), 1–12. doi:10.100713127-015-0234-6

Reuter, J. A., Spacek, D., & Snyder, M. P. (2015). High-Throughput Sequencing Technologies. *Molecular Cell*, *58*(4), 586–597. doi:10.1016/j.molcel.2015.05.004

Riya, M. P., Antu, K. A., Vinu, T., Chandrakanth, K. C., Anikumar, K. S., & Raghu, K. G. (2013). An in vitro study reveals neutraceutical properties of *Ananas comosus* (L.) Merr. Var. Mauritius fruit residue beneficial to diabetes. *Journal of the Science of Food and Agriculture*, *94*(5), 943–950. doi:10.1002/jsfa.6340

Roche, H. M., de Roos, B., & Brennan, L. (2015). Application of 'omics ' technology. In J. A. Lovegrove, L. Hodson, S. Sharma, & S. A. Lanham-New (Eds.), *Nutrition Research Methodologies* (pp. 198–211). John Wiley and Sons Ltd.; doi:10.1002/9781119180425.ch13

Roemer, T., & Boone, C. (2013). Systems-level antimicrobial drug and drug synergy discovery. *Nature Chemical Biology*, *9*(4), 222–231. doi:10.1038/nchembio.1205

Rozin, P. (2005). The meaning of food in our lives: A cross-culural perspective on eating and well-being. *Journal of Nutrition Education and Behavior*, *37*, S107–S112. doi:10.1016/S1499-4046(06)60209-1

Sabree, Z. L., Rondon, M. R., & Handelsman, J. (2009). *Metagenomics. Genetics, Genomics*. Elsevier Inc.

Sauer, U. G., Deferme, L., Gribaldo, L., Hackermuller, J., Tralau, T., van Ravenzwaay, B., ... Gant, T. W. (2017). The challenge of the application of 'omics technologies in chemical rick assessment: Background and outlook. *Regulatory Toxicology and Pharmacology*, *91*(supplement 1), S14–S26. doi:10.1016/j.yrtph.2017.09.020

Scelfo, C., Galeone, C., Bertolini, F., Caminati, M., Ruggiero, P., Facciolongo, N., & Menzella, F. (2018). Towards precision medicine: The application of omics technologies in asthma management. *F1000 Research*, *7*, 423. doi:10.12688/f1000research.14309.1

Schanes, K., Dobernig, K., & Gozet, B. (2018). Food waste matters - A systematic review of household food waste practices and their policy implications. *Journal of Cleaner Production*, *182*, 978–991. doi:10.1016/j.jclepro.2018.02.030

Shi, X. (2014). *Analysis of challenges to food security and its recommended strategies*. Retrieved from: https://www.eng.mcmaster.ca/sites/default/files/uploads/analysis_of_challenges_to_food_security.pdf

Stajcic, N. (2013). Understanding culture: Food as a means of communication. *Hemisphere*, *28*, 5–14.

Strotmann, C., Göbel, C., Friedrich, S., Kreyenschmidt, J., Ritter, G., & Teitcheid, P. (2017). A participatory approach to minimized food waste in the food industry-a manual for managers. *Sustainability*, *9*(1), 66. doi:10.3390u9010066

Tang, Y. (2015). Non-genomic omic techniques. In Y. Tang, M. Sussman, D. Liu, I. Poxton, & J. Schwartzman (Eds.), *Molecular Medical Microbiology* (pp. 399–406). London: Academic Press.

Tarazona, S., García-Canas, F., Dopazo, J., Ferrer, A., & Conesa, A. (2011). Differential expression in RNA-seq: A matter of depth. *Genome Research*, *21*(12), 2213–2223. doi:10.1101/gr.124321.111

Tarazona-Dıaz, M. P., & Aguayo, E. (2013). Assessment of by-products from fresh-cut products for reuse as bioactive compounds. *Food Science & Technology International*, *19*(5), 439–446. doi:10.1177/1082013212455346

Thi, N. B. D., Lin, C. Y., & Kumar, G. (2016). Waste-to-wealth for valorization of food waste to hydrogen and methane towards creating a sustainable ideal source of bioenergy. *Journal of Cleaner Production*, *122*, 29–41. doi:10.1016/j.jclepro.2016.02.034

Thomas, T., Gilbert, J., & Meyer, F. (2012). Metagenomics-a guide from sampling to data analysis. *Microbial Informatics and Experimentation*, *2*(1), 3. doi:10.1186/2042-5783-2-3

Tonini, D., Albizzati, P. F., & Astrup, T. F. (2018). Environmental impacts of food waste: Learning and challenges from a case study on UK. *Waste Management (New York, N.Y.)*, *76*, 744–766. doi:10.1016/j.wasman.2018.03.032

Vailati-Riboni, M., Palombo, V., & Loor, J. J. (2017). What are omics sciences? In B. Ametaj (Ed.), *Periparturient Diseases of Dairy Cows*. Cham: Springer. doi:10.1007/978-3-319-43033-1_1

Varzakas, T., Zakynthinos, G., & Verpoort, F. (2016). Plant food residues as a source of nutraceuticals and functional foods. *Foods*, *5*(4), 88–120. doi:10.3390/foods5040088

Vlaanderen, J., Moore, L. E., Smith, M. T., Lan, Q., Zhang, L., Skibola, C. F., ... Vermeulen, R. (2010). Application of Omics Technologies in Occupational and Environmental Health Research; Current Status and Projections. *Occupational and Environmental Medicine*, *67*(2), 136–143. doi:10.1136/oem.2008.042788

von Schneidemesser, E., Monks, P. S., Allan, J. D., Bruhwiler, L., Forster, P., Fowler, D., ... Sutton, M. A. (2015). Chemistry and the Linkages between Air Quality and Climate Change. *Chemical Reviews*, *115*(10), 3856–3897. doi:10.1021/acs.chemrev.5b00089

Walsh, A. M., Crispie, F., Claesson, M. J., & Cotter, P. D. (2017). Translating Omics to Food Microbiology. *Annual Review of Food Science and Technology*, *8*(1), 113–134. doi:10.1146/annurev-food-030216-025729

Whitworth, J. J. (2013). *Is Omics technology the future of food risk assessment?* Retrieved from: https://www.foodnavigator.com/Article/2013/10/14/Omics-technologies-and-applications-in-food-safety

WRAP. (2013). *Household food and drink waste in the UK 2012*. Retrieved from: http://www.wrap.org.uk/content/household-food-and-drink-waste-uk-2012

Chapter 12
Together We Will Reduce the Food Loss

Celin Tennis Raju
St. Xavier's College, India

Mahimaidoss Baby Mariyatra
St. Xavier's College, India

ABSTRACT

In September 2015, the United Nations General Assembly adopted a set of 17 Sustainable Development Goals (SDGs), which include Target 12.3's call for halving food waste and reducing food losses worldwide by 2030. One-third of all food produced each year in the world is never eaten, while at the same time nearly 800 million people go hungry. This situation urges us to take immediate resolutions and steps towards reduction of food waste and food loss. This cannot be done by one person or overnight. This requires systematic analysis in various layers and collective and appropriate effort. This target can be achieved altogether by various sectors including government organizations; non-government organizations; and private companies in collaboration with schools; colleges; universities; research institutes; religion-based organizations such as temples, churches, etc.; and charity-based organizations. This book chapter will discuss the various steps that can possibly be adopted and implemented to address the serious issue of reduction of food waste and food loss.

DOI: 10.4018/978-1-5225-7706-5.ch012

INTRODUCTION

Even though, the world produces enough food to feed twice the world's present population, every year it is estimated that about 72 billion pounds (32658650.64 tons=3265865084 kg) of food is wasted all over the world, which is enough to build a mountain 2 miles across and almost 8000 feet high (Shea, 2018; Stuart, 2009). These 72 billion pounds of foods are unutilized or goes as a waste in the processing, packaging or transporting. One-thirds of the world's production gets lost or wasted every year. About half the food being wasted or lost could have been utilized. "Food waste and food loss" range between 30 and 50% for both developed and emerging countries. The country that is leading in food wastage is USA (Tielens & Candel, 2014). The countries, all over the world, like Germany, Malaysia, Singapore, Finland, UK, Australia, Denmark, Canada, Norway, India are also a part of this catastrophe. Intensified famine prevails in countries such as Yemen, northern Nigeria, the Democratic Republic of Congo, South Sudan and Myanmar.

Our country India wastes about 40% of the food produced in its land. The food loss equals approximately $75 billion loss in income of our country. According to global hunger index survey in 2016, India ranked 97th among 118 countries. About 20 crores go into bed hungry and 7000 people die of hunger every day. It is estimated that India wastes as much food as UK consumes. India wastes more food than China. Especially India and China cause loss of 1.3 billion tons of food every year.

In 2017, around 124 million people in 51 countries face crisis of food insecurity or worse (World Food Programme, 2018). They require urgent humanitarian action to save lives, protect livelihoods, and reduce hunger and malnutrition. The world's worst food crises were in north-eastern Nigeria, Somalia, Yemen and South Sudan, where nearly 32 million people were food-insecure and in need of urgent assistance. Global report on food crisis identified 108 million people in Crisis food security or worse across 48 countries. Over the years, there is an 11 percent rise – in the number of food-insecure people needing urgent humanitarian action in Asia (Food and Agriculture organisation of United Nations, 2018).

In September 2015, the United Nations General Assembly adopted a set of 17 Sustainable Development Goals (SDGs), which include Target 12.3's call for halving food waste and reducing food losses worldwide by 2030. This requires systematic analysis in various layers and collective and appropriate effort. This target can be achieved altogether by various sectors including government organisations, Non-government organisations and private companies in collaboration with schools, colleges, universities, research institutes, religion based organisations such as temples, churches etc. and charity based organisations.

WHERE IS FOOD LOST?

Food loss and waste occurs in harvesting, storage, packing, transport, infrastructure or market / price mechanisms, as well as institutional and legal frameworks. Food waste starts in the farms. Planting more than the demands, food not harvested due to weather, price of production, transportation costs, labor shortages, etc. About 8% of fish caught in the world's fisheries are discarded. It is about 7.3 million tons per year. About 16-32% of by catch are thrown away into the sea. In the "food value chain", because of the tight specifications (such as size, blemishes and appearances) in supermarkets, about half of the harvests are thrown away. Most waste are produced at manufacturing and processing because of trimming off of edible portions such as skin, fats, crusts from food. Other reasons are over production, product damage and technical difficulties (Bloom, 2011).

MAJOR CAUSES OF FOOD LOSS

Persistent drought has also played a major role, causing consecutive poor harvests in 23 countries like eastern and southern Africa. Drought that prevails leads to less food production and less food availability. Conflict and insecurity continued to be the primary drivers of food insecurity in 18 countries (74 million people). Deforestation, unsustainable agricultural practices and excessive ground water extraction are also key factors of drought causing food loss or food wastage (Jaeger, 1992).

A large problem that occurs is rejection of perishable foods. The shorter shelf lives are a major problem (Stuart, 2009). Approximately 4% of food purchased by restaurants is wasted before reaching consumers. According to the Cornell University Food and Brand lab, on average, diners leave 17% of meals uneaten and 55% of edible leftovers are left. Food spoilage, over-preparing, date-label confusion, overbuying, poor planning are some major issues on food loss or waste. In India, the bigger the wedding, the larger the party and the more massive the waste is expected to be. About 20 to 25 per cent of food is wasted in weddings.

India's cold storage requirement is 66 million tons, and the national storage capacity currently available is approximately 30 million tons. Instituting community refrigerators in each and every block of the town would significantly help public awareness and would reduce food wastage.

COMMUNITY REFRIGERATOR

Community refrigerators can be a way out in food loss; the main aim of the community fridge is to curb food wastage and help feed the people in need. This is a platform where people can bring out their generosity and motivates young minds to help others and brings selflessness out of people.

Each set up is likely to have two fridges and a freezer. The latter and a fridge will hold food that would have gone to waste from local businesses and restaurants and the food will go to people who are struggling through a local network. The remaining fridge will work like an honesty box where anyone can leave and take food. The fridges will have set opening times, be monitored by CCTV and meet strict hygiene standards. The following list includes some of the governmental and non-governmental programs working towards reducing food loss and food waste.

- UK is the pioneer of the community fridge network. The People's fridge in Brixton that was launched and has been highly successful, with other community fridges having been established in Spain and Germany over the last few years (Wilson-Powell, 2017).
- NORTHERN Ireland's first ever `community fridge' has opened its door in Antrim village in an effort to tackle food waste. The initiative, by Cloughmills Community Action Team and environmental charity Hubbub, will allow anyone to help themselves to quality food that would otherwise be wasted. Despite increasing awareness of the issue of food waste, there are limited options for households and food businesses to get surplus perishable food to those who need it safely and within the `use-by' date.
- In Chennai, community fridge, installed outside Besant Nagar by Ayyamittu Unn project by a trust named The Public Foundation, is looking at anyone who is hungry with access to fresh, free food (Rekhi, 2017). The other community fridge is at Elliot's beach in Chennai which facilitates People to donate books, clothes, toys and footwear. Children often donate food for safekeeping in the fridge. The 'community fridge' idea could be implemented at more areas.
- This concept was first conceived in Germany and Spain, it has reached Indian cities like Kochi and Mumbai where community fridges have been installed and now, Gurugram has one too! The fridge gets loaded with 35-40 food packages every day including not just leftovers but also freshly cooked food (Bisaria, 2017).
- With an intention of reaching out to the needy and providing food to the hungry, members of the Versova Welfare Association began the concept of a community fridge. The fridge which is located outside a small temple in Versova does not fail to put a smile on the face of those deprived of food. The

concept also known as a 'food bank' has been receiving a lot of attention and becoming popular among the locals.

- Other associations like freedge, hubbub, versova welfare association and social activist all around the world has installed many fridges all over the world in abolishing poverty. Feeding India, Robin Hood, gift a meal India are some organizations in India helping the needy.

Wefood is a new supermarket in Denmark that sells surplus food other grocers have rejected.

Community fridges can be started with a just a fridge. The habit of sharing must be seeded in the young minds for a better future. Community fridge is a great platform to pin humanity in the hearts of our future pillars. As said before, there are people dying every day due to famine and hunger. They need to be cared about.

Anyone can start a community fridge because all they need is a fridge. Schools, colleges, places of worship, a shop, super market, etc. are the places where food is constantly wasted. There are plenty of young and potential minds that live in poverty and hunger in these places. The food that is wasted in supermarkets or restaurants or shops can be put up in the community fridge for an offer price or for free to the needy. In schools and colleges, this can be activated to the poor and the needy by influencing children and youngsters to donate at least a pack of biscuits or chocolates to be given as a donation for initiation. This makes the young minds to participate in social activities and to care about the people around them. In the place of worships in India, there are plenty of people in need of good food and other accessories for their living. So, in places of worship, these initiatives will be at its most beneficial form. In the places of worship, an initiative like the wall of kindness would also help the poor and the needy. This concept proves that everybody can be a social activist even though if they are not one. In places of worship, schools, colleges, supermarkets that has a large space can also install a root cellar that keeps the food for longer than it needed to be. A dark corner of garage in house, storage spaces in shops, super markets, a small room in places of worship will suffice to start a root cellar (Newton, 2003).

ACKNOWLEDGMENT

The informations and various activities to reduce food loss are taken from the following internet sources: www.fao.org, www.pebblemag.com, www.hubbub.org.uk, http://www.sustainabletable.org/5664/food-waste, www.theguardian.com, www.ballymenatimes.com, www.foo.ndtv.com, http://www.dnaindia.com. The authors of the above websites are acknowledged.

REFERENCES

Bisaria, A. (2017). *Now Gurgaon Gets Its First Community Fridge So There Are More Free Meals To Go Around*. Retrieved from www.indiatimes.com

Bloom, J. (2011). American Wasteland: How America Throws Away Nearly Half of Its Food (and What We Can Do About It). Hachette UK.

Food and Agriculture Organisation of United Nations (FAO). (2018). *Food security & nutrition around the world*. Retrieved from http://www.fao.org/3/CA1354EN/ca1354en.pdf

Jaeger, W. K. (1992). The causes of Africa's food crisis. *World Development*, *20*(11), 1631–1645. doi:10.1016/0305-750X(92)90019-R

Newton, S. (2003). Root cellars: types and storage tips. In *The Old Farmer's Almanac*. Retrieved from www.almanac.com

Rekhi, D. (2017, August 21). Chennai: Community fridge helps save surplus food, caters to the needy in Besant Nagar. *The New Indian Express*.

Shea, M. (2018). *Food waste & hunger: Solutions to the paradox*. Retrieved from www.rockefellerfoundation.org

Sturart, T. (2009). *Waste: Uncovering the global food scandal*. London: Penguin.

Tielens, J., & Candel, J. J. L. (2014). Reducing food wastage, improving food security? *Food & Business Knowledge Platform*. Retrieved from www.knowledge4food.net

Uton & Ostegard. (n.d.). *If we don't get help we will all die*. Retrieved from www.danchurchaid.org

Wilson-Powell, G. (2017, August 18). *UK to get world's first community fridge network*. Retrieved fromwww.pebblemag.com

World Food Programme (WFP). (2018). *Global report on food crisis*. Retrieved from https://www.wfp.org/content/global-report-food-crises-2018

Chapter 13
The Pursuit of Zero Food Waste:
A Gamified Approach Promoting Avoidance of Dormitory Mess Food Wastage in Educational Institutions

Bhavika Jain
Dr. B. R. Ambedkar National Institute of Technology, India

Arun Khosla
Dr. B. R. Ambedkar National Institute of Technology, India

Kulbhushan Chand
https://orcid.org/0000-0001-6502-0748
Dr. B. R. Ambedkar National Institute of Technology, India

Kiran Ahuja
DAV Institute of Engineering and Technology, India

ABSTRACT

Games are ordered activities, generally undertaken for recreation. The design elements of these games are being used by people all around the planet to make this world a better place. The opportunities for gamification are being discussed in this chapter along with the use of a decision-making method as both have been applied to the study using a local running mobile application as platform for encouraging students enrolled in various educational institutions to promote avoidance of mess food wastage and in gaining confidence to integrate to use this approach to fight the cause of this global malady in their everyday life. The overarching issue of student mentality about food wastage is being discussed along with how to merge gamification with digital technology in this aspect and its participatory design. This provides the background for addressing points of using a gameful system to foster empowerment and connection among the students of NIT Jalandhar where this case was studied and the proposed approach was implemented.

DOI: 10.4018/978-1-5225-7706-5.ch013

INTRODUCTION

The central theme of this chapter is to identify and analyze the problems that lead to food wastage in NIT Jalandhar mess halls. The focus is on discerning the framework within the students' and mess staff's environment that could be molded in a way to drive them to generate minimum food waste. The proposition primarily deals with game elements, introducing some variable factors for both students and mess staff and some additional prospects to make food accessible to the needy. The work primarily demonstrates the use of game theory to optimize the food distribution system within NIT Jalandhar messes. The objective was to bolster students' confidence in engaging with **gamification** through a meaningful integration of game elements and dynamics into their lives. Another objective of the research was to be quantifying the service satisfaction levels of the students who are using the services of mess in the college which involved some decision-making techniques.

The relationship between the students who enrolled for mess facility, mess workers and the non-profit organization, **Prayaas** provided a platform for building a gaming experience within the campus. Prayaas is a noble initiative by NIT Jalandhar students for underprivileged children. Under this initiative, a school is run that is solely run by the students of National Institute of Technology, Jalandhar (NITJ) to educate underprivileged children as well as children living in slum areas to provide them better education for a better future. More than 300 children are currently enrolled under the program from various downtrodden locations of Punjab. Free tuition classes, tutorial classes and guidance from NITJ students are provided by the NITJ students. These classes are being conducted in the form of evening school, so as to make it convenient for the children.

Major challenges were involved in the process of setting objectives, planning, implementing and executing the process of avoiding food wastage in NIT Jalandhar mess halls. The primary challenge was to convince the immediate hostel administration to give the permission to foster such modifications in the existing mess structure system. Also, the students in general do not use digital technology (internet, smart phones) for such cause. So, another hurdle was to promote the approach for the success of the study. This damper provided an opportunity to evolve a gamefic system that will support the regulators and participants relationship and assists in integrating the digital technology in a student's lifestyle in a meaningful way. In this way, the real purpose of introducing the technology in the first place, can be fulfilled.

The project is necessary to implement to know and understand the psychology of students. This assisted the authors to get familiar with lot of facts and helped them to know how to change the existing culture and mentality of students. Also, it opened new horizons for the possibilities of how to generalize this proposed structure in other college campuses or institutions. This aspect of the research demanded review of literature to get familiar with the present situation, listing down of different parameters, proposition of a questionnaire to get feedback, recording and segregating the data and finally the analysis followed by final conclusions.

BACKGROUND

As per the study, between 2004 and 2014, household food waste in Europe doubled to an estimated 30 million tons per year. Plastic packaging waste increased by half fraction over the same period, reaching over 15 million tones. It is to be noted here that the study was based on the Life Cycle Assessment Tool, the most comprehensive framework used in analyzing and evaluating the environmental impact of goods and services (G. Rebitzer Life Cycle Assessment, 2004). The study shows how annual per capita use of plastic packaging has extended simultaneously with levels of food waste since 1950s. (FUSIONS, Recommendations and guidelines for a common European food waste policy framework. 2016).

While the growing food devastation is acute, a recent pioneering research says that rise in food plastic packaging is failing to reduce widening of food waste problems, and in some cases may even be intensifying it (FAO,2018). Global significant nutriment losses and waste per year are roughly 30% for cereals, 40-50% for root crops, fruits and vegetables, 20% for oil seeds, meat and dairy plus and 35% for fish. The amount of food lost or wasted every year is equivalent to more than 50% of the world's yearly cereals crop (2.3 billion tons in 2009 and 2010) (FAO,2018). Food and Agriculture Organization, United States report states that roughly one third of the total food produced in the world for human consumption annually, approximately 1.3 billion tons — gets lost, devastated or wasted. The seriousness of the situation is evident from the fact that food loss amounts to roughly US$ 680 in industrialized countries and US$ 310 billion in developing countries. Fruits and vegetables, roots and tubers have the highest wastage rate of any food (FAO, 2018).

Food Wastage is any food substance, raw or cooked, which is discarded or required to be discarded according to legal definition of food waste by European Union (EU) Commission (Food Safety. 2018). The study says that the biggest wasters of food are young consumers (18-24 years) or families with children. Therefore, the research is conducted in mess halls of NIT Jalandhar identify and analyze the problems with youngsters that lead to food wastage and the goal was to implement a standard gamified approach for cutting the food loss in dormitory mess of educational institutions.

PROPOSED FRAMEWORK

The focus of the primary aspect of the chapter is on discerning the framework within the students' and mess staff's environment that could be molded in a way to drive them to generate minimum food waste. The proposition primarily deals with game elements, introducing some variable factors for both students and mess staff and some additional prospects to make food accessible to the needy. Another aspect involved decision making techniques to quantify the satisfaction levels of the mess facility availers regarding mess service quality, mess workers behavior, cleanliness required and maintained.

The variables included in this aspect of research are opinion about mess staff and facility/services provided by mess staff, food quality and service quality as independent variables and customer satisfaction as the dependent variable. The work primarily demonstrates the use of game theory to optimize the food distribution system within NIT Jalandhar messes. The background study analyzes the existing structure of the hostels and the respective mess halls that serve them along with the mess management that looks after the regular functioning of dormitory messes.

Mess Structure: There are three hostels (Saraswati hostel – GH1, Girls Hostel 2 – GH2 and Mega Girls Hostel - MGH) allocated to girls (refer Diagram 1) and eleven hostels (Boys Hostel 1 – BH1, Boys Hostel 2- BH2, Boys Hostel 3 - BH3, Boys Hostel 4 - BH4, Boys Hostel 5 - BH5, Boys Hostel 6 - BH6, Boys Hostel 7 - BH7, Boys Hostel 7E - BH7E, Mega Boys Hostel Block A, Mega Boys Hostel Block B, Mega Boys Hostel Block F) are allocated to boys (refer Diagram 2). Mess No.1 caters to Saraswati Hostel (GH1) and Mess No.2 cater to both GH2 and MGH. Mess No. 3 and 4 serves to BH1 and BH2 respectively. Mess No. 5 caters to both BH3 and BH4. Similarly, Mess No. 6 and 7 provide food to BH5 and BH6 respectively. Mess No.8 caters to BH7 and BH7E simultaneously and Mess No.9 serves to the three mega blocks named A, B and F. This implies that there are in total **fourteen hostels** and **nine mess halls** in NIT Jalandhar.

Mess Management Structure: Mess management system in NITJ follows a cooperative mess approach to cater the needs of the students. The system follows a tender being given to a mess manager who is answerable to the administration as well as the student body. The administration comprises of the hostel attendant and clerk who are answerable to the hostel warden. All the hostel wardens are further monitored by the Chief Warden which helps in an efficient approach to manage the mess. The student body of formally known as Mess committee comprises of student volunteers residing in that hostel who are shortlisted by the honorable Warden through a short interview. The student body is led by a Mess Secretary who is answerable to the warden for all the activities. The role of student body is to address the grievances of the students pertaining to the domain of hostel mess and to voice their opinion to the higher authorities.

The system, designed with the best of intentions is somehow not able to fulfill the expectations to standardize the performance of all the hostels and therefore definitely needs a review. Moreover, the difference in the number of students in each mess and different managers for different mess has affected the following things such as quality of food being offered, service time at each mess, variety in the menu, cleanliness and hygiene of the premises, ambience of the dining hall, behavior of the mess staff, cost of services being charged i.e. Mess Bill etc.

Gamification Approach: It can be defined as, the art to take out the entertaining, relaxing and the exciting elements of games and pouring them into boring, daily life non-gaming contexts. A game gives an opportunity to earn points or badges but because a game challenges a person, makes him or her to use his or her creativity and to make use of that inventiveness to add value to the surroundings. It makes a person more than he or she is today. Werbach, a leading expert in gamification describes the key elements of games by organizing them into three distinct categories: **Dynamics, Mechanics** and **Components**. Structured as a pyramid, Components are the base element, with Mechanics lie in the middle, and Dynamics are situated at thc top. Dynamics circumscribc most of the aspects of a gamified system. At the top of the pyramid, they are the most high-level conceptual elements in a game or gamified system. These are factors that must be considered, even if they don't enter directly into the game itself.

Dynamics: There are five dynamics known as Constraints, Emotions, Narrative, Progression, and Relationships. (Werbach and Hunter, 2012). Together, the five dynamics form the abstract notion of a game. When these five elements are considered in the design process, gamification occurs organically.

- ***Constraints*** are the limitations or trade-offs that are required in the design process. It will never be possible to do everything while making the prototype of a game. Therefore, trade-offs are an inherent part of the design process. It is important to examine the balance of elements in the gamification.
- ***Emotions*** are a driving part of gamification. It is something that varies from person to person. When considering how emotions drive engagement with games, it is important to realize that people, who are interacting with the system, are unique. What may frustrate one player may be the impetus for the other.
- ***Narrative*** refers to the storyline that thread its way through a game. Most such activities have some type of story or theme that frames how a user interacts with the game. The narrative can help the designer to capture interest and promote engagement.

- ***Progression*** considers the growth and development that occurs as a contestant navigates a game. The understanding, skills and abilities of the player progress, as he/ she involves in a game. This growth and development can be a motivating factor that increases engagement.
- ***Relationships*** comprise the last dynamic element. This element focuses on the social synergy that happens when the games are played by the participants. These interactions often generate feelings of comradeship, a sense of fraternity and brotherhood and selflessness. Even when playing a game alone, users may interact with virtual characters that are either human-driven or machine-driven.

Mechanics: The second group of elements is the Mechanics. These drive the contributors to engage with the sections and work for it. Following mentioned are the methods of Mechanics. (Wisconsin-Platteville, 2018).

- **Challenges** are the tasks presented that prompt the user to generate a solution. Typically, they are a rich scenario or intriguing objective that engages a user.
- **Chance** defines the element(s) of randomness or ambiguity in a game. A random event may trigger a user's curiosity or keep a player engaged as they anticipated the next occurrence of the random event.
- **Competition** is a common, intuitive mechanic associated with games. One player or group wins while another, or even others lose.
- **Cooperation** is closely associated with competition. The players who play together win together.
- **Feedback,** as components of mechanics in gamification, refers to providing information as to how a contender is doing. This concept is much the same as in learning environments.
- **Resource Acquisition** is a unique characteristic of games. As the participant progress, they often obtain useful or collectible items. The process of acquiring these items is often as valued as having the items themselves.
- **Rewards** are those benefits a user gets for completing some action or reaching some achievement.
- **Transactions** are deals between users, either directly or through an intermediary.
- **Turns** are the accustomed mechanics of games. The sequential participation of alternating contender keeps a game moving forward.
- **Win states** represent an important element of games that describes the goals that make one competent the winner.

Components: The final group of elements comprises of fifteen important Components (Werbach and Hunter, 2012):

- **Achievements** (defined goals)
- **Avatars** (visual portrayal of a contestant's character)
- **Badges** (visual depiction of achievements)
- **Boss Figures** (especially difficult challenges at the climax of a level)
- **Collections** (set of badges or points to accumulate)
- **Combat** (a defined short-lived battle)
- **Content** Unlocking (aspects available only when players reach the destination)
- **Gifting** (opportunities to share resources with others)
- **Leader boards** (visual representations of player progression and achievement)
- **Levels** (defined steps in player progression)
- **Points** (numerical displays of game progression)
- **Quests** (predefined challenges with objectives and rewards
- **Social Graphs** (presentation of players' social network within the game)
- **Teams** (defined groups of players working hard together for a common objective)
- **Virtual Goods** (gamc assets with pcrccivcd or rcal-moncy value)

Every game has some game elements; still most of the games are boring and not engaging. So, one can conclude that the "Good Gamification" doesn't start with game elements but starts with how it motivates a person's "Core Drives". This can be studied by a framework called Octalysis, proposed by Yu Kai Chou in 2010, who is popularly known as the "Guru" of gamification. It is designed as an octagon shape with 8 Core Drives representing each side.

Core Drives: The eight core drives consist of the following-

- **Epic Meaning and Calling:** The first core drive says that a person is motivated when he feels that he is a part of something bigger than himself. In a game, the world is about to end and somehow, he is the only person qualified to save it from demolishing.
- The second core drive, **Development and Accomplishment** says that the user gets motivated because he is leveling up, achieving something or mastering a field. "Nike Fuel Band" has very well exploited this core drive in engaging the customer base by focusing on earning short term health accomplishments.
- The third core drive, **Empowerment of Creativity of Feedback** motivates a user by making him use his creativity, try infinite number of different permutations and combinations to solve a problem see feedback and adjust to reduce the error.

- The fourth core drive, **Ownership and Possession** says that a person tends to improve something, be a part of it because he feels that he owns it. This core drive powers numerous of virtual goods, virtual currency etc.
- The fifth core drive, **Social Influence and Relatedness** makes use of human psychology. It postulates that a human being gravitates to do what the other people in his company are doing. "O Power", a utility SAS Company has used this core drive to try to get people lowers their utility bills by showing them how well their neighbors are doing.
- The sixth core drive, **Scarcity and Impatience** says that human beings tend to want something, just because it is scarce, or they cannot have effortlessly. For instance, if grapes are placed in front of you on a table you will not care about them. But if the same bunch is placed on a shelf just beyond your reach, you're always thinking about the grapes, about their color, whether they are sweet or not.
- The seventh core drive, **Unpredictability and Curiosity**, says that because a user doesn't know what's going to happen next, he wants to keep on getting engaged in it. This core drive is heavily utilized in gambling industry and because of the same core drive; we urge to finish a book or a movie.
- The eighth and final core drive, **Loss and Avoidance** is very straightforward. It refers to the action of human beings to avoid a loss. Video games like "Zombies Run" have used this core drive while weaving the game story.

These eight core drives are the foundation of every single thing a person do. Now, these core drives can be differentiated as:

Left Brain Drives and Right Brain Drives

- The Left-Brain Core Drives are on the left of octagon and are called Left Brain Core Drives, being more associated to logic, calculations and ownership.
- The Right Brain Core Drives are on the right of octagon and are called Right Brain Core Drives, being more related to creativity, self-expression and social aspects.

Interestingly, Left Brain Core Drives are **Extrinsic Motivators**. For instance, a person is motivated because he wants to obtain something, whether it be a goal, a good, or anything you cannot obtain; on the other hand, Right Brain Core Drives are **Intrinsic Motivators**. For instance, when there's an earning of a goal or reward to use your creativity or feel the suspense of unpredictability, the activity itself is rewarding on its own.

White Hat Gamification and Black Hat Gamification

The technique that utilizes the upper core drives involves White Hat Gamification. Activities that engage people because they feel that they are a part of something bigger than themselves or they get an opportunity to creatively use their existing skill set. On the other hand, the technique that utilizes the lower core involves Black Hat Gamification. Activities that utilizes a factor of ambiguity or fear of losing something dear, involve Black Hat Gamification (Yu-kai Chou: Gamification & Behavioral Design. 2018).

Octalysis Score

To calculate Octalysis score, a number between 0-10 is assigned based on the personal judgment, survey data, group experiences and then, square that number to get the core drive score. All eight core drive scores are added up, and the resultant gives the final Octalysis Score. i.e. Octalysis Score = (Sum of square of individual core drive scores) out of 800 (Yukaichou.com, 2018).

Level 2 Octalysis

This level emphasis on what player feels across his game journey. There are four phases of Level 2 Octalysis.

- **Discovery** refers to evaluate the reasons why would people even want to start the journey.
- **Onboarding** refers to the methods that how would to you teach the rules and the pros and the tolls to play the game.
- **Scaffolding** refers to the regular journey of repeated actions towards the goal.
- **Endgame** refers to the techniques used to retain and improve the status.

Plan and Implementation

The planning and the implementation phase include the setting up the survey questionnaire, analysis of data obtained based on survey and the credible solutions were accordingly proposed. Some of the solutions were based on gamification along with the additional prospects. Within these mess halls a drive was conducted for a period of three weeks to scrutinize the overall food wastage. A survey questionnaire was constructed based on items and factors determined from different observations. The data was analyzed, and it was estimated that in a year, around 15,000 kgs. of food was getting wasted within the institute (considering factors like winter camps, summer courses etc.). The description of the process is discussed below.

Proposition of the Questionnaire

These things were kept in mind while designing the questionnaire:

- The form made is short and precise
- Readily understandable
- Maximum information could be acquired in least number of questions
- Repeating of a few questions to confirm the authenticity of the person filling the survey
- Number of factors to be more than the alternatives
- A few miscellaneous questions to form the part of basic report online and easily accessible

Analysis of Data Considering Co-Relation of Different Factors: The calculation worked by considering the average amount of food wasted, per mess, per day. We also factored in differences for girl's messes and boy's messes, as it was often observed that girls tended to waste less food than boys. The data in all was averaged out for a period of 3 weeks and the average data for a day scaled to match the wastage in a year. Through survey forms, a broad base of information was being collected from students. Students were surveyed on issues such as most wasted meal, majorly wasted food item, influence of awareness posters/quotes on student psyche, while they are eating, etc. Students were interrogated on subjects like whether fines and stricter rules could be an effective method for curbing food wastage. It was identified that the size of ladle or serving spoon also had a bearing on the amount of food being wasted in a day.

Total no. of Mess = 9

Food wasted on a normal day in a single mess (Average of all) = 5.8 Kgs.

If all the mess work for almost 9 months, a year and four mess work for 2 and a half extra months (for summer courses, winter camps etc.) having food waste of approx. 1.6 Kgs per mess;

The total food wasted on an average in 1 year = (5.8 * 266 * 9) + (1.6 * 76 * 9) = 13885.2 + 109.4 = 14,979.6 Kgs. (approx.)

As part of survey feedback, students were asked whether they felt that the unconsumed meals could be donated to children from less privileged section of the society. For this, the children of construction workers living within the institute premises were identified. Moreover, the base of information was further used for categorizing the primary problems or data blocks that could be determined.

From the above questionnaires, some of the problems identified were:

- Poor quality of food being served
- The food prepared was being kept at some distance from the dining tables. Most people would thus avoid going for second helpings and fill their plates with extra food in one go
- A lack of sense of duty among the students and awareness on the perils of food wastage

Some key data highlights within our chosen sample set:

- **67.6%** of the surveyed audience highlighted the fact the vegetables and the dinner are the most wasted food item and the most wasted meal respectively
- **90.2%** said that it was because of the bad quality, why they waste mess food.
- **59.8%** think that specific dishes like curry, potato beans etc. which are less popular among youngsters are more wasted while **40.2%** suggested that the overall food quality is a dominant reason for food wastage.
- **57.8%** suggested that the size of service ladle should be reduced to minimize food wastage.
- **91.2%** were willing to donate the unconsumed meal to the underprivileged.

When asked about the techniques to be adopted apart from improving the food quality to reduce food wastage, most of the sample offered the idea of taking both the "quality of food" and the "frequency" at which a person wastes food while imposing fine for wasting food.

Additional Insights:

The work provides an important insight about the types of food wastage.

1. Avoidable Food Wastage

This category includes the food which either gets wasted because more food is being cooked or is out of date before we use it or gets wasted because more than required food has been shopped already.

2. Unavoidable Food Wastage

This category includes the food that isn't edible anymore and nothing can be done with it, except to dispose it off or to make compost out of it.

A concrete approach had to be followed as **69.6%** made it clear that they are somewhat or not at all influenced by the posters or inspirational quotes pasted on the mess walls. Therefore, "Gamification" must come into play.

SOLUTIONS AND RECOMMENDATIONS

The aim of the work covers the following bullet points:

- Improve the Quality of food and the current Mess System to avoid the "Avoidable Wastage" of food.
- To come up with an idea to effectively channelize the excessive food already cooked.
- To create an interface which make people donate the non-consumed food to the needy.

Based on the sample set, some of the proposed solutions include:

Building a Gamified Mobile Application: Enabling the students with the choice to donate their meal, a dedicated mobile application called "The Wheat Bowl" was designed that could manage the activities within the proposed system. Wheat Bowl is a gamified data collection platform to collect the food and its wastage associated data. The Wheat Bowl app, can create user profiles, allow users to upload multimedia, incorporates the game elements like points and badges and virtual money and a leader board. As such, the app is based on a collection of technologies that can be molded to fit the needs of this specific project. Various game elements are incorporated at various stages that manage wastage by individual and effective distribution of the surplus food. As per the institution's hostel rules the meal charges apply per student, at a rate of a minimum of 55 Meals per month (consumed or not). One has to inform on a notice of at least 2 days, if one wishes to have his or her mess shut off. A solution was proposed that with the help of this mobile application, hostel residents who avail the mess food facility can easily donate their unconsumed meals to the needy children quota by a single tap. Following the idea, a tie-up was made with an organization, Prayaas, a non-profit body engaged on educating the children of the local workers. These students can earn vouchers by participating in quizzes and competitions that concern with their educational curriculum and those vouchers can help them reserve meals for the upcoming days. All the relevant information is provided to the needy children about the vouchers. Different vouchers are described, that expounded in the mobile application which are equivalent to different food items. These vouchers are earned as per the difficulty levels, a tot clears. For instance: Answering two correct synonyms will award them a voucher for a chocolate, a good performance in a quiz can earn them more rewards with which, they can buy a better meal. This takes care of two birds with one arrow. Children get access to quality food, and learn in the process and hence, gamification plays its role well in the process.

The proposed model is placed within the Octalysis Framework (Yukaichou.com, 2018). The credibility of the proposed system was checked on eight parameters, Epic meaning and calling, Accomplishment, Empowerment of creativity and feedback, Ownership, Social Influence, Scarcity and impatience, Unpredictability and Avoidance of loss. This very scheme focuses more on development, accomplishment and empowers the needy children with creativity and the mess facility availers, with a call of epic meaning.

Introduction of Variable Salary for Mess Workers: We propose to introduce a concept of variable salary in the payment of mess workers.

Suppose, a mess worker is being paid Rs. 14,000 per month, the idea is to pay the mess worker:

- Fixed Salary = Rupees 11,000 per month
- Variable Salary = Rs. 5,000 per month,

This directly depends on the feedback given by the students about his performance (cooking, serving etc.) via the mobile application. Generally, the criteria of rating should be majorly based on cook's cooking skills. This very scheme focuses more on the core drive of unpredictability and curiosity of the results of their monthly performance, development of their cooking skills and a sense of avoiding the loss of the additional variable pay.

Competitive Environment: Developing an environment among those who make use of the mess facilities to measure the food wastage and incentives. This is done to motivate people by showing them on daily basis (on the mobile application) the cumulative food wastage and the account of food wastage done by them on a day with the anatomization of how many people can be fed by that food. It helps to plant in them a sense of loss in them.

Introduction of Points/Credit Based System: In the existing mess food distribution structure can create a huge change. The points earned by the user can be used as an investment to purchase the "Extra Refreshments". The track of the points credit/debit from one's account will be taken care by the algorithm embedded in the mobile application itself. Now, the question arises how a user can earn points? There are numerous ways like if the user shows the zero indicator on the weight scale (fixed on the counter), that he has not wasted the food in a group or individually for three consecutive times, it can earn him the credits which can be used to buy the extra mess refreshments. For Instance, User can buy a Lays Packet (Rs.20) for free at the end of the week if he has not wasted the food for a single day.

These proposed schemes, as a complete framework scored an octal score of 528 on a scale of 800 (which is discussed later in this chapter). The model works to demonstrate, how game principles can be effectively employed to modify student behavior in a way so as to reduce food wastage within the setup of the institute.

Additional Prospects Including the Self-Voluntary Activities

- An initiative by mess committee was taken in Mess No.2 so as to stress more on the ***portioning of food items***. Self-serve menu items are wasted more than portioned menu items by consumers. Therefore, this change was implemented in Mess No.2 and is proposed to impose in every mess of the institute.
- An easy and obvious step taken to reduce food waste in the mess halls was using ***Consumer Education Tools*** like posters or writing eye- catchy self-realization writings behind the serving dishes. This came out to be a fruitful initiative.
- ***Frequent Monitored Segregation Waste Audits*** were included in the record of duties and responsibilities of mess committee.
- It was suggested that the increase in communication between mess staff and students will highly help in mess food waste reduction, as the priorities and opinions of students will be clearer to the mess staff.
- Introduction of ***LED Lights*** in mess which will reduce power consumption and would further affect the service charges of hostel.
- A lot of complaints were observed for mess food menu in the additional comments sections and due considerations shall be taken from all the students while designing the menu.
- ***Biogas plants*** or ***manure compost plant*** should be setup for biodegradable waste being produced in mess (Wisconsin-Platteville, 2018).
- Night mess is not followed in all the hostels which come as a great relief to the students; efforts shall be made to instill night mess in all the hostel messes.

Calculation of Octalysis Score

- 200 students who were associated with the course and dynamics of gamification, were asked to rate this model's eight different core drives of gamification on the scale of 0 to 10, so as to calculate the overall octalysis score, as shown in Figure 1. The results came out to be:

- **Epic meaning and Calling:** The craze among the people to do something for their own educational institute attaches an epic meaning with this initiative. The overall score given to this core drive by the sample came out to be **8**.
- **Development and Accomplishment:** The campaign would be led by the points and incentives in terms of free snacks and goodies. Moreover, through a leader board, the performance of various students who enjoy the mess facility is tracked down and accordingly the incentives are provided. In this way, it becomes a routine task for students to accomplish the requirements to gain benefits and hence a sense of development and accomplishment is attached. This core drive was given an overall score of **9**.
- **Empowerment of Creativity and Feedback:** Like any sandbox or Nintendo game, participants are provided with gears/ tools and freedom to accomplish the goals they want. On the top of that, a continuous feedback mechanism is deployed to continuously figure out loop holes in the model. This gives them the power of creativity and feedback and as a result, this core drive scored **9** out of 10.
- **Ownership and Possession:** Participants including the students, the mess workers, the children of construction workers and the whole mess management are being awarded as per the goals accomplished by them and their consequent scores. The idea of "Helping yourself before others" was never delivered in a more effective manner. This core drive scored **8** on the scale of 10.
- **Social Influence and Relatedness:** Competition, Envy, Group pursuits and regular treats contribute to this core drive and make this core drive dominant of all. This is the sole reason that this core drive was the most popular of all and managed to score a full **10**.
- **Scarcity and Impatience:** The impatience to grab the special meal puts this core drive a special place and has earned this core drive 7 out of 10.
- **Curiosity and Unpredictability:** The idea of connecting the virtual world solutions with the real-world life problems has opened a brand-new domain and there, this core drive comes into play. Hence, this core was rated as **5** in the scale of 10.
- **Loss and Avoidance:** The promise of the provision of food is itself so convincing that it would persuade the students and the children of the construction workers to break the streak. The mess workers would be elated with the idea of the variable salary which is like the cherry on the top of cake and hence, they would not resist working on their existing cooking skill set. No wonder, this core drive was scored **8** on the scale of 10 by the sample.

And hence the overall Octalysis score can be calculated as:

Octalysis Score = (Sum of square of individual core drive scores) out of 800

= 64+81+81+64+100+49+25+64

= 528 out of 800

For quantifying the service satisfaction levels of the students who are using the services of mess in this college, some rating questions were asked. 200 students from all the hostels were selected as a sample. The data analysis shows that quantity, quality, services, hygiene environment and cost are significant factor that decide students' selection and satisfaction based on which the best mess of all the mess that students eat in, was rated.

Theoretical Foundation and Approach for Improving Service Quality

The research is to suggest a program to bring the service quality of all the mess at par. A method known as "Multiple criteria decision making" is used in which the weight determination is done by "Analytical Hierarchy Process" (AHP) method. The broad parameters based on which we based our judgment to rate the mess are Mess service quality, Special event management, Seating Arrangement, Waste Management, Cleanliness and hygiene of the premises, Ambience of the dining hall,

Figure 1.

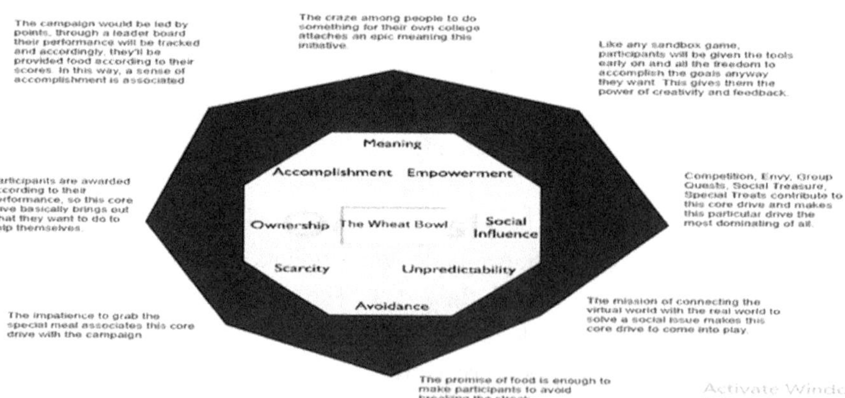

Behavior of the mess staff, Cost of services being charged i.e. Mess bill, Availability of food items like pickles, ketchup, salt etc.

Multiple criteria decision making (MCDM) is concerned with structuring and solving decision and planning problems involving multiple criteria. (Benita M. Beamon, 1999) The purpose is to support decision makers facing such problems. The MCDM process follows a common working principle as described below:

- **Selection of Criteria:** Selected criteria must be (refer Figure 1.):
    - Coherent with the decision
    - Independent of each other
    - Represented in same scale
    - Measurable
    - Not Unrelated with the alternatives

In this case, the various criterions were hygiene, cleanliness, quality of food etc.

- **Selection of Alternatives:** Selected alternatives must be (refer Figure 1.):
    - Available
    - Comparable
    - Real not Ideal
    - Practical/Feasible

In this case, the alternatives include all the nine mess halls.

- **Selection of the Weighing Methods to Represent Importance:** In this case, the weighing method used is Analytical Hierarchy Process (AHP). It is a type of Compensatory Weighing Method. The Analytic Hierarchy Process is a structured technique for organizing and analyzing complex decisions. (Benita M. Beamon, 1999)

Method of Aggregation: It can be a product or an average or a function. The result of this aggregation will separate the best alternative from the available options.

AHP Methodology

- **Development of a Model for the Decision:** The decision was broken down into a hierarchy of goals (Identification of the "Best Mess", criteria (Food Quality, Hygiene, cleanliness etc.), and alternatives (all the nine mess halls).
- **Derivation of Priorities (Weights) for the Criteria:** The importance of criteria is compared pair wise with respect to the desired goal to derive their

weights. The consistency of judgments was then checked; that is, a review of the judgments is done to ensure a reasonable level of consistency in terms of proportionality and transitivity.

- **Derivation of Local Priorities (Preferences) for the Alternatives:** The priorities or the alternatives were derived with respect to each criterion separately (following a similar process as in the previous step, i.e., compare the alternatives pair wise with respect to each criterion). Then, the consistency was checked and adjusted as required.
- **Derive Overall Priorities (Model Synthesis):** All alternative priorities obtained were combined as a weighted sum to consider the weight of each criterion and to establish the overall priorities of the alternatives. The alternate option with the highest gross priority represent the best option.
- **Performing Sensitivity Analysis:** This was performed to understand the rationale behind the obtained results by taking into consideration changes in the weights of the criteria.
- **Final Decision:** Based on the synthesis results and sensitivity analysis, a decision was made. (Saaty's 2012)

Questionnaire and Conduction of Survey for Improving Service Quality

The questionnaire consists of three sections:

- Personal Details
- Rating Questions
- Descriptive Questions

In this survey, full confidentiality was maintained about the personal details and views of the participants. Following is the set of the rating enquiry that was made as another aspect of this research work. The rating varies on a scale of 0 to 5 which refers to "Very Less"/ "None at all" and "Lots of Variety"/ "Excellent" respectively. The rating questions were divided into nine sorts:

Table 1.

Expert Opinion Matrix	Quality	Cleanliness	Dining Area	Staff Behavior	Special Event	Seating	Mess Bill	Waste Management	Additional

- **Mess Service Quality:** This includes a detailed inspection of variety of meals to choose from, extent of green vegetables and level of nutrient values content in menu, promptness of service, level of concern of student's choice in menu making, availability of extra items, average waiting time, overall value for money etc.
- **Cleanliness:** In this section, the subjects like cleanliness and décor in dining area, cleanliness of utensils and cleanliness of kitchen were broached.
- **Dining Area:** This group includes the questions like comfortability of dining area seating and the lightening of halls.
- **Mess Staff Behavior:** This group lets the user to rate the mess staff behavior.
- **Special Event Management:** This group covers the enquiry about sufficient funds and resources to organize any special event, whether hostel staff has adequate management skills to organize any special event, willingness of hostel staff to work for a special event like hostel farewells.
- **Seating Arrangement:** This includes questions regarding the convenience level and availability of seats, space provided between two seats and aisle width, maintenance of furniture, seating furniture condition etc.
- **Mess Bill:** This section questions the students regarding the transparency in the mess bill, whether the mess bill is presented in an understandable format or not and on right time or not.
- **Waste Management:** In this section, the subjects like extent of food wastage (including all dairy products, vegetables, dals, chapattis etc.), reasons and the source of food wastage and effective ways for the reduction of food wastage were introduced. This group also lets user to rate the functionality of the waste management system employed in the existing mess structure.
- **Additional Items Availability:** The objective of this group was to gather data about the availability of different items like spices, pickles, ketchup etc. in mess.

Expert Opinion on Co-Relation of Different Factors

To find the co-relation of different factors on each other, three experts were asked on the matter to fill out a form for the same. The results from this form were used to conclude the following expert opinion matrix and are described in Table 2.

Using the above matrix, priority vector (weight of each factor) is calculated (refer Table 3). The consistency of the weights is checked using CI/RI approach. It comes out to be <0.1, hence consistent.

Survey Analysis: After the calculation of priority matrix and checking for consistency ratio, the data was to be analyzed for proceeding further to define the ranks of hostel mess. The procedure involved the following steps:

Table 2. Expert opinion matrix

Expert Opinion Matrix	Quality	Cleanliness	Dining Area	Staff Behavior	Special Event	Seating	Mess Bill	Waste Management	Additional
Quality	1	2	2	3	5	3	1	5	4
Cleanliness	0.5	1	1	3	5	2	0.3333	5	4
Dining Area	0.5	1	1	2	4	2	0.25	4	3
Staff Behaviour	0.3333	0.3333	0.5	1	3	0.5	0.25	3	3
Special Event	0.2	0.2	0.25	0.3333	1	0.3333	0.2	0.3333	0.5
Seating	0.333	0.5	0.5	2	3	1	0.25	3	2
Mess Bill	1	3	4	4	5	4	1	5	4
Waste Management	0.2	0.2	0.25	0.3333	3	0.3333	0.2	1	0.5
Additional	0.25	0.25	0.3333	0.3333	2	0.5	0.25	2	1
Column Sum	4.3167	8.4833	9.833	16	31	13.667	3.7333	28.333	22

- The data was bifurcated hostel wise and plotted against the respective factors on excel sheet.
- The average of the recorded entries was taken factor wise and their respective weight was multiplied which was decided in priority matrix.
- The cumulative sum of all the factors was taken after step 2.
- The result was compared in an ascending or descending order and the ranks were awarded accordingly.
- The graphs were plotted factor wise for ease of comparison.

Inferences

- The rankings of the mess are as follows: (refer Figure 2.)
    - Boys Hostel 7/7E (ii) Boys Hostel 1 (iii) Mega Boys Hostel (iv) Mega Girls Hostel (v) Girls Hostel 1 (vi) Boys Hostel 5 (vii) Boys Hostel 2 (viii) Boys Hostel 6 (ix) Boys Hostel 3/4
- In terms of food quality, cleanliness, dining area Boys Hostel 7 is declared the best.
- In terms of staff behavior Boys Hostel 1 is best
- Seating area at Boys hostel 7 and Mega Boys Hostel are at par.
- In terms of mess bill Mega Girls Hostel leads.
- Boys Hostel 3/4 is rated as the worst mess.

Table 3. Calculation of priority vector and consistency along with calculation of weightage of each question

Normalized Column Sums									Priority Index	AX	AX/X	Weight for each question
0.2317	0.2358	0.2034	0.1875	0.1613	0.2195	0.2679	0.1765	0.1818	0.20725054	1.990282	9.603263	0.025906318
0.1158	0.1179	0.1017	0.1875	0.1613	0.1463	0.0893	0.1765	0.1818	0.14201217	1.369742	9.645245	0.028402433
0.1158	0.1179	0.1017	0.125	0.129	0.1463	0.067	1412	0.1364	0.12003126	1.158412	9.650921	0.04001042
0.0772	0.0393	0.0508	0.0625	0.0968	0.0366	0.067	0.1059	0.1364	0.7471446	0.702426	9.4011469	0.074714456
0.0463	0.0236	0.0254	0.0208	0.0323	0.0244	0.0536	0.0118	0.0227	0.02898627	0.269577	9.300151	0.009662091
0.0772	0.0589	0.0508	0.125	0.0968	0.0732	0.067	0.1059	0.0909	0.0825637	0.794879	9.593453	0.016571273
0.2317	0.3536	0.4068	0.25	0.1613	0.2927	0.2679	0.1765	0.1818	0.25802151	2.59927	9.805101	0.086007172
0.063	0.0236	0.0254	0.0208	0.0968	0.0244	0.0536	0.053	0.0227	0.03876911	0.353395	9.115384	0.019384556
0.0579	0.0295	0.0339	0.0208	0.0645	0.0366	0.067	0.0706	0.0455	0.04735831	0.441034	9.312696	0.0738312
										^max	9.491965	
										CI	0.061496	
										RI	1.45	
										CR	0.042411	

Figure 2.

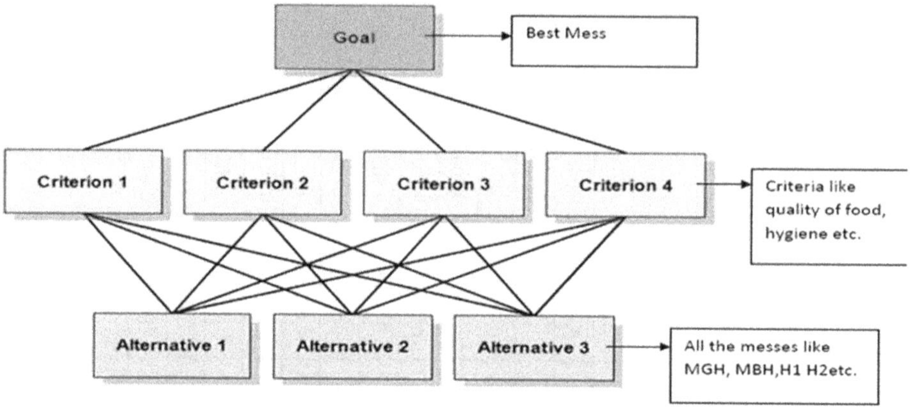

Outcomes

- Diet rate in **Boys Hostel 7/7E** is Rs. 22 per diet which when compared to the diet rate of Hostel 3 and 4 (Rs. 23.5 per diet) is way low. Diet rate in **Boys hostel 3/4** is the prime factor for worst rank of being awarded to **Boys hostel 3/4** and efforts must be made to reduce the diet rate in **Boys hostel 3-4**.
- Students eating in **Boys Hostel 3/4** have complained about mess bill not displayed periodically and transparent system is not followed in while adding the extra cost in the final bill and efforts must be made to change it.
- Seats to no. of students' ratio are way too less in **Boys Hostel 2** which causes trouble for students to find seats while having their food. If the budget constraints allow, an expansion shall be made for the same.
- Additional items (like ketchup, pickles etc.) is less in **Boys hostel 2** and regular stock depending on the consumption should be maintained.
- It is observed from the data that the staff behavior is very poor in **Mega Girls Hostel** which should be discouraged, and a proper grievance readdressing committee shall be formed. Training should be provided for the same.
- Special event management is poor in **Boys Hostel 5** and suggestions from the students shall be taken to address to this problem.
- Problems like **Boys hostel 3/4** are observed in **Hostel 6** in terms of diet rate and mess bill and similar actions shall be taken to reprimand the same.

Challenges: It is very difficult to motivate the students to use the android phone application for this specific cause which is required for feedback and to donate the unconsumed food. Another issue is to make the biggest wasters of food realize how important this is, as the generation have its own priorities and opinions. Another deal is to be convincing enough to make the age group between 12-30 years aware about the importance and the benefits of donation.

CONCLUSION

After analyzing food waste at NIT Jalandhar's nine mess halls, several recommendations were devised for the institute based on the data collection followed by the analysis. Waste audits were conducted in these nine mess halls, focus groups were run with mess consumers, mess management and campus administrators were interviewed, and this is how a set of improvements was proposed to implement in the future. Some of these suggestions regard systematic changes, like Introduction of Points/Credit Based System in the existing mess food distribution structure and track the unconsumed meal status using a mobile application called "Wheat Bowl". Other recommendations involve introducing competitive environment for the mess staff by introducing a factor of variable salary which makes them improve the quality of food they cook, aware them about portion sizes, food waste reduction as well as introducing more variety to the menu. These recommendations were simultaneously presented to members of the Mess Committee and mess managers, in hope that they will take further action to reduce mess hall food waste. The provided solution is a ready to use solution which can be implemented in any educational institution very easily with some basic support. This system if applied successfully is able to reduce the food wastage drastically. Plus, it's a win-win situation for every single individual associated with this new proposed system.

Prospects: The work proposes the extension for the involvement to the banquet halls hosting parties or marriages to contribute to this cause. Furthermore, the aim is to involve a local NGO in the time to come for the effective implementation of the solutions. Also, an agenda can be brought to fruition to cater to the children in the slums located in the suburbs of Jalandhar.

ACKNOWLEDGMENT

We are deeply indebted to Dr. Arun Khosla, our supervisor in this project, for consistently providing us with the required guidance to help us in the timely and successful completion of this project. Despite his extremely busy schedules, he was always available to share with us his deep insights, wide knowledge and extensive experience. We are appreciative of the "Prayaas" and their community for their selfless efforts to participate in the development of this research. This work is supported by the mess committees of every mess in the institute. We would also especially like to thank the mess workers who went out of their way to help us audit waste and test the proposed model. The students who make use of the mess facility of the institute played a major role in the successful implementation of this proposed approach in Mess No.2. We also appreciate of the support given by the Girls Hostel-2 and Mega Girls Hostel Wardens and the Chief Warden. They have helped us in building the vision and made critical judgment about the project. In this way, we were able to rectify the errors that may otherwise have gone unnoticed and guided us the way to success.

REFERENCES

Beamon, B. (1999). Measuring supply chain performance. *International Journal of Operations & Production Management*, *19*(3), 275–292. doi:10.1108/01443579910249714

Biswas, A. K. (2014). *India must tackle food waste*. Retrieved from https://www.weforum.org/agenda/2014/08/india-perishable-food-waste-population-growth/

Chou, Y. (2018). *Octalysis: Complete Gamification Framework - Yu-kai Chou.* Available at: https://yukaichou.com/gamification-examples/octalysis-complete-gamification-framework/

Dash, A., Sankaran, K., Shrimali, P., Iyer, P., Upadhyay, S., & Javadekar, S. (2015, November). Food for thought. *Insight: IIT Bombay's Student Media Body*, 6-7.

FAO. (2011). Global food losses and food waste – Extent, causes and prevention. Rome: Food &Agriculture Organization of the United Nations.

Fijnheer, J. D., & Van Oostendorp, H. (2016). Steps to design a household energy game. *International Journal of Serious Games*, *3*(3). doi:10.17083/ijsg.v3i3.131

FUSIONS. (2016). Recommendations and guidelines for a common European food waste policy framework. FUSIONS EU FP7 Research Project.

Klüppel, H. (1998). ISO 14041: Environmental management — life cycle assessment — goal and scope definition — inventory analysis. *The International Journal of Life Cycle Assessment*, *3*(6), 301–301. doi:10.1007/BF02979337

Learn Gamification with Yu-kai Chou: cheat codes to win the game of life. (2018). Retrieved from https://yukaichou.com/

Werbach, K., & Hunter, D. (2012). *For the Win: How game thinking can revolutionize your business*. Philadelphia: Wharton Digital Press.

Whitehair, K., Shanklin, C., & Brannon, L. (2013). Written messages improve edible food waste behaviors in a university dining facility. *Journal of the Academy of Nutrition and Dietetics*, *113*(1), 63–69. doi:10.1016/j.jand.2012.09.015 PMID:23260724

Compilation of References

Abdalla, A. M., Darwish, S. M., Ayad, E. E., & El-Hamahmy, R. M. (2007). Egyptian mango by-product 1. Compositional quality of mango seed kernel. *Food Chemistry*, *103*(4), 1134–1140. doi:10.1016/j.foodchem.2006.10.017

Abdel-Rahman, M. A., Tashiro, Y., & Sonomoto, K. (2013). Recent advances in lactic acid production by microbial fermentation processes. *Biotechnology Advances*, *31*(6), 877–902. doi:10.1016/j.biotechadv.2013.04.002 PMID:23624242

Abdel-Rahman, M. A., Xiao, Y., Tashiro, Y., Wang, Y., Zendo, T., Sakai, K., & Sonomoto, K. (2015). Fed-batch fermentation for enhanced lactic acid production from glucose/xylose mixture without carbon catabolite repression. *Journal of Bioscience and Bioengineering*, *119*(2), 153–158. doi:10.1016/j.jbiosc.2014.07.007

Achaya, K. T. (1994). *Indian food: a historical companion*. Delhi: Oxford University Press.

ACR+. (2009). Municipal Waste in Europe – Towards a European Recycling Society. *Victoires Editions*.

Adame, Samino, & Sanchez, , & Gonzalez. (2012). In vitro estimation of the antibacterial activity and antioxidant capacity of aqueous extracts from grape seeds. *Food Control*, *24*, 136 – 141.

Adams, M. R. (1990). Topical aspects of fermented foods. *Trends in Food Science & Technology*, *1*, 140–144. doi:10.1016/0924-2244(90)90111-B

Adams, M. R. (1998). Fermented weaning foods. In J. B. Wood (Ed.), *Microbiology of fermented foods* (pp. 790–811). London: Blackie Academic. doi:10.1007/978-1-4613-0309-1_25

Adams, M. R., & Dougan, J. (1981). Waste Products. In R. J. Clarke & R. Macrae (Eds.), *Coffee Technology*. New York: Elsevier Applied Science Publishers Ltd.

Afrc, R. F. (1989). Probiotics in man and animals. *The Journal of Applied Bacteriology*, *66*(5), 365–378. doi:10.1111/j.1365-2672.1989.tb05105.x PMID:2666378

Agarwal, K. N., & Bhasin, S. K. (2002). Feasibility studies to control acute diarrhea in children by feeding fermented milk preparation Actimel and Indian Dahi. *European Journal of Clinical Nutrition*, *56*(4), S56–S59. doi:10.1038j.ejcn.1601664 PMID:12556949

Agrahar-Murugkar, D., & Subbulakshmi, G. (2006). Preparation techniques and nutritive value of fermented foods from the Khasi tribes of Meghalaya. *Ecology of Food and Nutrition*, *45*(1), 27–38. doi:10.1080/03670240500408336

Agte, V. V., Gokhale, M. K., & Chiplonkar, S. A. (1997). Effect of natural fermentation on in vitro zinc bioavailability in cereal–legume mixtures. *International Journal of Food Science & Technology*, *32*(1), 29–32. doi:10.1046/j.1365-2621.1997.00372.x

Ahamed, A., Yin, K., Ng, B. J. H., Ren, F., Chang, V. W.-C., & Wang, J.-Y. (2016). Life cycle assessment of the present and proposed food waste management technologies from environmental and economic impact perspectives. *Journal of Cleaner Production*, *131*, 607–614. doi:10.1016/j.jclepro.2016.04.127

Ahlawat, S., Dhiman, S. S., Battan, B., Mandhan, R. P., & Sharma, J. (2009). Pectinase production by *Bacillus subtilis* and its potential application in biopreparation of cotton and micropoly fabric. *Process Biochemistry*, *44*(5), 521–526. doi:10.1016/j.procbio.2009.01.003

Ahmed, I., Muhammad, A., Muhammad, A., Zain, A., & Muhammad, T. (2015). Bioprocessing of citrus waste peel for induced pectinase production by *Aspergillus niger* ; Its purification and characterization. *Journal of Radiation Research and Applied Sciences*, 1–7.

Aidoo, K. E., Nout, N. J. R., & Sarkar, P. K. (2006). Occurrence and function of yeasts in Asian indigenous fermented foods. *FEMS Yeast Research*, *6*(1), 30–39. doi:10.1111/j.1567-1364.2005.00015.x PMID:16423068

Ajila, C. M., Aalami, M., Leelavathi, K., & Rao, U. P. (2010). Mango peel powder: A potential source of antioxidant and dietary fibre in macaroni preparations. *Innovative Food Science & Emerging Technologies*, *11*(1), 219–224. doi:10.1016/j.ifset.2009.10.004

Aldridge, B. B., & Rhee, K. Y. (2014). Microbial metabolomics: Innovation, application, insight. *Current Opinion in Biotechnology*, *19*, 90–96.

Alibardi, L., Muntoni, A., & Polettini, A. (2014). Hydrogen and waste: Illusions, challenges and perspectives. *Waste Management (New York, N.Y.)*, *34*(12), 2425–2426. doi:10.1016/j.wasman.2014.09.001 PMID:25442106

Aliyu, S., & Bela, M. (2011). Brewer's spent grain: A review of its potential and applications. *African Journal of Biotechnology*, *10*, 324–331.

Ambardar, S., Gupta, R., Trakroo, D., Lal, R., & Vakhlu, J. (2016). High Throughput Sequencing: An Overview of Sequencing Chemistry. *Indian Journal of Microbiology*, *56*(4), 394–404. doi:10.100712088-016-0606-4

Amore, A., Giacobbe, S., & Faraco, V. (2013). Regulation of cellulase and hemicellulase gene expression in fungi. *Current Genomics*, *14*(4), 230–249. doi:10.2174/1389202911314040002

Anaerobic Digestion. (n.d.). Retrieved from https://www.epa.gov/anaerobic-digestion

Andjelković, U., Šrajer Gajdošik, M., Gašo-Sokač, D., Martinović, T., & Josić, D. (2017). Foodomics and Food Safety: Where We Are. *Food Technology and Biotechnology*, *55*(3), 290–307. doi:10.17113/ftb.55.03.17.5044

Ángel Siles López, J., Li, Q., & Thompson, I. P. (2010). Biorefinery of waste orange peel. *Critical Reviews in Biotechnology*, *30*(1), 63–69. doi:10.3109/07388550903425201 PMID:20148755

Anto, H., Trivedi, U. B., & Patel, K. C. (2006). Glucoamylase production by solid- state fermentation using rice flake manufacturing waste products as substrate. *Bioresource Technology*, *97*(10), 1161–1166. doi:10.1016/j.biortech.2005.05.007 PMID:16006122

Antonio, A. G., Moraes, R. S., Perrone, D., Maia, L. C., Santos, K. R. N., I'orio, N. L. P., & Farah, A. (2010). Species, roasting degree and decaffeination influence the antibacterial activity of coffee against *Streptococcus mutans*. *Food Chemistry*, *118*, 782–788.

Antony, U., & Chandra, T. S. (1997). Microbial population and biochemical changes in fermenting finger millet (*Eleusine coracana*). *World Journal of Microbiology & Biotechnology*, *13*(5), 533–537. doi:10.1023/A:1018561224777

Antony, U., George, M. L., & Chandra, T. S. (1998). Inhibition of *Salmonella typhimurium* and *Escherichia coli* by fermented Finger millet (*Eleusine coracana*). *World Journal of Microbiology & Biotechnology*, *14*(6), 883–886. doi:10.1023/A:1008871412183

Anu Appaiah, K. A., & Ganesh, D. (2006). Coffee cherry husk, a feedstock for methane generation. In *ASIC – Proceedings of 20th ASIC Conference, Association Scientifique* (pp.540–545). Internationale Du café.

Appaiah Anu, K. A. (2017). Fruit and vegetable wastes: An alternative feed stock for alcohol production. *Indian Food Industry*, *36*(6), 30–41.

Arvanitoyannis, S., Ladas, D., & Mavromattes, A. (2006). Potential Uses and application of treated wine waste: A Review. *International Journal of Food Science & Technology*, *41*(5), 475–487. doi:10.1111/j.1365-2621.2005.01111.x

Arvind, K., Nikhlesh, K. S., & Pushpalata, R. S. (2010). Inhibition of 1,2-dimethylhydrazine induced colon genotoxicity in rats by the administration of probiotic curd. *Molecular Biology Reports*, *37*(3), 1373–1376. doi:10.100711033-009-9519-1 PMID:19330535

Aryana, K. J., & Olson, D. W. (2017). A 100-Year Review: Yogurt and other cultured dairy products. *Journal of Dairy Science*, *100*(12), 9987–10013. doi:10.3168/jds.2017-12981 PMID:29153184

Asad, M. J., Asghar, M., Sheikh, M. A., & Sultan, J. I. (2001). Production of α-amylase by *Aspergillus niger* and its partial purification. *Pakistan Journal of Agricultural Sciences*, *38*, 3–4.

Ashraf, R., Shahid, F., & Ali, T. A. (2007). Association of fungi, bacteria and actinomycetes with different composts. *Pakistan Journal of Botany*, *39*(6), 2141–2151.

Aslanzadeh, S., & Özmen, P. (2009). *Biogas production from municipal waste mixed with different portions of orange peel*. Academic Press.

Asli, I., Jennifer, N. H., & Anthony, L. (2008). Aqueous ammonia soaking of switchgrass followed by simultaneous saccharification and fermentation. *Applied Biochemistry and Biotechnology*, *144*(1), 69–77. doi:10.100712010-007-8008-z PMID:18415988

Autrel, E., Betrhier, F., Lutze Zajac, A., & Nicolas, F. (2007). Inceneration of municipal and material recovery performances. *Journal of Hazardous Materials*, *139*, 569. doi:10.1016/j.jhazmat.2006.02.065 PMID:16707217

Aysun, H., Mercimek T., & Filiz U. T. (2016). Extracellular Pectinase Production and Purification from a Newly Isolated *Bacillus subtilis* Strain Extracellular Pectinase Production and Purification from a Newly Isolated *Bacillus subtilis* Strain. *International Journal of Food Properties, 19*(11), 2443–50.

Babu, P. D., Bhakyaraj, R., & Vidhyalakshmi, R. (2009). A low cost nutritious food "tempeh"-a review. *World J Dairy Food Sci*, *4*(1), 22–27.

Bachert, C., Bidlingmaier, W., & Surapong, W. (2008). Open windrow composting manual. *Orbit (Amsterdam, Netherlands)*, V.

Backes, M. (2014). *Cannabis Pharmacy: The Practical Guide to Medical Marijuana*. New York: Black Dog & Leventhal.

Bairagi, S., & Rohtak, H. (2016). Optimization of Cellulase Enzyme from Vegetable Waste by Using *Trichoderma atroviride* in Solid State Fermentation. *Journal of Environmental Science, Toxicology and Food Technology*, *10*(5), 68–73.

Bajón Fernández, Y., Soares, A., Villa, R., Vale, P., & Cartmell, E. (2014). Carbon capture and biogas enhancement by carbon dioxide enrichment of anaerobic digesters treating sewage sludge or food waste. *Bioresource Technology*, *159*, 1–7. doi:10.1016/j.biortech.2014.02.010 PMID:24632434

Bajželj, B., Richards, K. S., Allwood, J. M., Smith, P., Dennis, J. S., Curmi, E., & Gilligan, C. A. (2014). Importance of food-demand management for climate mitigation. *Nature Climate Change*, *4*(10), 924–929. doi:10.1038/nclimate2353

Baker, D., Fear, J., & Denniss, R. (2009). *What a waste: An analysis of house hold expenditure on food, in policy Brief no. 6*. Canberra: The Australia Institute.

Bansal, N., Rupinder, T., Raman, S., & Sanjeev, K. (2012). Production of Cellulases from Aspergillus niger NS-2 in Solid State Fermentation on Agricultural and Kitchen Waste Residues. Waste Management, 32(7), 1341–46.

Bansal, N., & Kanwar, S. S. (2013). Peroxidase (s) in environment protection. *The Scientific World Journal*, *2013*, 714639. doi:10.1155/2013/714639 PMID:24453894

Barbulova, A., Colucci, G., & Apone, F. (2015). New trends in cosmetics: By-products of plant origin and their potential use as cosmetic active ingredients. *Cosmetics*, *2*(2), 82–92. doi:10.3390/cosmetics2020082

Barton, S. (2010, September 17). *Retailers halve waste to landfill since 2005*. Retrieved from https://www.letsrecycle.com/news/latest-news/retailers-halve-waste-to-landfill-since-2005/

Battacharya, S., & Bhat, K. K. (1997). Steady shear rheology of rice blackgram suspensions and suitability of rheological models. *Journal of Food Engineering*, *32*(3), 241–250. doi:10.1016/S0260-8774(97)00027-7

Battestin, V., & Macedo, G. A. (2007). Tannase production by *Paecilomyces variotii*. *Bioresource Technology*, *98*(9), 1832–1837.

Bazzoffi, P., Pellegrini, S., Rocchini, A., & Morandi, M. (1973). Municipal waste in a plantation of young slash pine: Effects on soil and trees. *Journal of Environmental Quality*, *2*(4), 441–444. doi:10.2134/jeq1973.00472425000200040006x

Beamon, B. (1999). Measuring supply chain performance. *International Journal of Operations & Production Management*, *19*(3), 275–292. doi:10.1108/01443579910249714

Belewu, M. A., & Babalola, F. T. (2009). Nutrient enrichment of some waste agriculture residue after solid state fermentation: Application to animal nutrition. *Animal Feed Science and Technology*, *14*(4), 122.

Ben Rebah & Miled. (2013). Fish processing wastes for microbial enzyme production: A review. *Biotech*, *3*, 255–265.

Ben Rebah, F., Frikha, F., Kammoun, W., Belbahri, L., Gargouri, Y., & Miled, N. (2008). Culture of Staphylococcus xylosus in fish processing by-product-based media for lipase production. Lett. *Applied Microbiology*, *47*(6), 549–554. doi:10.1111/j.1472-765X.2008.02465.x PMID:19120924

Beretta, C., Stoessel, F., Baier, U., & Hellweg, S. (2013). Quantifying food losses and the potential for reduction in Switzerland. *Waste Management (New York, N.Y.)*, *33*(3), 764–773. doi:10.1016/j.wasman.2012.11.007 PMID:23270687

Bernstad, A., & Jansen, J. (2012). Review of Comparative LCAs of Food Waste Management Systems - Current Status and Potential Improvements. *Waste Management (New York, N.Y.)*, *32*(12), 2439–2455. doi:10.1016/j.wasman.2012.07.023 PMID:22922048

Bhoite, Navya, & Murthy. (2013a), Statistical optimization, partial purification and characterization of coffee pulp β-glucosidase and its application in ethanol production. *Food Science and Biotechnology*, *22*(S), 205-212.

Bhoite, Navya, & Murthy. (2013b). Statistical optimization of Bioprocess parameters for enhanced Gallic acid production from coffee pulp tannins by *Penicillium verrucosum*. *Preparative Biochemistry & Biotechnology*, *43*(4), 350–363.

Bilbao, C., Ezequiel, M. E. I., Benavent, M. A., Calleja, M. H., del Moral, M. P., & Ariz, M. V. (2014). *Application of products of coffee silverskin in anti-ageing cosmetics and functional food*. Patent Number: EP2730171 A1.

Billings, T. (1998). *On fermented foods*. Available: http://www.livingfoods.com

Bisaria, A. (2017). *Now Gurgaon Gets Its First Community Fridge So There Are More Free Meals To Go Around*. Retrieved from www.indiatimes.com

Biswas, A. K. (2014). *India must tackle food waste*. Retrieved from https://www.weforum.org/agenda/2014/08/india-perishable-food-waste-population-growth/

Blandinob, A., Al-Aseeria, M. E., Pandiellaa, S. S., Canterob, D., & Webba, C. (2003). Review: Cereal-based fermented foods and beverages. *Food Research International*, *36*(6), 527–543. doi:10.1016/S0963-9969(03)00009-7

Bloom, J. (2011). American Wasteland: How America Throws Away Nearly Half of Its Food (and What We Can Do About It). Hachette UK.

Bokulich, N. A., Lewis, Z. T., Boundy-Mills, K., & Mills, D. A. (2016). A new perspective on microbial landscapes within food production. *Current Opinion in Biotechnology*, *37*, 182–189. doi:10.1016/j.copbio.2015.12.008

Bol, J., & de Vos, W. M. (1997). Fermented foods: an overview. In J. Green (Ed.), *Biotechnological innovations in food processing* (pp. 45–76). Oxford, UK: Butterworth-Heinemann.

Bos, A., & Hamelinck, C. (2014). *Green House Impact of marginal fossil fuel use*. Project no. BIEN 14973.

Bouallagui, H., Torrijos, M., Godon, J., Moletta, R., Cheikh, R. B., & Touhami, Y. (2004). Two phases anaerobic digestion of fruit and vegetable wastes: Bioreactors performance. *Biochemical Engineering Journal*, *21*(2), 193–197. doi:10.1016/j.bej.2004.05.001

Boye, J., Zare, F., & Pletch, A. (2010). Pulse proteins: Processing, characterization, functional properties and applications in food and feed. *Food Research International*, *43*(2), 414–431. doi:10.1016/j.foodres.2009.09.003

Braconi, D., Bernardini, G., Millucci, L., & Santucci, A. (2018). Foodomics for human health: Current status and perspectives. *Expert Review of Proteomics*, *15*(2), 153–164. doi:10.1080/14789450.2018.1421072

Bragg, P. C., Patricia Bragg, N. D., & Bragg, P. C. (2003). *Apple Cider Vinegar Miracle Health System*. Health Science Publications, Inc.

Braun, S. (2012). *Food Waste Report on the situation and recent activities in Germany, Stuttgart University*. Retrieved from http://ec.europa.eu/dgs/health_foodsafety/dgs_consultations/docs/ag/summary_ahac_05102012_3_susanne_braun_en.pdf

Brigidi, P., Bolognani, F., Rossi, M., Cerre, C., & Matteuzzi, D. (1993). Cloning of the gene for cholesterol oxidase in Bacillus spp., Lactobacillus reuteri and its expression in Escherichia coli. *Letters in Applied Microbiology*, *17*(2), 61–64. doi:10.1111/j.1472-765X.1993.tb00371.x PMID:7763933

Brook Lyndhurst Blog. (2009). *Enhancing Participation in Kitchen Waste Collections*. Retrieved from http://www.brooklyndhurst.co.uk/enhancing-participation-in-kitchen-waste-collections-_119

Brown, M. E., & Chang, M. C. (2014). Exploring bacterial lignin degradation. *Curr. opin. Inchem. biol., 19*, 1-7.

Brown, K. A., & David, M. H. (1994). Using landfill gas: A UK perspective. *Renewable Energy, 5*(5-8), 774–781. doi:10.1016/0960-1481(94)90086-8

Bryant, M. (1979). Microbial Methane Production—Theoretical Aspects 1, 2. *Journal of Animal Science, 48*(1), 193–201. doi:10.2527/jas1979.481193x

Burkhardt, M., Koschack, T., & Busch, G. (2015). Biocatalytic methanation of hydrogen and carbon dioxide in an anaerobic three-phase system. *Bioresource Technology, 178*, 330–333. doi:10.1016/j.biortech.2014.08.023 PMID:25193088

Butella, C., Diaz, A., Ory, I. W., Webb, C., & Blandino, A. (2007). Xylanase and pectinase production by Aspergillus awamori on grape pomace in solid state fermentation. *Process Biochem., 42*, 98-101.

Campbell-Platt, G. (1987). *Fermented foods of the world. A dictionary and guide*. Butterworths.

Canakci, M. (2007). The potential of restaurant waste lipids as biodiesel feedstocks. *Bioresource Technology, 98*(1), 183–190. doi:10.1016/j.biortech.2005.11.022 PMID:16412631

Canteri, M. G., Scheer, A., Wosiacki, G., Ginies, C., Reich, M., & Renard, C. G. (2010). A comparative study of pectin extracted from passion fruit rind flours. *Journal of Polymers and the Environment, 18*(4), 593–599. doi:10.100710924-010-0206-z

Caplice, E., & Fitzgerald, G. F. (1999). Food fermentations: Role of microorganisms in food production and preservation. *International Journal of Food Microbiology, 50*(1-2), 131–149. doi:10.1016/S0168-1605(99)00082-3 PMID:10488849

Capozzi, F., & Bordoni, A. (2013). Foodomics: A new comprehensive approach to food and nutrition. *Genes & Nutrition, 8*(1), 1–4. doi:10.100712263-012-0310-x

Carucci, A., Dionisi, D., Majone, M., Rolle, E., & Smurra, P. (2001). Aerobic storage by activated sludge on real wastewater. *Water Research, 35*(16), 3833–3844. doi:10.1016/S0043-1354(01)00108-7 PMID:12230166

Cerda, A., Adriana, A., Xavier, F., Raquel, B., Teresa, G., Antoni, S., & Alejandra, C. (2017). Composting of Food Wastes : Status and Challenges. *Bioresource Technology, 248*(Part A), 57-67.

Chakona, G., & Shackleton, C. M. (2017). Local setting influences the quantity of household food waste in mid-sized South African towns. *PLoS One, 12*(12), e0189407. doi:10.1371/journal.pone.0189407

Chalak, A., Abou-Daher, C., Chaaban, J., & Abiad, M. G. (2016). The global economic and regulatory determinants of household food waste generation: A cross-country analysis. *Waste Management (New York, N.Y.), 48*, 418–422. doi:10.1016/j.wasman.2015.11.040 PMID:26680687

Chandrasekaran. (2012). *Valorization of food processing by-products*. CRC Press.

Chavan, J. K., Kadam, S. S., & Beuchat, L. R. (1989). Nutritional improvement of cereal fermentation. *CRC Critical Reviews in Food Science and Nutrition*, *28*(5), 349–400. doi:10.1080/10408398909527507 PMID:2692608

Chaves-López, C., Serio, A., Paparella, A., Martuscelli, M., Corsetti, A., Tofalo, R., & Suzzi, G. (2014). Impact of microbial cultures on proteolysis and release of bioactive peptides in fermented milk. *Food Microbiology*, *42*, 117–121. doi:10.1016/j.fm.2014.03.005 PMID:24929726

Chen, K., & Pachter, L. (2005). Bioinformatics for Whole-Genome Shotgun Sequencing of Microbial Communities. *PLoS Computational Biology*, *1*(2), e24. doi:10.1371/journal.pcbi.0010024

Chiari, B. G., Trovatti, E., Pecoraro, E., Corrêa, M. A., Cicarelli, R. M. B., Ribeiro, S. J. L., & Isaac, V. L. B. (2014). Synergistic effect of green coffee oil and synthetic sunscreen for health care application. *Industrial Crops and Products*, *52*, 389–393.

Chilakala, R., & Ahn, A. (2017). Environmental Effect of the Coffee Waste and Anti-Microbial Property of Oyster Shell Waste Treatment. *Journal of Energy Engineering*, *26*(2), 39–49.

Chompreeda, P. T., & Fields, M. L. (1984). Effects of heat and fermentation on the extractability of minerals from soybean meal and corn meal blends. *Journal of Food Science*, *49*(2), 566–568. doi:10.1111/j.1365-2621.1984.tb12469.x

Chonhenchob, V., & Singh, S. P. (2003). A comparison of corrugated boxes and reusable plastic containers for mango distribution. *Packaging Technology & Science*, *16*(6), 231–237. doi:10.1002/pts.630

Chonhenchob, V., & Singh, S. P. (2005). Packaging performance comparison for distribution and export of papaya fruit. *Packaging Technology & Science*, *18*(3), 125–131. doi:10.1002/pts.681

Chou, L. (2003, Fall). Chinese and Other Asian Pickles. *Flavor and Fortune*.

Chou, Y. (2018). *Octalysis: Complete Gamification Framework - Yu-kai Chou*. Available at: https://yukaichou.com/gamification-examples/octalysis-complete-gamification-framework/

Choudhury, D., Sahu, J. K., & Sharma, G. D. (2011). Bamboo shoot based fermented food products: A review. *Journal of Scientific and Industrial Research (New Delhi, India)*, *70*, 199–203.

Ćilerdžić, Stajić, Vukojević, Duletić-Laušević, & Knežević. (2011). Potential of *Tramete Hirsuta* to Produce Ligninolytic Enzymes during Degradation Of Agricultural Residues. *BioResources*, *6*(3), 2885–2895.

Clarke, W. P., & Alibardi, L. (2010). Anaerobic digestion for the treatment of solid organic waste: What's hot and what's not. *Waste Management (New York, N.Y.)*, *30*(10), 1761–1762. doi:10.1016/j.wasman.2010.06.019 PMID:20638829

Cleveland, J., Montville, T. J., Nes, I. F., & Chikindas, M. L. (2001). Bacteriocins: Safe, natural antimicrobials for food preservation. *International Journal of Food Microbiology*, *71*(1), 1–20. doi:10.1016/S0168-1605(01)00560-8 PMID:11764886

Coker, C. (2006). Environmental remediation by composting. *BioCycle, 47*(12), 18–23.

Cournoyer, M. (1996). Sanitation and stabilization of slaughterhouse sludges through composting. *Proceedings of the Canadian meat research institute technology symposium.*

Couto, S. R., & Sanromán, M. A. (2006). Application of solid-state fermentation to food industry—a review. *Journal of Food Engineering, 76*(3), 291–302. doi:10.1016/j.jfoodeng.2005.05.022

Cruz, A. G., Ana, A. S. J., Marchione, M. M., Teixeira, A. M., & Schmaltz, F. I. (2009). Milk drink using whey butter, cheese and acerola juice as a potential source of Vit C. *Food and Bioprocess Technology, 28*, 368–373. doi:10.100711947-008-0059-9

da Silva, M. C. S., Naozuka, J., da Luz, J. M. R., de Assunção, L. S., Oliveira, P. V., & Vanetti, M. C. D. (2012). Enrichment of *Pleurotus ostreatus* mushrooms with selenium in coffee husks. *Food Chemistry, 131*(2), 558–563.

Dahiya, N., Tewari, R., & Hoondal, G. S. (2006). Biotechnological aspects of chitinolytic enzymes: A review. *Applied Microbiology and Biotechnology, 71*(6), 773–782. doi:10.100700253-005-0183-7 PMID:16249876

Dahlén, L., & Lagerkvist, A. (2010). Evaluation of recycling programmes in household waste collection systems. *Waste Management & Research, 28*(7), 577–586. doi:10.1177/0734242X09341193 PMID:19748961

Dash, A., Sankaran, K., Shrimali, P., Iyer, P., Upadhyay, S., & Javadekar, S. (2015, November). Food for thought. *Insight: IIT Bombay's Student Media Body*, 6-7.

Daubresse, P., Ntibashirwa, S., Gheysen, A., & Meyer, J. A. (1987). A process for protein enrichment of cassava by solid substrate fermentation in rural conditions. *Biotechnology and Bioengineering, 29*(8), 962–968. doi:10.1002/bit.260290807 PMID:18576545

De Gioannis, G., Muntoni, A., Polettini, A., & Pomi, R. (2013). A review of dark fermentative hydrogen production from biodegradable municipal waste fractions. *Waste Management (New York, N.Y.), 33*(6), 1345–1361. doi:10.1016/j.wasman.2013.02.019 PMID:23558084

De Las Fuentes, L. B., Sanders, B., Lorenzo, A., & Aber, S. (2004). Awareness Agrofood wastes minimization and reduction network. Total Food Proceedings, 233 – 244.

De Simone, C. (1986). Microflora yogurt and the immune system. *International Journal of Immunotherapy*, 19–23.

Delgado-Andrade, C., Rufia´n-Henares, J.A., & Morales, F.J. (2005). Assessing the antioxidant activity of melanoidins from coffee brews by different antioxidant methods. *Journal of Agricultural and Food Chemistry, 53*, 7832–7836.

Dewan, S., & Tamang, J. P. (2007b). Microbial and analytical characterization of Chhu-A traditional fermented milk product of the Sikkim Himalayas. *Journal of Scientific and Industrial Research (New Delhi, India), 65*, 747–752.

DEWHA, National Waste Policy. (2009). Less waste more resources. Department of Environment, Heritage and the Arts, Editor, Common Wealth of Australia.

DeWitt, C. A. M., & Morrissey, M. T. (2002). Pilot plant recovery of catheptic proteases from surimi wash water. *Bioresource Technology*, *82*(3), 295–301. doi:10.1016/S0960-8524(01)00178-X PMID:11991080

Dhanapal, A.P., & Govindaraj, M. (2015). Unlimited Thirst for Genome Sequencing, Data Interpretation, and Database Usage in Genomic Era: The Road towards Fast-Track Crop Plant Improvement. *Genetics Research International*. doi:10.1155/2015/684321

Dirar, H. A. (1993). *The indigenous fermented foods of the Sudan: a study in African food and nutrition*. CAB international.

Djilas, S., Canadanovic-Brunet, J., & Cetkovic, G. (2009). By products of fruit processing as source of phytochemicals. *Chem Ind.Chem. Eng.*, *15*(4), 191–202. doi:10.2298/CICEQ0904191D

Dobbs, R., Oppenheim, J., Thompson, F., Brinkman, M., & Zornes, M. (2011). *Resource Revolution: Meeting the world's energy, materials, food, and water needs*. Academic Press.

Dorman, H. J. D., & Deans, S. G. (2008). Antimicrobial agents from plants: Antibacterial activity of plant volatile oils. *Journal of Applied Microbiology*, *88*(2), 308–316. doi:10.1046/j.1365-2672.2000.00969.x

Dos Santos, B. S., da Silva, L. C. N., da Silva, T. D., Rodrigues, J. F. S., Grisotto, M. A. G., Correia, M. T. dos S., ... Paiva, P. M. G. (2016). Application of Omics Technologies for Evaluation of Antibacterial Mechanisms of Action of Plant-Derived Products. *Frontiers in Microbiology*, *7*, 1466. doi:10.3389/fmicb.2016.01466

Dung, T. N. B., Sen, B., Chen, C. C., Kumar, G., & Lin, C. Y. (2014). Food waste to bioenergy via anaerobic processes. *Energy Procedia*, *61*, 307–312. doi:10.1016/j.egypro.2014.11.1113

Duszkiewicz-Reinhard, W., Gujska, E., & Khan, K. (1994). Reduction of stachyose in legume flours by lactic acid bacteria. *Journal of Food Science*, *59*(1), 115–117. doi:10.1111/j.1365-2621.1994.tb06911.x

Duval, A., & Lawoko, M. (2014). A review on lignin-based polymeric, micro- and nano-structured materials. *Reactive & Functional Polymers*, *85*(0), 78–96. doi:10.1016/j.reactfunctpolym.2014.09.017

Ebringer, L., Ferenčík, M., & Krajčovič, J. (2008). Beneficial health effects of milk and fermented dairy products--review. *Folia Microbiologica*, *53*(5), 378–394. doi:10.100712223-008-0059-1 PMID:19085072

Edjabou, V. M. E., Pivnenko, K., Petersen, C., Scheutz, C., & Astrup, T. F. (2015). *Compositional data analysis of household food waste in Denmark*. Abstract from 6th International Workshop on Compositional Data Analysis, Spain. Retrieved from http://orbit.dtu.dk/files/113896186/Vincent_Edajbou.pdf

Edjabou, M. E., Petersen, C., Scheutz, C., & Astrup, T. F. (2016). Food waste from Danish households: Generation and composition. *Waste Management (New York, N.Y.)*, *52*, 256–268. doi:10.1016/j.wasman.2016.03.032 PMID:27026492

Efraim, P., Pezoa-García, N. H., Jardim, D. C. P., Nishikawa, A., Haddad, R., & Eberlin, M. N. (2010). Influência da fermentação e secagem damêndoas de cacau no teor de compostos fenólicos e na aceitação sensorial. *Food Science and Technology (Campinas)*, *30*(1), 142–150.

Eka, O. U. (1980). Effect of fermentation on the nutrient status of locust beans. *Food Chemistry*, *5*(4), 303–308. doi:10.1016/0308-8146(80)90051-5

Eller, F. J., Mosser, J. K., Kenar, J. A., & Taylor, S. L. (2010). Extraction and analysis of tomato seed oil. *Journal of the American Oil Chemists' Society*, *87*(7), 755–762. doi:10.100711746-010-1563-4

Enden, J.C., & Calvert, K.C. (2002). *Limit environmental damage by basic knowledge of coffee wastewaters*. GTZ-PPP Project-improvement of coffee quality and sustainability of coffee production in Vietnam.

Ercolini, D. (2013). High-throughput sequencing and metagenomics: Moving forward in the culture-independent analysis of food microbial ecology. *Applied and Environmental Microbiology*, *79*(10), 3148–3155. doi:10.1128/AEM.00256-13

Ermgassen, E., Phalan, B., Green, R., & Balmford, A. (2016). Reducing the land use of EU pork production: Where there's will, there's a way. *Food Policy*, *58*, 35–48. doi:10.1016/j.foodpol.2015.11.001 PMID:26949285

Esakkiraj, P., Austin Jeba Dhas, G., Palavesam, A., & Immanuel, G. (2010a). Media preparation using tuna-processing wastes for improved lipase production by shrimp gut isolate *Staphylococcus epidermidis* CMST Pi2. *Applied Biochemistry and Biotechnology*, *160*(4), 1254–1265. doi:10.100712010-009-8632-x PMID:19430738

Esakkiraj, P., Rajkumarbharathi, M., Palavesam, A., & Immanuel, G. (2010b). Lipase production by *Staphylococcus epidermidis* CMST-Pi 1 isolated from the gut of shrimp *Penaeusindicus*. *Annals of Microbiology*, *60*(1), 37–42. doi:10.100713213-009-0003-x

Escobar-Zepeda, A., Vera-Ponce de Leon, A., & Sanchez-Flores, A. (2015). The road to metagenomics: From microbiology to DNA sequencing technologies and bioinformatics. *Frontiers in Genetics*, *6*, 348. doi:10.3389/fgene.2015.00348

Essien, J. P., Akpan, E. J., & Essien, E. P. (2005). Studies on mould growth and biomass production using waste banana peel. *Bioresource Technology*, *96*(13), 1451–1456. doi:10.1016/j.biortech.2004.12.004 PMID:15939272

Exodus Market Research. (2006). *A quantitative assessment of the nature, scale and origin of post consumer food waste arising in Great Britain 2006.* Author. (unpublished)

Fankhauser, D. B. (2007). Fankhauser's Cheese Page. *Taken To, 9*(23), 2007.

Fan, L., Pandey, A., & Soccol, C. R. (2001). *Flammulina velutipes* on coffee husk and coffee spentground. *Brazilian Archives of Biology and Technology*, *44*, 205–212.

FAO. (1981). Food loss prevention in perishable crops. FAO Agricultural Services Bulletin, 43, 72.

FAO. (1998). *Fermented Fruits and Vegetables-A Global Perspective* (Vol. 134). Rome, Italy: FAO Agricultural Services Bulletin.

FAO. (1999). Fermented Cereals—A Global Perspective. FAO Agricultural Services Bulletin, 138.

FAO. (2001). *The state of Food Insecurity in the World 2001*. FAO.

FAO. (2002). *Guidelines for evaluation of probiotics in food*. Report of a Joint FAO/WHO Working Group on Drafting Guidelines for the Evaluation of Probiotics in Food.

FAO. (2011). Global food losses and food waste – Extent, causes and prevention. FAO.

FAO. (2011). *Global food losses and food waste - extent, causes and prevention*. FAO. Retrieved from http://www.fao.org/docrep/014/mb060e/mb060e.pdf

FAO. (2011). Global food losses and food waste – Extent, causes and prevention. Rome: Food &Agriculture Organization of the United Nations.

FAO. (2011). *Global food losses and waste, extent, causes and prevention*. Rome: FAO.

FAO. (2013). Food and Agriculture Organization of the United Nations. Food wastage footprint – Impacts on natural resources, Summary Report. FAO.

FAO. (2013). *Food wastage footprint: impacts on natural resources. Rome*. Retrieved from http://www.fao.org/docrep/018/i3347e/i3347e.pdf

FAO. (2014). *Definitional framework of food loss. Save Food: Global Initiative on Food Loss and Waste Reduction*. Food and Agriculture Organization of the United Nations. Retrieved from http://www.fao.org/fileadmin/user_upload/save-food/PDF/FLW_Definition_and_Scope_2014.pdf

FAO. (2015). *The state of food insecurity in the world*. Rome: FAO.

FAO. (2017). *The future of food and agriculture - trends and challenges*. FAO. Retrieved from http://www.fao.org/3/a-i6583e.pdf

FAO. (2018). *Sustainable pathways-foot wastage footprint*. Retrieved from http://www.fao.org/nr/sustainability/food-loss-and-waste/en/

FAO/WHO Expert Consultation Report. (2001). *Evaluation of health and nutritional properties of powder milk and live lactic acid bacteria*. Author.

Farah, A., & Donangelo, C. M. (2006). Phenolic compounds in coffee. *Brazilian Journal of Plant Physiology*, *18*, 23–36.

Favaro, L., Alibardi, L., Lavagnolo, M. C., Casella, S., & Basaglia, M. (2013). Effects of inoculum and indigenous microflora on hydrogen production from the organic fraction of municipal solid waste. *International Journal of Hydrogen Energy*, *38*(27), 11774–11779. doi:10.1016/j.ijhydene.2013.06.137

Fernandes, A. S. (2017). Impacts of discarded coffee waste on human and environmental health. *Ecotoxicology and Environmental Safety*, *141*, 30–36.

Fia, F.R.L., Matos, A.T., Borges, A.C., Moreira, D.A., Fia, R., & Eustaquio, V.Junior. (2010). "Remoção, "decompostos fenólicos em reatores anaeróbios de leito fixo com diferentes materiais suporte. *Revista Brasileira de Engenharia Agrícola e Ambiental*, *14*(10), 1079–1086.

Fierer, N., Leff, J. W., Adams, B. J., Nielsen, U. N., Bates, S. T., Lauber, C. L., ... Caporaso, J. G. (2012). Cross-biome metagenomic analyses of soil microbial communities and their functional attributes. *Proceedings of the National Academy of Sciences of the United States of America*, *109*(52), 21390–21395. doi:10.1073/pnas.1215210110

Fijnheer, J. D., & Van Oostendorp, H. (2016). Steps to design a household energy game. *International Journal of Serious Games*, *3*(3). doi:10.17083/ijsg.v3i3.131

Flores, R. A., Shanklin, C. W., Loza-Garay, M., & Wie, S. H. (1999). Quantification and characterization of food processing wastes/ residues. *Compost Science & Utilization*, *7*(1), 63–71. doi:10.1080/1065657X.1999.10701954

FLWP. (2016). *Food loss and waste accounting and reporting standard*. Retrieved from http://flwprotocol.org/

Fogarty, W. M., & Kelly, C. T. (1983). Pectic enzymes. In W. M. Fogarty (Ed.), *Microbial Enzymes and Biotechnology*. London: Applied Science Publisher.

Fogarty, W. M., & Kelly, C. T. (1990). Recent Advances in Microbial Amylases. In W. M. Fogarty & C. T. Kelly (Eds.), *Microbial Enzymes and Biotechnology*. Dordrecht: Springer. doi:10.1007/978-94-009-0765-2_3

Food and Agriculture Organisation of United Nations (FAO). (2018). *Food security & nutrition around the world.* Retrieved from http://www.fao.org/3/CA1354EN/ca1354en.pdf

Food Standards Agency. (2006). *Consumer Attitudes to Food Safety 2005*. Author.

Franca, A. S., Oliveira, L. S., & Ferreira, M. E. (2009). Kinetics and equilibrium studies of methylene blue adsorption by spent coffee grounds. *Desalination*, *249*, 267–272.

Franzosa, E. A., Hsu, T., Sirota-Madi, A., Shafquat, A., Abu-Ali, G., Morgan, X. C., & Huttenhower, C. (2015). Sequencing and beyond: Integrating molecular 'omics' for microbial community profiling. *Nature Reviews. Microbiology*, *13*(6), 360–372. doi:10.1038/nrmicro3451

Fratamico, P. M. (2008). The application of "Omics" technology for food safety research. *Foodborne Pathogens and Disease*, *5*(4), 369–370. doi:10.1089/fpd.2008.9994

Freire, E. S., Mororó, R. C., & Schwan, R. F. (1996). The cacao-pulp agroindustry and the uses of its residues in Bahia: Progress achieved in the last ten years. *Proc. 12th International Cocoa Research Conference*.

Frias-Lopez, J., Shi, Y., Tyson, G. W., Coleman, M. L., Schuster, S. C., Chiholm, S. W., & Delong, D. F. (2008). Microbial community gene expression in ocean surface waters. *Proceedings of the National Academy of Sciences of the United States of America*, *105*(10), 3805–3810. doi:10.1073/pnas.0708897105

Friedman, M. (1996). Nutritional value of proteins from different food sources. A review. *Journal of Agricultural and Food Chemistry*, *44*(1), 6–29. doi:10.1021/jf9400167

Friends of the Earth. (2006). *Briefing: food waste collections*. Retrieved from http://www.foe.co.uk/resource/briefings/food_waste.pdf

Fuessl, A., Yamamoto, M., & Schneller, A. (2012). *Opportunities in bi-based building blocks for polycondensates and vinyl polymers. In Polymer Science: A Comprehensive Reference* (Vol. 5, pp. 49–70). Elsevier.

FUSIONS. (2016). *EU FUSIONS website*. Retrieved from www.eu-fusions.org

FUSIONS. (2016). Recommendations and guidelines for a common European food waste policy framework. FUSIONS EU FP7 Research Project.

Galanakis, C. M. (2012). Recovery of high added-value components from food wastes: Conventional, emerging technologies and commercialized applications. *Trends in Food Science & Technology*, *26*, 68–87.

Ganders, D. (2012). *Wasted: How America is losing up to 40 percent of its food from farm to fork to landfill*. New York: Natural Resources Defense Council.

Gandhi, D. N., Marwaha, S. S., & Arora, J. K. (Eds.). (2000). *Food Processing: Biotechnological Applications*. New Delhi: Asiatech Publishers Inc.

Gan, S., Lau, E., & Ng, H. (2009). Remediation of soils contaminated with polycyclic aromatic hydrocarbons (PAHs). *Journal of Hazardous Materials*, *172*(2-3), 532–549. doi:10.1016/j.jhazmat.2009.07.118 PMID:19700241

Gaoa, A. (2017). Comparison between the technologies for food waste treatment. *Energy Procedia*, *105*, 3915–3921. doi:10.1016/j.egypro.2017.03.811

García-Canas, V., Simó, C., Herrero, M., Ibáñez, E., & Cifuentes, A. (2012). Present and future challenges in food analysis: Foodomics. *Analytical Chemistry*, *84*(23), 10150–10159. doi:10.1021/ac301680q

Garcia-Garcia, Woolley, & Rahimifard, Colwill, White, & Needham. (2017). A Methodology for Sustainable Management of Food Waste. *Waste and Biomass Valorization*, *8*, 2209–2227.

Garnett, T. (2014). *What is a sustainable healthy diet? A discussion paper*. Food Climate Research Network. Retrieved from https://www.fcrn.org.uk/sites/default/files/fcrn_what_is_a_sustainable_healthy_diet_final.pdf

Gasparatos, A., Stromberg, P., & Takeuchi, K. (2011). Biofuels, ecosystem services and human wellbeing: Putting biofuels in the ecosystem services narrative. *Agriculture, Ecosystems & Environment*, *142*(3-4), 111–128. doi:10.1016/j.agee.2011.04.020

Gerardi, M. H. (2003). *The microbiology of anaerobic digesters*. John Wiley & Sons. doi:10.1002/0471468967

Ghanavati, H., Nahvi, I., & Karimi, K. (2015). Organic fraction of municipal solid waste as a suitable feedstock for the production of lipid by oleaginous yeast Cryptococcus aerius. *Waste Management (New York, N.Y.)*, *38*, 141–148. doi:10.1016/j.wasman.2014.12.007 PMID:25595390

Gibson, G. R., Probert, H. M., Van Loo, J. A. E., & Roberfroid, M. B. (2004). Dietary modulation of the human colonic microbiota: Updating the concept of prebiotics. *Nutrition Research Reviews*, *17*(2), 257–259.

Gigras, P., Vikram, S., & Rani, G. (2002). Statistical Media Optimization and Production of ITS-Amylase from *Aspergillus oryzae* in a Bioreactor. *Current Microbiology*, *45*(3), 203–208. doi:10.100700284-001-0107-4 PMID:12177743

Girotto, F., Luca, A., & Raffaello, C. (2015). Food Waste Generation and Industrial Uses : A Review. *Waste Management (New York, N.Y.)*, *45*, 32–41. doi:10.1016/j.wasman.2015.06.008 PMID:26130171

Giusquiani, P., Pagliai, M., Gigliotti, G., Businelli, D., & Benetti, A. (1995). Urban waste compost: Effects on physical, chemical, and biochemical soil properties. *Journal of Environmental Quality*, *24*(1), 175–182. doi:10.2134/jeq1995.00472425002400010024x

Goldin, B. R. (1983). The effect of oral administration of Lactobacillus and antibiotics on intestinal bacterial activity and chemical induction of large bowel tumors. *Developments in Industrial Microbiology*, *25*, 139–150.

Gopalan, C., Rama Sastri, B. V., & Balasubramanian, S. C. (2014). *Nutritive value of Indian Foods* (p. 161). Hyderabad: NIN.

Gormley, R. (2013). *Fish as a functional food: some issues and outcomes, Sea Health*. UCD.

Gouda, M. K., Swellam, A. E., & Omar, S. H. (2001). Production of PHB by a *Bacillus megaterium* strain using sugarcane molasses and corn steep liquor as sole carbon and nitrogen source. *Microbiological Research Journal*, *156*(3), 201–204. doi:10.1078/0944-5013-00104 PMID:11716209

Gouvea, B. M., Torres, C., Franca, A. S., Oliveira, L. S., & Oliveira, E. S. (2009). Feasibility of ethanol production from coffee husks. *Biotechnology Letters*, *31*, 1315–1319.

Graf, W. (1983). Studies on the therapeutic properties of acidophilus milk. In *Symposia of the Swedish Nutrition Foundation*. Almqvist och Wiksell International.

Graminha, Goncalves, Pirota, Balsalobre, Silva, & Gomes. (2008). Enzyme production by solid feed. *Sci. Technol, 144*, 1–22.

Graziani, V., Scognamiglio, M., Belli, V., Esposito, A., D'Abrosca, B., Chambery, A., ... Fiorentino, A. (2018). Metabolomic approach for a rapid identification of natural products with cytotoxic activity against human colorectal cancer cells. *Scientific Reports, 5309*. doi:10.103841598-018-23704-9

Griffin, M. E., McMahon, K. D., Mackie, R. I., & Raskin, L. (1998). Methanogenic population dynamics during start-up of anaerobic digesters treating municipal solid waste and biosolids. *Biotechnology and Bioengineering*, *57*(3), 342–355. doi:10.1002/(SICI)1097-0290(19980205)57:3<342::AID-BIT11>3.0.CO;2-I PMID:10099211

Grobe, K. (1994). Composter links up with food processor. *BioCycle*, *34*(40), 42–43.

Grunewald, K. K. (1992). Serum Cholesterol levels in rats fed skim milk fermented by *Lactobacillus acidophilus*. *Journal of Food Science*, *47*(6), 2078–2079. doi:10.1111/j.1365-2621.1982.tb12955.x

Guermani, L., Villaume, C., Bau, H. W., Chandrasiri, V., Nicolas, J. P., & Mejan, L. (1992). *Composition and nutritional value of okara fermented by Rhizopus oligosporus*. Sciences des Aliments.

Guil-Guerrero, J. L., Ramos, L., Moreno, C., Zúñiga-Paredes, J. C., Carlosama-Yepez, M., & Ruales, P. (2016). Plant Foods By-Products as Sources of Health-Promoting Agents for Animal Production: A Review Focusing on the Tropics. *Agronomy Journal*, *108*(5), 1759–1774. doi:10.2134/agronj2015.0555

Gul, K., Yousuf, B., Singh, A. K., Sing, P., & Wane, A. (2015). Rice Bran, Nutritional value and its emerging potential for development of functional food: A review. *J of Bioactive Carbohydrates and Dietary Fibers*, *6*(1), 24–30. doi:10.1016/j.bcdf.2015.06.002

Gupta, U., Rudramma, Rati, E. R., & Joseph, R. (1998). Nutritional quality of lactic fermented bitter gourd and fenugreek leaves. *International Journal of Food Sciences and Nutrition*, *49*(2), 101–108. doi:10.3109/09637489809089389 PMID:9713580

Guru, M., Bilgesu, A. Y., & Pamuk, V. (2001). Production of oxalic acid from sugar beet molasses by formed nitrogen oxides. *Bioresource Technology*, *77*(1), 81–86. doi:10.1016/S0960-8524(00)00122-X PMID:11211079

Gustafsson, B. E. (1983). Introduction to the ecology of the intestinal microflora and its general characteristics. *Nutrition and the Intestinal Flora*, 11-16.

Gustavsson, J., Cederberg, C., Sonesson, U., van Otterdijk, R., & Meybeck, A. (2011). Global food losses and food waste. Extent, causes and prevention. Rome: Academic Press.

Gustavsson, J., Cederberge, Sonesson, U., Van O Hedrick, R., & Me beck, A. (2011). *Global Food issues and food waste: Extent causes and prevention*. Rome: FAO, UNO.

Gustavsson, J., Cederberg, C., Sonesson, U., Van Otterdijk, R., & Meybeck, A. (2011). *Global food losses and food waste: extent, causes and prevention*. Rome: FAO.

Haberer, P., Holzapfel, W. H., & Wagner, H. (1997). Moegliche Rolle von Milchsaeurebakterien bei der Cholesterinsenkung im Blutserum. *Mitteilungsblatt-Bundesanstalt Fur Fleischforschung Kulmbach, 1*(136), 202–207.

Hachemi, N., Nouani, A., & Benchabane, A. (2015). Bioconversion of Oranges Wastes for Pectinase Production Using *Aspergillus niger* under Solid State Fermentation. *International Journal of Bioengineering and Life Sciences, 9*(9), 1004–1009.

Hafuka, A., Sakaida, K., Satoh, H., Takahashi, M., Watanabe, Y., & Okabe, S. (2011). Effect of feeding regimens on polyhydroxybutyrate production from food wastes by *Cupriavidus necator. Bioresource Technology, 102*(3), 3551–3553. doi:10.1016/j.biortech.2010.09.018 PMID:20870404

Haile, M. (2014). Integrated volarization of spent coffee grounds to biofuels. *Biofuel Research Journal, 2*, 65–69.

Hall, K. D., Quo, J., Dore, M., & Chow, C. (2009). The progressive increase of food waste in America and its environmental impact. National Institute of Diabetes and Digestive and Kidney Diseases.

Handelsman, J., Rondon, M. R., Brady, S. F., Clardy, J., & Goodman, R. M. (1998). Molecular biological access to the chemistry of unknown soil microbes: A new frontier for natural products. *Chemistry & Biology, 5*(10), R245–R249. doi:10.1016/S1074-5521(98)90108-9

Harland, B. F., & Harland, J. (1980). Fermentative reduction of phytate in rye, white and whole wheat breads. *Cereal Chemistry, 57*(3), 226–229.

Harlander, S. (1992). Food biotechnology. In J. Lederberg (Ed.), *Encyclopedia of microbiology* (pp. 191–207). New York: Academic Press.

Harrison, V. C., & Peat, G. (1975). Serum cholesterol and bowel flora in the newborn. *The American Journal of Clinical Nutrition, 28*(12), 1351–1355. doi:10.1093/ajcn/28.12.1351 PMID:45573

Harun-ur-Rashid, M., Togo, K., Useda, M., & Miyamoto, T. (2007). Probiotic characteristics of lactic acid bacteria isolated from traditional fermented milk 'Dahi' in Bangladesh. *Pakistan Journal of Nutrition, 6*(6), 647–652. doi:10.3923/pjn.2007.647.652

Hasin, Y., Seldin, M., & Lusis, A. (2017). Multi-omics approaches to disease. *Genome Biology, 18*(1), 83. doi:10.118613059-017-1215-1

Hawkins, K. (2007). *Allotment Cookbook*. New Holland Publishers.

Hebeish, & Ibrahim, N.A. (2007). The impact of frontier sciences on textile industry. *Colourage, 54*, 41–55.

Helkar, P. B., Sahoo, A. K., & Patil, N. J. (2016). Review: Food industry by-products used as a functional food ingredients. *J. of Waste Resource, 6*(3), 1-6.

Heller, K. J. (2001). Probiotic bacteria in fermented foods: Product characteristics and starter organisms. *The American Journal of Clinical Nutrition, 73*(2), 374s–379s. doi:10.1093/ajcn/73.2.374s PMID:11157344

Hepner, G., Fried, R., St Jeor, S., Fusetti, L., & Morin, R. (1979). Hypocholesterolemic effect of yogurt and milk. *The American Journal of Clinical Nutrition, 32*(1), 19–24. doi:10.1093/ajcn/32.1.19 PMID:581636

Hesseltine, C. W., & Wang, H. L. (1980). The importance of traditional fermented foods. *Bioscience, 30*(6), 402–404. doi:10.2307/1308003

Hettich, R. L., Sharma, R., Chourey, K., & Giannone, R. J. (2012). Microbial metaproteomics: Identifying the repertoire of proteins that microorganisms use to compete and cooperate in complex environmental communities. *Current Opinion in Biotechnology, 15*, 373–380.

Hirahara, T. (1998). Functional food science in Japan. In T. Mattila-Sandholm & K. Kauppila (Eds.), Functional food research in Europe (pp. 19–20). Julkaisija-Utgivare.

HLPE. (2014). *Food losses and waste in the context of sustainable food systems.* A report by the High-Level Panel of Experts on Food Security and Nutrition of the Committee on World Food Security. Retrieved from: http://www.fao.org/fileadmin/user_upload/hlpe/hlpe_documents/HLPE_Reports/HLPE-Report-8_EN.pdf

Hmidet, N., Ali, N. E. H., Haddar, A., Kanoun, S., Alya, S. K., & Nasri, M. (2009). Alkaline proteases and thermostable α-amylase co-produced by *Bacillus licheniformis* NH1: Characterization and potential application as detergent additive. *Biochemical Engineering Journal, 47*(1-3), 71–79. doi:10.1016/j.bej.2009.07.005

Hogg, D., Barth, J., Schleiss, K., & Favoino, E. (2007). *Dealing with food waste in the UK.* Eunomia Research and Consulting Limited. WRAP. Retrieved from http://www.wrap.org.uk/sites/files/wrap/Dealing with Food Waste - Final 2 March_07.pdf

Homayouni, A., Alizadeh, M., Alikhan, H., & Zijah, V. (2012). *Functional dairy probiotic development trends, concepts and products.* Intech. doi:10.5772/48797

Hoornweg, D., Bhada-Ta-ta, P., & Kennedy, C. (2013). Environment: waste production must peak this century. *Nature, 502*, 615 – 617.

Horton, B. S. (1995). Whey processing and utilization. *IDF Bulletin, 308*, 2–6.

Hosono, K. T., Kashina, T., & Kada, T. (1986). Antimutagenic properties of lactic acid-cultured milk on chemical and fecal mutagens. *Journal of Dairy Science, 69*(9), 2237–2242. doi:10.3168/jds.S0022-0302(86)80662-2 PMID:3097092

Hossain, M. F., Akhtar, S., & Anwar, M. (2015). Nutritional value and medicinal benefits of pineapple. *Int. J. Nutri. Food Sci., 4*(1), 84–88. doi:10.11648/j.ijnfs.20150401.22

Hrdlickova, R., Toloue, M., & Tian, B. (2017). RNA-Seq methods for transcriptome analysis. *Wiley Interdisciplinary Reviews. RNA, 8*(1), e1364. doi:10.1002/wrna.1364

Hudler, P., Kocevar, N., & Komel, R. (2014). Proteomic Approaches in Biomarker Discovery: New Perspectives in Cancer Diagnostics. *The Scientific World Journal*.. doi:10.1155/2014/260348

Hussein, A. M. S., Kamil, M. M., Hegazy, N. A., Mahmoud, K. F., & Ibrahim, M. A. (2015). Utilization of some fruits and vegetables by-products to produce high dietary fibre jam. *Food Sci. Quality and Mgmt*, *37*, 39–45.

Hutkins, R. W. (2008). *Microbiology and technology of fermented foods* (Vol. 22). John Wiley & Sons.

Ibrahim, M. F., Abd-Azizi, S., Razak, M. A., Phang, L. Y., & Hassan, M. A. (2012). Oil palm empty fruit bunch as alternative substrate for acetone-butanol-ethanol production by *Clostridium butyricum* EB6. *Applied Biochemistry and Biotechnology*, *166*(7), 1615–1625. doi:10.100712010-012-9538-6 PMID:22391689

Imran, M., Anwar, Z., Irshad, M., Asad, M. J., & Ashfaq, H. (2016). Cellulase, production from species of fungi and bacteria from agricultural wastes and its utilization in industry: A review. *Advances in Enzyme Research*, *4*(02), 44–55. doi:10.4236/aer.2016.42005

Institution of Mechanical Engineers (IME). (2013). *Global food waste not want not*. Author.

Ismali,, A. E., & Abdellader,, M. O., & Ali, A. (2011). Microbial and chemical evolution of whey based mango beverage, Advances. *Journal of Food Science and Technology*, *38*, 250–253.

Iyer, B. K., Singhal, R. S., & Ananthanarayan, L. (2013). Characterization and in-vitro probiotic evaluation of lactic acid bacteria isolated from idli batter. *Food Sci. Technol.*, *50*, 1114–1121. PMID:24426023

Jaeger, W. K. (1992). The causes of Africa's food crisis. *World Development*, *20*(11), 1631–1645. doi:10.1016/0305-750X(92)90019-R

Jagavati, S., Adivikatla, V., Paritala, N., & Linga, V. (2012). Cellulase Production by co-culture of *Trichoderma* sp. and *Aspergillus* sp. under submerged fermentation. *Dynamic Biochemistry, Process Biotechnology and Molecular Biology*, *6*(S1), 79-83.

Jahan, N., Faiza, S., Afsheen, A., Talat, Y. M., Shah, A., & Qader, U. I. (2017). Utilization of Agro Waste Pectin for the Production of Industrially Important Polygalacturonase. *Heliyon (London)*, *3*(6), e00330. doi:10.1016/j.heliyon.2017.e00330 PMID:28664192

Jasani, H., Nimita, U., Darshan, D., & Jagdish, P. (2016). Isolation, Optimization and Production of Cellulase by *Aspergillus niger* from Agricultural Waste Isolation. *Journal of Pure & Applied Microbiology*, *10*(2), 1159–1166.

Jayachandra, C., & Venugopal, K.A., & Appaiah, A. (2011). Utilization of phytotoxic agro waste – coffee cherry husk through pretreatment by the ascomycetes fungi *Mycotypha* for biomethanation. *Energy for Sustainable Development*, *15*(1), 104–108.

Jerez, C. A. (2008). The use of genomics, proteomics and other OMICS technologies for the global understanding of biomining microorganisms. *Hydrometallurgy*, *94*(1-4), 162–169. doi:10.1016/j.hydromet.2008.05.032

Jeyaram, K., Mohendro Singh, W., Premarani, T., Devi, A. R., Chanu, K. S., Talukdar, N. C., & Singh, M. R. (2008). Molecular identification of dominant microflora associated with 'Hawaijar'da traditional fermented soybean (*Glycine max* (L.)) food of Manipur. India. *International Journal of Food Microbiology*, *22*(3), 259–268. doi:10.1016/j.ijfoodmicro.2007.12.026 PMID:18281117

Jeyaram, K., Singh, A., Romi, W., Devi, A. R., Singh, W. M., & Dayanithi, H. (2009). Traditional fermented foods of Manipur. *Indian Journal of Traditional Knowledge*, *8*(1), 115–121.

Jin, F., Zhou, Z., Moriya, T., Kishida, H., Higashijima, H., & Enomoto, H. (2005). Controlling hydrothermal reaction pathways to improve acetic acid production from carbohydrate biomass. *Environmental Science & Technology*, *39*(6), 1893–1902. doi:10.1021/es048867a PMID:15819253

Johnvesly, B., Manjunath, B. R., & Naik, G. R. (2002). Pigeon Pea Waste as a Novel Inexpensive Substrate for Production of a Thermostable Alkaline Protease from Thermoalkalophilic *Bacillus* Sp. JB-99. *Bioresource Technology*, *82*(1), 61–64. doi:10.1016/S0960-8524(01)00147-X PMID:11848379

Joshi, V. K., Chauhan, S. K., & Lal, B. B. (1991). Extraction of juices from peaches, plums and apricots by pectinolytic treatment. *Journal of Food Science and Technology*, *28*, 64–65.

Josic, D., & Giacometti, J. (2013). Foodomics-Use of Integrated Omics in Nutrition, Food Technology and Biotechnology. *Journal of Data Mining in Genomics & Proteomics*, *4*(02), e106. doi:10.4172/2153-0602.1000e106

Josic, D., Sokac, D. G., Gajdosik, M. S., & Clifton, J. (2016). Microbial omics for food safety. *Journal of Hygienic Engineering and Design*, *67*, 116–129.

Juliano, B. O. (1985). *Rice Chemistry and Technology*. American Association of Cereal Chemistry.

Kadam, S., & Prabha Sankar, P. (2010). Marine foods as functional ingredients in bakery and pasta products. *Food Research International*, *3*(8), 1975–1980. doi:10.1016/j.foodres.2010.06.007

Kadir, A. A., Azhari, N. W., & Jamaludin, S. N. (2017). Study on Composition and Generation of Food Waste in Makanan Ringan Mas Industry. *MATEC Web of Conferences, 103*. DOI: 10.1051/matecconf/20171030

Kalaichelvan, P. T. (2014). Production and Optimization of Pectinase from *Bacillus* Sp. MFW7 Using Cassava Waste. *Asian Journal of Plant Science and Research*, *2*(3), 369–375.

Kandasamy, S., Govarthanan, M., & Senthilkumar, B. (2016). Optimization of Protease Production from Surface-Modified Coffee Pulp Waste and Corncobs Using *Bacillus* Sp . by SSF. *3 Biotech*, *6*(2), 1–11.

Kanwar, S. S., Gupta, M. K., Katoch, C., Kumar, R., & Kanwar, P. (2007). Traditional fermented foods of Lahaul and Spiti area of Himachal Pradesh. *Indian Journal of Traditional Knowledge*, *6*, 42–45.

Karakashev, D., Batstone, D. J., & Angelidaki, I. (2005). Influence of environmental conditions on methanogenic compositions in anaerobic biogas reactors. *Applied and Environmental Microbiology*, *71*(1), 331–338. doi:10.1128/AEM.71.1.331-338.2005 PMID:15640206

Karmakar, M., & Ray, R. R. (2011). Current trends in research and application of microbial cellulases. *Res J Microbiol*, *6*(1), 41–53. doi:10.3923/jm.2011.41.53

Kashyap, D. R., Vohra, P. K., Chopra, S., & Tewari, R. (2001). Applications of Pectinases in the Commercial Sector : A Review. *Bioresource Technology*, *77*(3), 215–227. doi:10.1016/S0960-8524(00)00118-8 PMID:11272008

Katami, T., & Gifu, P. (2004). Formation of Dioxins from Incineration of Foods Found in Domestic Garbage. *Environmental Science & Technology*, *38*(4), 1062–1065. doi:10.1021/es030606y PMID:14998019

Kaur, R., Kapoor, S., & Sharma, S. (2017). Utilization potential of fruit and vegetable pomace. *Indian Food Industry*, *36*(2), 24 – 30.

Kawaguchi, H., Vertes, A. A., Okino, S., Inui, M., & Yukawa, H. (2006). Engineering of a xylose metabolic pathway in *Corynebacterium glutamicum*. *Applied and Environmental Microbiology*, *72*(5), 3418–3428. doi:10.1128/AEM.72.5.3418-3428.2006

Keuth, S., & Bisping, B. (1994). Vitamin B12 production by Citrobacter freundii or Klebsiella pneumoniae during tempeh fermentation and proof of enterotoxin absence by PCR. *Applied and Environmental Microbiology*, *60*(5), 1495–1499. PMID:8017933

Khaliq, A., Abbasi, M. K., & Hussain, T. (2006). Effects of integrated use of organic and inorganic nutrient sources with effective microorganisms (EM) on seed cotton yield in Pakistan. *Bioresource Technology*, *97*(8), 967–972. doi:10.1016/j.biortech.2005.05.002 PMID:16023343

Khetarpaul, N., & Chauhan, B. M. (1990). Fermentation of pearl millet flour with yeasts and lactobacilli: In vitro digestibility and utilisation of fermented flour for weaning mixtures. *Plant Foods for Human Nutrition (Dordrecht, Netherlands)*, *40*(3), 167–173. doi:10.1007/BF01104139 PMID:2217082

Khetarpaul, N., & Chauhan, B. M. (1991). Biological utilisation of pearl millet fluor fermented with yeasts and lactobacilli. *Plant Foods for Human Nutrition (Dordrecht, Netherlands)*, *41*(4), 309–319. doi:10.1007/BF02310625 PMID:1796088

Khetarpaul, N., & Chauhan, B. M. (1991). Sequential fermentation of pearl millet by yeasts and lactobacilli—effect on the antinutrients andin vitro digestibility. *Plant Foods for Human Nutrition (Dordrecht, Netherlands)*, *41*(4), 321–327. doi:10.1007/BF02310626 PMID:1796089

Kim, J., Kim, H., Baek, G., & Lee, C. (2017). Anaerobic co-digestion of spent coffee grounds with different waste feedstocks for biogas production. *Waste Management (New York, N.Y.)*, *60*, 322–328.

Kim, M. H., Han, B. S., Yuleum, S., In, T. J., & Jung, W. K. (2013). Evaluation of Food Waste Disposal Options in Terms of Global Warming and Energy Recovery : Korea. *International Journal of Energy and Environmental Engineering*, *4*(1), 1. doi:10.1186/2251-6832-4-1

Kingston, J. J., Radhika, M., Roshini, P. T., Raksha, M. A., Murali, H. S., & Batra, H. V. (2010). Molecular characterization of lactic acid bacteria recovered from natural fermentation of beet root and carrot Kanji. *Indian Journal of Microbiology*, *50*(3), 292–298. doi:10.100712088-010-0022-0 PMID:23100843

Kiran, E. U., Antoine, P. T., & Yu, L. (2014). Glucoamylase Production from Food Waste by Solid State Fermentation and Its Evaluation in the Hydrolysis of Domestic Food Waste. *Biofuel Research Journal*, *3*, 98–105. doi:10.18331/BRJ2015.1.3.7

Kiran, E. U., Trzcinski, A. P., Ng, W. J., & Liu, Y. (2014). Bioconversion of food waste to energy: A review. *Fuel*, *134*, 389–399. doi:10.1016/j.fuel.2014.05.074

Kiran, Trzcinski, & Liu. (2014). Glucoamylase production from food waste by solid state fermentation and its evaluation in the hydrolysis of domestic food waste. *Biofuel Research Journal*, *3*, 98–105.

Klüppel, H. (1998). ISO 14041: Environmental management — life cycle assessment — goal and scope definition — inventory analysis. *The International Journal of Life Cycle Assessment*, *3*(6), 301–301. doi:10.1007/BF02979337

Kneifel, W., & Mayer, H. K. (1991). Vitamin profiles of kefirs made from milks of different species. *International Journal of Food Science & Technology*, *26*(4), 423–428. doi:10.1111/j.1365-2621.1991.tb01985.x

Kojima, R., & Ishikawa, M. (2013). Prevention and recycling of food wastes in Japan: Policies and achievements. Kobe University.

Koller, R., Chiellini, B. E., Fernandes, E. G., Horvat, P., Kutschera, C., Hesse, P., & Braunegg, G. (2008). Polyhydroxy alkanoate production from whey by *Pseudomonas hydrogenovora*. *Bioresource Technology*, *99*(11), 4854–4863. doi:10.1016/j.biortech.2007.09.049 PMID:18053709

Kroyer, G. T., Kretschmer, L., & Washiittl, J. (1989). Antioxidant properties of tea and coffee extracts. *Proceedings of 5th European Conference of Food Chemistry*, 2, 433–437.

Kubicek, C. P., Mikus, M., Schuster, A., Schmoll, M., & Seiboth, B. (2009). Metabolic engineering strategies for the improvement of cellulose production by Hypocreajecorina. *Biotechnology for Biofuels*, *2*(1), 19. doi:10.1186/1754-6834-2-19

Kuboye, A. O. (1985, October). Traditional fermented foods and beverages of Nigeria. In *Proceedings of the IFA/UNU workshop on the development of indigenous fermented foods and food technology in Africa, Douala, Cameroon. Stockholm: International Foundation for Science* (pp. 224-36). Academic Press.

Kuhad, R. C., Gupta, R., & Singh, A. (2011). Microbial cellulases and their industrial applications. *Enzyme Research, 2011*, 1–10. doi:10.4061/2011/280696 PMID:21912738

Kulasinski, K., Keten, S., Churakov, S. V., Derome, D., & Carmeliet, J. (2014). A comparative molecular dynamics study of crystalline, paracrystalline and amorphous states of cellulose. *Cellulose (London, England), 21*(3), 1103–1116. doi:10.100710570-014-0213-7

Kumar, R. S. (2005). Whey beverage: A review. *Beverage and Food World*, 58-60.

Kumar, R., Singh, S., & Singh, O. V. (2008). Bioconversion of lignocellulosic biomass: Biochemical and molecular perspectives. *Journal of Industrial Microbiology & Biotechnology, 35*(5), 377–391. doi:10.100710295-008-0327-8

Kummu, M., de Moel, H., Porkka, M., Siebert, S., Varis, O., & Ward, P. (2012). Lost food, wasted resources: Global food supply chain losses and their impact on freshwater, cropland and fertilizer use. *The Science of the Total Environment, 43*, 477–489. doi:10.1016/j.scitotenv.2012.08.092 PMID:23032564

Kwatia, S., & Isaac, W. O. (2017). Optimization of Amylase Production by *Aspergillus niger* Cultivated on Yam Peels in Solid State Fermentation Using Response Surface Methodology. *African Journal of Biochemistry Research, 11*(7), 34–42. doi:10.5897/AJBR2017.0941

Lacey, J. (1989). Pre-and post-harvest ecology of fungi causing spoilage of foods and other stored products. *The Journal of Applied Bacteriology, 67*, 11s–25s. doi:10.1111/j.1365-2672.1989.tb03766.x PMID:2508232

Lafarga, T., & Teagase, M. H. (2014). Bioactive peptides from meat muscle and by products: Generation functionality and application as functional ingredients. *Meat Science, 98*(2), 227–239. doi:10.1016/j.meatsci.2014.05.036 PMID:24971811

Lal, G., Siddappa, G. S., & Tandon, G. L. (1988). *Preservation of Fruits and Vegetables*. New Delhi: ICAR Pub.

Lambing, E. F., & May Froidt, P. (2011). Global land use change, deglobalization and booming land scarcity. *Proceedings of the National Academy of Sciences of the United States of America, 108*(9), 3465–3472. doi:10.1073/pnas.1100480108 PMID:21321211

Langille, M. G. I., Zaneveld, J., Caporaso, J. G., McDonald, D., Knights, D., Reyes, J. A., ... Huttenhower, C. (2013). Predictive functional profiling of microbial communities using 16S rRNA marker gene sequences. *Nature Biotechnology, 31*(9), 814–821. doi:10.1038/nbt.2676

Langley, R. J., & Wong, H. R. (2017). Early diagnosis of sepsis: Is an integrated omic approach the way forward. *Molecular Diagnosis & Therapy, 21*(5), 525–537. doi:10.100740291-017-0282-z

Laufenberg, G., Kunz, B., & Nystroem, M. (2003). Transformation of vegetable waste into value-added products: (A) the upgrading concept; (B) practical implementations. *Bioresource Technology*, *87*(2), 167–198. doi:10.1016/S0960-8524(02)00167-0 PMID:12765356

Lauwers, J. L., Appels, P., Thompson, I., Degr'eve, J., Impe, J. F. V., & Dewil, R. (2013). Mathematical modelling of anaerobic digestion of biomass and waste: Power and limitations. *Progress in Energy and Combustion Science*, *39*(4), 383–402. doi:10.1016/j.pecs.2013.03.003

Lay, M. M. G., & Fields, M. L. (1981). Nutritive value of germinated corn and corn fermented after germination. *Journal of Food Science*, *46*(4), 1069–1073. doi:10.1111/j.1365-2621.1981.tb02993.x

Leach, B., & Swannell, R. (2017). Food loss and waste: A key issue for our generation. *Environmental Scientist*, *26*(1), 47–51.

Learn Gamification with Yu-kai Chou: cheat codes to win the game of life. (2018). Retrieved from https://yukaichou.com/

Lee, C., & Clin Chim, A. (2000). Antioxidant ability of caffeine and its metabolites based on the study of oxygen radical absorbing capacity and inhibition of LDL peroxidation. *Clinica Chimica Acta*, *295*, 141–154.

Lee, S. G., & Xu, X. (2004). A simplified life cycle assessment of reusable and single -use bulk transit packaging. *Packaging Technology & Science*, *17*(2), 67–83. doi:10.1002/pts.643

Leroy, F., & De Vuyst, L. (2004). Lactic acid bacteria as functional starter cultures for the food fermentation industry. *Trends in Food Science & Technology*, *15*(2), 67–78. doi:10.1016/j.tifs.2003.09.004

Levis, J. W., Barlaz, M. A., Themelis, N. J., & Ulloa, P. (2010). Assessment of the state of food waste treatment in the United States and Canada. *Waste Management (New York, N.Y.)*, *30*(8-9), 1486–1494. doi:10.1016/j.wasman.2010.01.031 PMID:20171867

Lidbeck, A., Övervik, E., Rafter, J., Nord, C. E., & Gustafsson, J. Å. (1992). Effect of Lactobacillus acidophilus supplements on mutagen excretion in faeces and urine in humans. *Microbial Ecology in Health and Disease*, *5*(1), 59–67. doi:10.3109/08910609209141305

Liebler, D. C. (2002). Proteomic approaches to characterize protein modifications: New tools to study the effects of environmental exposures. *Environmental Health Perspectives*, *110*(Suppl 1), 3–9. doi:10.1289/ehp.02110s113

Liener, I. E. (1994). Implications of antinutritional components in soybean foods. *Critical Reviews in Food Science and Nutrition*, *34*(1), 31–67. doi:10.1080/10408399409527649 PMID:8142044

Liimatainen, H., Kuokkanen, T. K., & Ariainen, J. (2004). Development of bioethanol production from waste potatoes. In *Proceedings of the Waste Minimization and Resources Use Optimization Conference*. University of Oulu.

Lin, C. S. K., Koutinas, A. A., Stamatelatou, K., Mubofu, E. B., Matharu, A. S., Kopsahelis, N., & Luque, R. (2014). Current and future trends in food waste valorization for the production of chemicals, materials and fuels: A global perspective. *Biofuels, Bioproducts & Biorefining*, *8*(5), 686–715. doi:10.1002/bbb.1506

Lin, C. S. K., Pfaltzgraff, L. A., Herrero-Davila, L., Mubofu, E. B., Abderrahim, S., Clark, J. H., ... Thankappan, S. (2013). Food waste as a valuable resource for the production of chemicals, materials and fuels, Current situation and global perspective. *Energy & Environmental Science*, *6*(2), 426–464. doi:10.1039/c2ee23440h

Liu, H., Ma, X., Li, L., Hu, Z., Guo, P., & Jiang, Y. (2014). The catalytic pyrolysis of food waste by microwave heating. *Bioresource Technology*, *166*, 45–50. doi:10.1016/j.biortech.2014.05.020 PMID:24905041

Liu, J., & Jichu, Y. (2007). Cellulase Production by *Trichoderma koningii* AS3. 4262 in Solid-State Fermentation Using Lignocellulosic Waste from the Vinegar Industry. *Food Technology and Biotechnology*, *45*(4), 420–425.

Liu, Y. (2017). Direct transesterification of spent coffee grounds for biodiesel production. *Fuel*, *199*, 157–161.

Li, X., Chen, Y., Zhao, S., Chen, H., Zheng, X., Luo, J., & Liu, Y. (2015). Efficient production of optically pure l-lactic acid from food waste at ambient temperature by regulating key enzyme activity. *Water Research*, *70*, 148–157. doi:10.1016/j.watres.2014.11.049 PMID:25528545

Loman, N. J., & Pallen, M. J. (2015). Twenty years of bacterial genome sequencing. *Nature Reviews. Microbiology*, *13*(12), 787–794. doi:10.1038/nrmicro3565

Lopez, H. W., Leenhardt, F., Coudray, C., & Remesy, C. (2002). Minerals and phytic acid interactions: Is it a real problem for human nutrition? *International Journal of Food Science & Technology*, *37*(7), 727–739. doi:10.1046/j.1365-2621.2002.00618.x

Lorri, W., & Svanberg, U. (1994). Lower prevalence of diarrhoea in young children fed lactic acid-fermented cereal gruels. *Food and Nutrition Bulletin-United Nations University*, *15*, 57–57.

Lowe, R., Shirley, N., Bleackley, M., Dolan, S., & Shafee, T. (2017). Transcriptomics technologies. *PLoS Computational Biology*, *13*(5), e1005457. doi:10.1371/journal.pcbi.1005457

Luque, R., & Clark, J. H. (2013). Valorisation of food residues: Waste to wealth using green chemical technologies. *Sustainable Chemical Processes*, *1*(1), 10. doi:10.1186/2043-7129-1-10

Macfarlane, S., Macfarlane, G. T., & Cummings, J. H. (2006). Prebiotics in the gastrointestinal tract. *Alimentary Pharmacology & Therapeutics*, *24*(5), 701–714. doi:10.1111/j.1365-2036.2006.03042.x PMID:16918875

Machado, E. S. M. (2009). *Reaproveitamento de resíduos da indústria do café como matériaprimapara a produc¸ ão de etanol* (MSc thesis). Braga, Portugal: Department of Biological Engineering, University of Minho.

Machado, C. M. M., Soccol, C. R., Oliveira, B. H., & Pandey, A. (2002). Gibberellic acid production by solid-state fermentation in coffee husk. *Applied Biochemistry and Biotechnology*, *15*, 102–106.

Machado, E.M.S., & Rosa, M., Rodriguez-Jasso, J. A., & Teixeira, S. I. M. (2012). Growth of fungal strains on coffee industry residues with removal of polyphenolic compounds. *Biochemical Engineering Journal*, *60*, 87–90.

Madamwar, D., & Patel, S. (1992). Formation of cellulases by co-culturing of *Trichoderma reesei* and *Aspergillus niger* on cellulosic wastes. In V. S. Malik & P. Sridhar (Eds.), *Industrial biotechnology* (pp. 471–478). Oxford, UK: IBH.

Ma, E., Cervera, Q., & Sánchez, G. M. M. (1993). Integrated utilization of orange peel. *Bioresource Technology*, *44*(1), 61–63. doi:10.1016/0960-8524(93)90209-T

Maeno, M., Yamamoto, N., & Takano, T. (1996). Identification of an antihypertensive peptide rom casein hydrolysate produced by a proteinase from *Lactobacillus helveticus* CP790. *Journal of Dairy Science*, *79*(8), 1316–1321. doi:10.3168/jds.S0022-0302(96)76487-1 PMID:8880454

Makris, D. P. (2015). Recovery and applications of enzymes from food wastes. Food Waste Recovery, 361-379. doi:10.1016/B978-0-12-800351-0.00016-X

Ma, L., Cui, Y., Cai, R., Liu, X., Zhang, C., & Xiao, D. (2015). Optimization and evaluation of alkaline potassium permanganate pretreatment of corncob. *Bioresource Technology*, *180*(0), 1–6. doi:10.1016/j.biortech.2014.12.078

Mangala, S. L., Malleshi, N. G., Tharanathan, R. N., & Not Available, N. A. (1999). Resistant starch from differently processed rice and ragi (finger millet). *European Food Research and Technology*, *209*(1), 32–37. doi:10.1007002170050452

Mani, U., Prabhu, S., Damie, S., & Mani, I. (1993). Glycemic index of some commonly consumed foods in Western India. *Asia Pacific Journal of Clinical Nutrition*, *2*, 111–114. PMID:24352140

Mariana, M., Mahidin, M., Mulana, F., & Aman, F. (2018). Utilization of Activated Carbon Prepared from Aceh Coffee Grounds as Bio-sorbent for Treatment of Fertilizer Industrial Wastewater. *Materials Science and Engineering, 358*, 12-27.

Marín, F. R., Soler-Rivas, C., Benavente-García, O., Castillo, J., & Pérez-Alvarez, J. A. (2007). By-products from different citrus processes as a source of customized functional fibres. *Food Chemistry*, *100*(2), 736–741. doi:10.1016/j.foodchem.2005.04.040

Marin, M. L., Tejada-Simon, M. V., Murtha, J., Ustunol, Z., & Pestka, J. J. (1997). Effects of Lactobacillus spp. on cytokine production by RAW 264.7 macrophage and EL-4 thymoma cell lines. *Journal of Food Protection*, *60*(11), 1364–1370. doi:10.4315/0362-028X-60.11.1364

Marshall E. and Mejia D., (2011) Traditional fermented food and beverages for improved livelihoods. *FAO Diversification Booklets, 21.*

Marteau, P., & Rambaud, J. C. (1993). Potential of using lactic acid bacteria for therapy and immunomodulation in man. *FEMS Microbiology Reviews*, *12*(1-3), 207–220. doi:10.1111/j.1574-6976.1993.tb00019.x PMID:8398215

Martín, C., García, A., Schreiber, A., Puls, J., & Saake, B. (2015). Combination of water extraction with dilute sulphuric acid pretreatment for enhancing the enzymatic hydrolysis of *Jatrophacurcas* shells. *Industrial Crops and Products*, *64*(0), 233–241. doi:10.1016/j.indcrop.2014.09.040

Martinez-Saez, N. (2017). Use of spent coffee grounds as food ingredient in bakery products. *Food Chemistry*, *216*, 114–122.

Mason, L., Boyle, T., Fyfe, J., Smith, T., & Cordell, D. (2011). *National Food Waste Data Assessment: Final Report*. Prepared for the Department of Sustainability, Environment, Water, Population and Communities, by the Institute for Sustainable Futures, University of Technology, Sydney. Retrieved from https://www.environment.gov.au/system/files/resources/128a21f0-5f82-4a7d-b49c-ed0d2f6630c7/files/food-waste.pdf

Mathew, J. J., Prem, J. V., & Sajeshkumar, N. K. (2016). Amylase Production by *Aspergillus niger* through Submerged Fermentation Using Starchy Food Byproducts as Substrate Amylase Production by *Aspergillus niger* through Submerged Fermentation Using Starchy Food Byproducts as Substrate. *International Journal of Herbal Medicine*, *4*(6), 34–40.

Mathur, Navya, Basavaraj, & Murthy. (2015). Bioprocess of Robusta cherry coffee with polyphenol oxidase and quality enhancement. *European Journal of Food Research and Technology*, *240*, 319–325.

Matsakas, L., Kekos, D., Loizidou, M., & Christakopoulos, P. (2014). Utilization of household food waste for the production of ethanol at high dry material content. *Biotechnology for Biofuels*, *7*(1), 1–9. doi:10.1186/1754-6834-7-4 PMID:24401142

Matsuda, T., Yano, J., Hirai, Y., & Sakai, S. (2012). Life-cycle greenhouse gas inventory analysis of household waste management and food waste reduction activities in Kyoto, Japan. *The International Journal of Life Cycle Assessment*, *17*(6), 743–752. doi:10.100711367-012-0400-4

Mayer, J., Scheid, S., Widmer, F., Fließbach, A., & Oberholzer, H. R. (2010). How effective are 'Effective microorganisms® (EM)'? Results from a field study in temperate climate. *Applied Soil Ecology*, *46*(2), 230–239. doi:10.1016/j.apsoil.2010.08.007

McKay, L. L., & Baldwin, K. A. (1990). Applications for biotechnology: Present and future improvements in lactic acid bacteria. *FEMS Microbiology Reviews*, *87*(1-2), 3–14. doi:10.1111/j.1574-6968.1990.tb04876.x PMID:2271224

Melikoglu, M., Carol, S., Ki, L., & Colin, W. (2013). Analysing Global Food Waste Problem : Pinpointing the Facts and Estimating the Energy Content. *Central European Journal of Engineering*, *3*(2), 157–164.

Melikoglu, M., Lin, C., & Webb, C. (2013). Food and Bioproducts Processing Stepwise Optimisation of Enzyme Production in Solid State Fermentation of Waste Bread Pieces. *Food and Bioproducts Processing*, *91*(4), 638–646. doi:10.1016/j.fbp.2013.04.008

Mena, C. B., Adenso-Diaz, B., & Yurt, O. (2011). The causes of food waste in the supplier – retail interface: Evidence from the UK and Spain. *Resources, Conservation and Recycling*, *55*(6), 648–654. doi:10.1016/j.resconrec.2010.09.006

Menon, M. A. (2010). Integrated solid waste management based on the '3R'approach. *J. Motor Cycles Waste*, *12*(1), 30–40. doi:10.100710163-009-0274-0

Meyers, W. H., Ziolkowska, J. R., Tothova, M., & Goychuk, K. (2012). *Issues affecting the future of agriculture and food security for Europe and Central Asia*. FAO Regional Office for Europe and Central Asia. Policy Studies on Rural Transition No. 2012-3. Retrieved from http://www.fao.org/3/a-aq343e.pdf

Mirabella, N., Castellani, V., & Sala, S. (2014). Current options for the valorization of food manufacturing waste: A review. *Journal of Cleaner Production*, *65*, 28–41. doi:10.1016/j.jclepro.2013.10.051

Misra, R. V., Roy, R. N., & Hiraoka, H. (2003). *On-Farm Composting Methods*. Food and Agriculture Organization of the United Nations.

Mitall, B. K., & Garg, S. K. (1995). Anticarcinogenic, hypocholesterolemic, and antagonistic activities of Lactobacillus acidophilus. *Critical Reviews in Microbiology*, *21*(3), 175–214. doi:10.3109/10408419509113540 PMID:8845062

Mitidieri, S., Martinelli, A. H. S., Schrank, A., & Vainstein, M. H. (2006). Enzymatic detergent formulation containing amylase from *Aspergillus niger*: A comparative study with commercial detergent formulations. *Bioresource Technology*, *97*(10), 1217–1224. doi:10.1016/j.biortech.2005.05.022 PMID:16112858

Mitra, S., Chakrabartty, P. K., & Biswas, S. R. (2007). Production of Nisin Z by *Lactococcus lactis* Isolated from Dahi. *Applied Biochemistry and Biotechnology*, *143*(1), 41–53. doi:10.100712010-007-0032-5 PMID:18025595

Mittal, B. K., Prasad, R., & Singh, S. (1977). Effect of carbohydrates and phosphates on acid production by lactic acid bacteria in soy milk [India]. Research note. *Journal of Food Science and Technology*.

Mittal, K. M. (1997). Biogas Systems, Policies, Progress and Prospects. New Age International.

Moktan, B., Roy, A., & Sarkar, P. K. (2011). Antioxidant activities of cereal-legume mixed batters as influenced by process parameters during preparation of dhokla and idli, traditional steamed pancakes. *International Journal of Food Sciences and Nutrition*, *62*(4), 360–369. doi:10.3109/09637486.2010.532116 PMID:21142877

Moldes, D., Gallego, P. P., Rodriguez Couto, S., & Sanroman, M. A. (2003). Grape seeds: The best lignocellulosic waste to produce laccase by solid state cultures of Trameteshirsuta. *Biotechnology Letters*, *25*(6), 491–495. doi:10.1023/A:1022660230653 PMID:12882277

Monente, C., Bravo, J., Vitas, A.I., & Arbillaga, L., De Pen˜ a, M.P., & Cid, C. (2015). Coffee and spent coffee extracts protect against cell mutagens and inhibit growth of food-borne pathogen microorganisms. *Journal of Functional Foods*, *12*, 365–374.

Monier, V., Shailendra, M., Escalon, V. O., Connor, C., Gibbon, T., Anderson, G., . . . Reisinger, H. (2010). Preparatory study on food waste across EU 27. European Commission (DGENV) Directorate Industry, Final Report.

Monier, V., Shailendra, M., Escalon, V., O'Connor, C., Gibon, T., Anderson, G., . . . Reisinger, H. (2010). Preparatory Study on Food Waste across EU 27. European Commission (DG ENV) Directorate C-Industry. Final Report.

Moon, H. C., Song, I. S., Kim, J. C., Shirai, Y., Lee, D. H., Kim, J. K., ... Cho, Y. S. (2009). Enzymatic hydrolysis of food waste and ethanol fermentation. *International Journal of Energy Research*, *33*(2), 164–172. doi:10.1002/er.1432

Morales, F. J., Somoza, V., & Fogliano, V. (2012). Physiological relevance of dietary melanoidins. *Amino Acids*, *42*, 1097–1109.

Morita, M., & Sasaki, K. (2012). Factors influencing the degradation of garbage in methanogenic bioreactors and impacts on biogas formation. *Applied Microbiology and Biotechnology*, *94*(3), 575–582. doi:10.100700253-012-3953-z PMID:22395906

Moroney, P. (2016). Recycling, recovering and preventing food waste competing solutions for food systems substantially in the United States and France. Academic Press.

Moslehuddin, A. B. M., & Hang, Y. D. (1987). Effect of processing methods on the nutritional value of Lathyrus sativus seeds. Nutrition Reports International (USA).

Mourad, M. (2016). Recycling, recovering and preventing "food waste": Competing solutions for food systems sustainability in the United States and France. *Journal of Cleaner Production*, *126*, 461–477. doi:10.1016/j.jclepro.2016.03.084

Mukherjee, S. K., Alburry, M. N., Pederson, C. S., Vanveen, A. G., & Steinkraus, K. H. (1965). Role of *Leuconostoc mesenteroides* in leavening the batter of idli, a fermented food of India. *Applied Microbiology*, *13*, 227–231. PMID:14325884

Murdock, F. A., & Fields, M. L. (1984). B-Vitamin Content of Natural Lactic Acid Fermented Cornmeal. *Journal of Food Science*, *49*(2), 373–375. doi:10.1111/j.1365-2621.1984.tb12425.x

Murthy, P. S., Manjunatha, M. R., Sulochannama, G., & Naidu, M. M. (2012). Extraction, Characterization and Bioactivity of Coffee Anthocyanins. *European Journal of Biological Sciences*, *4*(1), 13–19.

Murthy, P. S., & Manonmani, H. K. (2008). Recycling of spent mushroom substrate to vermicompost. *Journal of Environmental Sciences (China)*, *3*(2), 212–216.

Murthy, P. S., & Naidu, M. M. (2010). Protease production by *Aspergillus oryzae* in solid-state fermentation utilizing coffee by-products. *World Applied Sciences Journal*, *8*(2), 199–205.

Murthy, P. S., & Naidu, M. M. (2011). Improvement of Robusta Coffee Fermentation with Microbial Enzymes. *European Journal of Applied Sciences*, *3*(4), 130–139.

Murthy, P. S., & Naidu, M. M. (2012a). Production and application of Xylanase from *Penicillium* sp. utilizing coffee by-products. *Food and Bioprocess Technology*, *5*(2), 657–664.

Murthy, P. S., & Naidu, M. M. (2012b). Sustainable management of coffee industry byproducts and value addition – A review. *Resources, Conservation and Recycling*, *66*, 45–58.

Murthy, P. S., & Naidu, M. M. (2012c). Recovery of phenolic antioxidants and functional compounds from coffee industry by-products. *Food and Bioprocess Technology*, *5*(3), 897–903.

Murthy, P. S., Naidu, M. M., & Srinivas, P. (2009). Production of amylase under solid–state fermentation utilizing coffee waste. *Journal of Chemical Technology and Biotechnology (Oxford, Oxfordshire)*, *84*, 1246–1249.

Murthy, P. S., Naidu, M. M., & Srinivas, P. (2009). Production of α -Amylase under Solid-State Fermentation Utilizing Coffee Waste. *Journal of Chemical Technology and Biotechnology (Oxford, Oxfordshire)*, *84*(8), 1246–1249. doi:10.1002/jctb.2142

Mushtaq, Q., Muhammad, I., Fouzia, T., & Javed, I. (2016). Potato Peels : A Potential Food Waste for Amylase Production. *Journal of Food Process Engineering*.

Mussatto, S. I., Carneiro, L. M., Silva, J. P. A., Roberto, I. C., & Teixeira, J. A. (2011). A study on chemical constituents and sugars extraction from spent coffee grounds. *Carbohydrate Polymers*, *83*, 368 374.

Muttalib, A., Aminah, S., Ismail, S., Norkhadijah, S., & Praveena, S. M. (2016). Application of effective microorganism (EM) in food waste composting: A review. *Asia Pacific Environmental and Occupational Health Journal*, *2*(2), 37–47.

Nagavallemma, K. P., Wani, S. P., Stephane, L., Padmaja, V. V., Vineela, C., & Babu Rao, M. (2006). Vermicomposting: Recycling wastes into valuable organic fertilizer. *International Crops Research Institute for the Semi-Arid Tropics*, *2*, 16.

Nagpal, R., Behare, P. V., Kumar, M., Mohania, D., Yadav, M., Jain, S., ... Henry, C. J. K. (2012). Milk, milk products, and disease free health: An updated overview. *Critical Reviews in Food Science and Nutrition*, *52*(4), 321–333. doi:10.1080/10408398.2010.500231 PMID:22332596

Nakayama, R., & Imai, M. (2013). Promising ultrasonic irradiation pretreatment for enzymatic hydrolysis of Kenaf. *Journal of Environmental Chemical Engineering*, *1*(4), 1131–1136. doi:10.1016/j.jece.2013.08.030

Namkoong, W., Hwang, E.-Y., Park, J.-S., & Choi, J.-Y. (2002). Bioremediation of diesel-contaminated soil with composting. *Environmental Pollution*, *119*(1), 23–31. doi:10.1016/S0269-7491(01)00328-1 PMID:12125726

Nanqi, R., Wanqian, G., Bingfeng, L., Guangli, C., & Jie, D. (2011). Biological hydrogen production by dark fermentation: Challenges and prospects towards scaled-up production. *Current Opinion in Biotechnology*, *22*(3), 365–370. doi:10.1016/j.copbio.2011.04.022 PMID:21612910

Nasir, I. M., Ghazi, T. I. M., & Omar, R. (2012). Production of biogas from solid organic wastes through anaerobic digestion: A review. *Applied Microbiology and Biotechnology*, *95*(2), 321–329. doi:10.100700253-012-4152-7 PMID:22622840

Navya, P. N., & Pushpa, S. M. (2013). Production, statistical optimization and application of endoglucanase from *Rhizopus stolonifer* utilizing coffee husk. *Bioprocess and Biosystems Engineering*, *36*, 1115–1123.

Navya, P.N., & Bhoite, R. N., & Murthy, P. S. (2012a). Improved β-glucosidase production from *Rhizopus stolonifer* utilizing coffee husk. *International Journal of Current Research*, *4*(8), 123–129.

Navya, P.N., & Bhoite, R.N., & Murthy, P. S. (2012b). Bioconversion of Coffee Husk Cellulose and Statistical Optimization of Process for Production of Exoglucanase by *Rhizopus stolonifer. World Applied Sciences Journal*, *20*(6), 781–789.

Nehal, N. (2013). Knowledge of traditional fermented food products harbored by the Tribal Folks of the Indian Himalayan Belt. *Int. J.Agri. Food Sci. Technol.*, *4*, 401–414.

Nerantzis, E. T., & Tetaridis, P. (2006). *Integrated enology: utilization of winery by-products into high value added products.* Academic Press.

Nestel, A. J., & Fernandez, B. (2018). Treatment of mining waste leachate by the adsorption process using spent coffee grounds. *Environmental Technology*, *15*, 1–15.

Newton, S. (2003). Root cellars: types and storage tips. In *The Old Farmer's Almanac*. Retrieved from www.almanac.com

Ngoc, T., Bao, D., Biswarup, S., & Chin-chao, C. (2014). Food Waste to Bioenergy via Anaerobic Processes. *Energy Procedia*, *61*, 307–312.

Nguyen, N. H., Song, Z., Bates, S. T., Branco, S., Tedersoo, L., Menke, J., ... Kennedy, P. G. (2016). FUNGuild: An open annotation tool for parsing fungal community datasets by ecological guild. *Fungal Ecology*, *20*, 241–248. doi:10.1016/j.funeco.2015.06.006

Nguyen, T. A. D., Kim, K. R., Han, S. J., Cho, H., Kim, J. W., Park, S. M., & Sim, S. J. (2010). Pretreatment of rice straw with ammonia and ionic liquid for lingo cellulose conversion to fermentable sugars. *Bioresource Technology*, *101*(19), 7432–7438. doi:10.1016/j.biortech.2010.04.053 PMID:20466540

Nigam, J. N. (1998). Single cell protein from pineapple cannery effluent. *World Journal of Microbiology & Biotechnology*, *14*(5), 693–696. doi:10.1023/A:1008853303596

Nishida, J. (2014) Reducing Food Waste and Promoting Food Recovery Globally. *EPA Connect. The official blog of the EPA Leadership*. Retrieved from https://blog.epa.gov/blog/2014/10/reducing-food-waste-and-promoting-food-recovery-globally/

Nishida, J. (2014). Reducing food waste and promoting two recovery, globally, EPA c connect. *The Official Blog of the EPA Leadership*. Retrieved from https:/blog.epa.gov/blog/2014/reducingfood-waste-and-promotingfood-recovery-gglobally

Nnam, N. M. (1995). Evaluation of nutritional quality of fermented cowpea (Vigna unguiculata) flours. *Ecology of Food and Nutrition, 33*(4), 273–279. doi:10.1080/03670244.1995.9991435

Nout, M. J. R., Rombouts, F. M., & Hautvast, J. G. A. J. (1989). Accelerated natural lactic fermentation of infant food formulations. *Food and Nutrition Bulletin, 11*(1), 65–73. doi:10.1177/156482658901100102

Nout, M. J. R., & Sarkar, P. K. (1999). Lactic acid food fermentation in tropical climates. *Antonie van Leeuwenhoek, 76*(1/4), 395–401. doi:10.1023/A:1002066306013 PMID:10532396

Nout, M. R. (2003). Traditional fermented products from Africa, Latin America and Asia. In *Yeasts in food-beneficial and detrimental aspects* (pp. 451–473). Behr's Verlag. doi:10.1533/9781845698485.451

Nout, M. R., Rombouts, F. M., & Havelaar, A. (1989). Effect of accelerated natural lactic fermentation of infant good ingredients on some pathogenic microorganisms. *International Journal of Food Microbiology, 8*(4), 351–361. doi:10.1016/0168-1605(89)90006-8 PMID:2701696

Novinscak, A., Filion, M., Surette, C., & Allain, A. (2008). Application of molecular technologies to monitor the microbial content of biosolids and composted biosolids. *Water Science and Technology, 57*(4), 471–477. doi:10.2166/wst.2008.019 PMID:18359983

Nwabueze, T. U., & Ogumtimein, G. B. (1987). Sweet orange (*Citrus sinensis*) residue as a substrate for single cell protein production. *Biological Wastes, 20*(1), 71–75. doi:10.1016/0269-7483(87)90085-1

Obizoba, I. C., & Atii, J. V. (1991). Effect of soaking, sprouting, fermentation and cooking on nutrient composition and some anti-nutritional factors of sorghum (Guinesia) seeds. *Plant Foods for Human Nutrition (Dordrecht, Netherlands), 41*(3), 203–212. doi:10.1007/BF02196388 PMID:1924184

Obizoba, I. C., & Egbuna, H. I. (1992). Effect of germination and fermentation on the nutritional quality of bambara nut (Voandzeia subterranea L. Thouars) and its product (milk). *Plant Foods for Human Nutrition (Dordrecht, Netherlands), 42*(1), 13–23. doi:10.1007/BF02196068 PMID:1546053

ODT. (2004). *Food: the new definition*. Retrieved from http://tzhcpas.com/wp-content/uploads/2011/01/streamlined_sales_tax_food_BR.pdf

Odunfa, S. A., & Komolafe, O. B. (1989). Nutritional characteristics of staphylococcus species from fermenting african locust bean (parkia-biglobosa). *Nahrung-food, 33*(7), 607–615.

Oki, K., Rai, A. K., Sato, S., Watanbe, K., & Tamang, J. P. (2011). Lactic acid bacteria isolated from Ethnic preserved meat products of the Western Himalayas. *Food Microbiology*, *28*(7), 1308–1315. doi:10.1016/j.fm.2011.06.001 PMID:21839380

OLIO. (2017). *The problem of food waste*. Retrieved from: https://olioex.com/food-waste/the-problem-of-food-waste/

Opatokun, S. A., Strezov, V., & Kan, T. (2015). Product based evaluation of pyrolysis of food waste and its digestate. *Energy*, *92*, 349–354. doi:10.1016/j.energy.2015.02.098

Östergren, K., Gustavsson, J., Bos-Brouwers, H., Timmermans, T., Hansen, O.-J., Møller, H., . . . Redlingshöfer, B. (2014). FUSIONS Definitional Framework for Food Waste. Full Report.

Otles, S., Despoudi, S., Bucatariu, C., & Kartal, C. (2015). Food waste management, valorization, and sustainability in the food industry. In C. M. Galanakis (Ed.), *Food Waste Recovery Processing Technologies and Industrial Techniques* (pp. 3–18). London: Academic Press, Elsevier. doi:10.1016/B978-0-12-800351-0.00001-8

Ouattara, D. H., Ouattara, H. G., Goualie, B. G., Kouame, L. M., & Niamke, S. L. (2014). Biochemical and functional properties of lactic acid bacteria isolated from Ivorian cocoa fermenting beans. *Journal of Applied Biosciences*, *77*, 6489–6499.

Padmaja, G., George, M., Moorthy, S. N., Bainbridge, Z., Plumb, V., Wood, J. F., & Powell, C. J. (1994). Nutritional evaluation of the starchy flour obtained from cassava tubers on fermentation with a mixed-culture inoculum. *Journal of Agricultural and Food Chemistry*, *42*(3), 766–770. doi:10.1021/jf00039a033

Pagni, J. (1998). *Demonstrating bacteria for health.* VTT Biotechnology and Food Research (Catalogue 5).

Pal, V., Jamuna, M., & Jeevaratnam, K. (2005). *Isolation and characterization of bacteriocin producing lactic acid bacteria from a south Indian special dosa (appam) batter*. J Cult Collect.

Pandey, A., Tripathi, P. H., Pandey, S. C., Pathak, V. M., & Nailwal, T. K. (2018). Removal of Toxic Pollutants From Soil Using Microbial Biotechnology. In Microbial Biotechnology in Environmental Monitoring and Cleanup (pp. 86-105). Academic Press. doi:10.4018/978-1-5225-3126-5.ch006

Pandey, A. (1992). Recent process developments in solid-state fermentation. *Process Biochemistry*, *27*(2), 109–117. doi:10.1016/0032-9592(92)80017-W

Pandey, A. (2003). Solid-state fermentation. *Biochemical Engineering Journal*, *13*(2-3), 81–84. doi:10.1016/S1369-703X(02)00121-3

Pandey, A., Soccol, C. R., Nigam, P., Brand, D., Mohan, R., & Roussos, S. (2000). Biotechnological potential of coffee pulp and coffee husk for bioprocesses. *Biochemical Engineering Journal*, *6*, 153–162.

Pandey, V. N., & Srivastava, A. K. (1993). A simple, low energy requiring method of coagulating leaf proteins for food use. *Plant Foods for Human Nutrition (Dordrecht, Netherlands)*, *43*(3), 241–245. doi:10.1007/BF01886225 PMID:8506239

Pandey, V. P., Awasthi, M., Singh, S., Tiwari, S., & Dwivedi, U. N. (2017). A Comprehensive Review on Function and Application of Plant Peroxidases. *Biochemistry and Analytical Biochemistry*, *6*(01), 308. doi:10.4172/2161-1009.1000308

Panesar, P. S. (2011). Fermented dairy products: Starter cultures and potential nutritional benefits. *Food Nutrit. Sci.*, *2*, 47.

Panesar, P. S., Kennedy, J. F., Gandhi, D. N., & Bunko, K. (2007). Bioutilisation of whey for lactic acid production. *Food Chemistry*, *105*(1), 1–14. doi:10.1016/j.foodchem.2007.03.035

Panesar, P. S., Panesar, R., Singh, R. S., Kennedy, J. F., & Kumar, H. (2006). Review Microbial production, immobilization and applications of β-D-galactosidase J. *Chem. Tech. Biotech.*, *81*(4), 530–543. doi:10.1002/jctb.1453

Papargyropoulou, E., Lozano, R., Steinberger, J. K., & Wright, N., & bin Ujang, Z. (2014). The food waste hierarchy as a framework for the management of food surplus and food waste. *Journal of Cleaner Production*, *76*, 106–115. doi:10.1016/j.jclepro.2014.04.020

Paranthaman, R., Alagusundaram, K., & Indhumathi, J. (2009). Production of Protease from Rice Mill Wastes by *Aspergillus niger* in Solid State Fermentation. *World Journal of Agricultural Sciences*, *5*(3), 308–312.

Parfitt, J., Barthel, M., & Macnaughton, S. (2010). Food waste within food supply chains: Quantification and potential for change to 2050. *Phil. Trans. R. Soc.*, *365*(1554), 3065–3081. doi:10.1098/rstb.2010.0126 PMID:20713403

Paritosh, K., Kushwaha, S. K., Yadav, M., Pareek, N., Chawade, A., & Vivekanand, V. (2017). Food waste to energy: an overview of sustainable approaches for food waste management and nutrient recycling. BioMed Research International.

Paritosh, K., Sandeep, K. K., Monika, Y., Nidhi, P., Aakash, C., & Vivekanand, V. (2017). Food Waste to Energy : An Overview of Sustainable Approaches for Food Waste Management and Nutrient Recycling. *BioMed Research International*, *2017*, 1–19. doi:10.1155/2017/2370927 PMID:28293629

Parry, A., Bleazard, P., & Okawa, K. (2015). *Preventing Food Waste: Case Studies of Japan and the United Kingdom. OECD Food, Agriculture and Fisheries Papers, No. 76*. Paris: OECD Publishing. doi:10.1787/5js4w29cf0f7-

Parvez, S., Malik, K. A., Ah Kang, S., & Kim, H. Y. (2006). Probiotics and their fermented food products are beneficial for health. *Journal of Applied Microbiology*, *100*(6), 1171–1185. doi:10.1111/j.1365-2672.2006.02963.x PMID:16696665

Patel, A., Shah, N., & Prajapati, J. B. (2013). Biosynthesis of vitamins and enzymes in fermented foods by lactic acid bacteria and related genera-A promising approach. *Croatian Journal of Food Science and Technology*, *5*(2), 85–91.

Patidar, S., & Prajapati, J. (1998). Standardization and evaluation of lassi prepared using *Lactobacillus acidophilus* and *Streptococcus thermophilus*. *Journal of Food Science and Technology*, *35*, 428–431.

Patring, J. D., Jastrebova, J. A., Hjortmo, S. B., Andlid, T. A., & Jagerstad, I. M. (2005, April). Development of a simplified method for the determination of folates in baker's yeast by HPLC with ultraviolet and fluorescence detection. *Journal of Agricultural and Food Chemistry*, *53*(7), 2406–2411. doi:10.1021/jf048083g PMID:15796570

Peñas, E., Martinez-Villaluenga, C., & Frias, J. (2016). Sauerkraut: production, composition, and health benefits. In Fermented Foods in Health and Disease Prevention (pp. 557-576). Academic Press.

Perdigon, G., Alvarez, S., Rachid, M., Agüero, G., & Gobbato, N. (1995). Immune system stimulation by probiotics. *Journal of Dairy Science*, *78*(7), 1597–1606. doi:10.3168/jds.S0022-0302(95)76784-4 PMID:7593855

Pfaltzgraff, L. A., De bruyn, M., Cooper, E. C., Budarin, V., & Clark, J. H. (2013). Food waste biomass: A resource for high-value chemicals. *Green Chemistry*, *15*(2), 307–3014. doi:10.1039/c2gc36978h

Pfotzer, G., & Schüler, C. (1997). Effects of different compost amendments on soil biotic and faunal feeding activity in an organic farming system. *Biological Agriculture and Horticulture*, *15*(1-4), 177–183. doi:10.1080/01448765.1997.9755192

Pham, T. P., Kaushik, R., Parshetti, G. K., Mahmood, R., & Balasubramanian, R. (2015). Food waste-to-energy conversion technologies: Current status and future directions. *Waste Management (New York, N.Y.)*, *38*, 399–408. doi:10.1016/j.wasman.2014.12.004 PMID:25555663

Phimsen, S. (2016). Oil extracted from spent coffee grounds for bio-hydrotreated diesel production. *Energy Conversion and Management*, *126*, 1028–1036.

Pillaiyar, P. (1988). *Rice post production Manual*. New Delhi: Willey Eastern Ltd.

Pirozzi, D., Ausiello, A., & Yousuf, A. (2012). Exploitation of Lignocellulosic Materials for the Production of II Generation Biodiesel. In *Proceeding Venice 2012, Fourth International Symposium on Energy from Biomass and Waste*. Cini Foundation.

Pourfarzad, A., Mahdavian-Mehr, H., & Sedaghat, N. (2013). Coffee silverskin as a source of dietary fiber in bread-making: Optimization of chemical treatment using response surface methodology. *Food Science and Technology (Campinas)*, *50*(2), 599–606.

Prashant, Tomar, Singh, Gupta, Arora, & Joshi. (2009). Phenotypic and genotypic characterization of *Lactobacilli* from Churpi cheese. *Dairy Sci Technol*, *89*, 531–540.

Prazeres, A. R., Carvalho, F., & Rivas, J. (2012). Cheese whey management: A review. *Journal of Environmental Management, 110*, 48–68. doi:10.1016/j.jenvman.2012.05.018 PMID:22721610

Premarani, T., & Chhetry, G. (2010). Evaluation of traditional fermentation technology for the preparation of hawaijar in Manipur. *Assam Univ. J. Sci. Technol., 6*, 82–88.

Priefer, C., Jörissen, J., & Bräutigam, K.-R. (2016). Food waste prevention in Europe – A cause-driven approach to identify the most relevant leverage points for action. *Resources, Conservation and Recycling, 109*, 155–165. doi:10.1016/j.resconrec.2016.03.004

Purna, M., & Reddy, K.V. (2016). Effects of the Culture Media Optimization on Pectinase Production by *Enterobacter* Sp . PSTB-1. *3 Biotech, 6*(2), 1–11.

Purushothaman, D., Dhanapal, N., & Rangaswami, G. (1977). *Microbiology and biochemistry of idli fermentation*. Symposium on Indigenous Fermented Food, Bangkok, Thailand.

Pushpangadan, P., Dan, V. M., Ijinu, T., & George, V. (2012). Food, nutrition and beverage. *Indian Journal of Traditional Knowledge, 11*, 26–34.

Qin, W., Wu, L., Zheng, Z., Dong, C., & Yang, Y. (2014). Lignin hydrolysis and phosphorylation mechanism during phosphoric acid–acetone pretreatment: A DFT study. *Molecules (Basel, Switzerland), 19*(12), 21335–21349. doi:10.3390/molecules191221335

Quested, T. E., Ingle, R., & Parry, A. D. (2013). *Household Food and Drink Waste in the UK 2012*. Retrieved from http://www.wrap.org.uk/sites/files/wrap/hhfdw-2012-main.pdf.pdf

Radhakrishnamurty, R., Desikachar, H. S., Srinivasan, M., & Subrahmanyan, V. (1961). Studies on idli fermentation. II. Relative participation of black gram flour and rice semolina in the fermentation. *Journal of Scientific and Industrial Research (New Delhi, India), 20*, 342–345. PMID:14038559

Rahmat, H., Hodge, R., Manderson, G., & Yu, P. (1995). Solid substrate fermentation of *Kloechera apiculata* and *Candida utilis* on apple pomace to produce an improved stock-feed. *World Journal of Microbiology & Biotechnology, 11*(2), 168–170. doi:10.1007/BF00704641 PMID:24414495

Rai, A. K., Palni, U., & Tamang, J. P. (2009). Traditional knowledge of the Himalayan people on the production of indigenous meat products. *Indian Journal of Traditional Knowledge, 8*(1), 104–109.

Ramakrishnan, C. (1980). Studies on Indian fermented foods. *Baroda J. Nutr., 6*, 1–57.

Ramanavičiene, A., Mostovojus, V., Bachmatova, I., & Ramanavičiene, A. (2003). Anti-bacterial effect of caffeine on *Escherichia coli* and *Pseudomonas fluorescens. Acta Medica Lithuanica, 10*, 185–188.

Ramchandra, Y. L., Padmalatha Rai, S., Sujan Ganapathy, P. S., Sudeep, H. V., & Krushnamurthy, N. B. (2008). Chitinase Production by Solid State Fermentation using Shrimp waste. *Asian Journal Of Microbiol.Biotech.Env.Sc., 10*(3), 615–620.

Ramos, O. L., Rodrigues, R. M., Texeira, J. A., & Vicentre, A. A. (2016). Whey and whey powders: Production and uses. Food Science Encyclopedia of Food and Health, 498-505.

Ramukhwatho, F. R., du Plessis, R., & Oelofse, S. (2014). Household Food Wastage in a Developing Country: A Case Study of Mamelodi Township in South Africa. *Proceedings of the 20th WasteCon Conference*. Retrieved from https://researchspace.csir.co.za/dspace/bitstream/handle/10204/7757/Oelofse_2014.pdf?sequence=1&isAllowed=y

Rao, D. G. (2010). *Fundamentals of food engineering*. New Delhi: PHI Learning Private Ltd.

Rasit & Kuan. (2018). Investigation on the Influence of Bio-catalytic Enzyme Produced from Fruit and Vegetable Waste on Palm Oil Mill Effluent. *IOP Conf. Series: Earth and Environmental Science, 140.*

Rathanivelu, R., & Graziosi, G. (2005). Potential alternative use of coffee waste and by-products. ICO Proceedings, ED 1967/05.

Raupauch, M. J., Amann, R., Wheeler, Q. D., & Roos, C. (2016). The application of "-omics" technologies for the classification and identification of animals. *Organisms, Diversity & Evolution, 16*(1), 1–12. doi:10.100713127-015-0234-6

Ravindran, R., & Jaiswal, A.K. (2015). A comprehensive review on pre-treatment strategy for lignocellulosic food industry waste: Challenges and opportunities. *Bioresource Technology*.

Ravindran, R., & Amit, J. (2016). Microbial Enzyme Production Using Lignocellulosic Food Industry Wastes as Feedstock: A Review. *Bioengineering, 3*(4), 30. doi:10.3390/bioengineering3040030 PMID:28952592

Razak, M. A., Ibrahim, M. F., Yee, P. L., Hassan, M. A., & Abd-Azizi, S. (2013). Statistical optimization of butanol Production from oil palm decantater cake hydrolysate by *Clostridium acetobutylicum* ATCC 824. *BioResources, 8*, 1758–1770.

Reddy, K. B. P. K., Raghavendra, P., Kumar, B. G., Misra, M. C., & Prapulla, S. G. (2007). Screening of probiotic properties of lactic acid bacteria isolated from Kanjika, an ayruvedic lactic acid fermented product: An *in-vitro* evaluation. *The Journal of General and Applied Microbiology, 53*(3), 207–213. doi:10.2323/jgam.53.207 PMID:17726302

Reddy, N. R., Sathe, S. K., Pierson, M. D., & Salunkhe, D. K. (1982). Idli, an Indian fermented food: A review. *Journal of Food Quality, 5*(2), 89–101. doi:10.1111/j.1745-4557.1982.tb00736.x

Reed, G. (1981). Use of microbial cultures-yeast products. *Food Technology, 35*(1), 89–94.

Refaat, A. A. (2012). Biofuels from Waste Materials. *J. Compr. Renew. Energy, 5*, 217–261. doi:10.1016/B978-0-08-087872-0.00518-7

Reis, M., Albuquerque, M., Villano, M., Majone, M. (2011). Mixed culture processes for Resource Futures. *Bournemouth Borough Council Waste Composition Analysis Phase 2 and Sciences,* (1), 17-26.

Reisch, L., Eberle, U., & Sylvia Lorek, S. (2013). Sustainable food consumption: an overview of contemporary issues and policies. *Sustainability: Science, Practice and Policy*, *9*(2), 7–25. doi:10.1080/15487733.2013.11908111

Rekhi, D. (2017, August 21). Chennai: Community fridge helps save surplus food, caters to the needy in Besant Nagar. *The New Indian Express*.

Resource Futures. (2009). *Bournemouth Borough Council Waste Composition Analysis Phase 2 and Comparative*. RF Project no.510.

Reuter, J. A., Spacek, D., & Snyder, M. P. (2015). High-Throughput Sequencing Technologies. *Molecular Cell*, *58*(4), 586–597. doi:10.1016/j.molcel.2015.05.004

Riggle, D. (1989). Revival time for composting food industry wastes. *BioCycle*, *29*, 35–37.

Riya, M. P., Antu, K. A., Vinu, T., Chandrakanth, K. C., Anikumar, K. S., & Raghu, K. G. (2013). An in vitro study reveals neutraceutical properties of *Ananas comosus* (L.) Merr. Var. Mauritius fruit residue beneficial to diabetes. *Journal of the Science of Food and Agriculture*, *94*(5), 943–950. doi:10.1002/jsfa.6340

Roche, H. M., de Roos, B., & Brennan, L. (2015). Application of 'omics ' technology. In J. A. Lovegrove, L. Hodson, S. Sharma, & S. A. Lanham-New (Eds.), *Nutrition Research Methodologies* (pp. 198–211). John Wiley and Sons Ltd.; doi:10.1002/9781119180425.ch13

Rodríguez Couto, S. (2008). Exploitation of biological waste for the production of value added products under solid state fermentation conditions. *Biotechnology Journal*, *3*(7), 859–870. doi:10.1002/biot.200800031 PMID:18543242

Rodriguez-Duran, L. V., Ramirez-Coronel, M. A., Aranda-Delgado, E., Nampoothiri, K. M., Favela-Torres, E., & Aguilar, C. N. (2014). Soluble and bound hydroxycinnamates in coffee pulp (*Coffea arabica*) from seven cultivars at three ripening stages. *Journal of Agricultural and Food Chemistry*, *62*(31), 7869–7876.

Rodr, S., & Sanrom, M. A. (2005). Application of Solid-State Fermentation to Ligninolytic Enzyme Production. *Biochemical Engineering Journal*, *22*(3), 211–219. doi:10.1016/j.bej.2004.09.013

Roemer, T., & Boone, C. (2013). Systems-level antimicrobial drug and drug synergy discovery. *Nature Chemical Biology*, *9*(4), 222–231. doi:10.1038/nchembio.1205

Rolfe, R. D. (2000). The role of probiotic cultures in the control of gastrointestinal health. *The Journal of Nutrition*, *130*(2), 396S–402S. doi:10.1093/jn/130.2.396S PMID:10721914

Roman, A., Vetal, G., Illzes, A., Kovacs, Z., & Czermak, P. (2012). Modeling of dia filteration process for determination of acid whey: an empirical approach. *J. of food Process Engg*, *35*(5), 708-714.

Romani, S., Silvia, G., Richard, P. B., & Ada, M. B. (2018). Domestic Food Practices: A Study of Food Management Behaviors and the Role of Food Preparation Planning in Reducing Waste. *Appetite*, *121*, 215–227. doi:10.1016/j.appet.2017.11.093 PMID:29155173

Romney, A. (1990). *CIP: cleaning in place* (2nd ed.). Society of Dairy Technology.

Roopali, Navya, & Murthy. (2013c). Purification and characterization of a Coffee Pulp Tannase Produced by *Penicillium verrucosum. Journal of Food Science and Engineering*, *3*, 323–331.

Roopashri, A. N., & Varadaraj, M. C. (2009). Molecular characterization of native isolates of lactic acid bacteria, *Bifidobacteria* and yeasts for beneficial attributes. *Applied Microbiology and Biotechnology*, *83*(6), 1115–1126. doi:10.100700253-009-1991-y PMID:19407995

Roos, U., Caneva, P., Escher, F., & Amadò, R. (1990). Influence of fermentation on the phytic acid content of conventionally produced bread. *Processing and Quality of Foods: Food Biotechnology: Avenues to Healthy and Nutritious Products*, *2*, 105–110.

Rosales, E., Rodriguez Couto, S., & Sanhromán, M. A. (2007). Increased laccase production by *Tramete shirsuta* grown on ground orange peelings. *Enzyme and Microbial Technology*, *40*(5), 1286–1290. doi:10.1016/j.enzmictec.2006.09.015

Rosales, S., Rodríguez Couto, S., & Sanromán, A. (2002). New uses of food waste: Application to laccase production by *Tramete shirsuta*. *Biotechnology Letters*, *24*(9), 701–704. doi:10.1023/A:1015234100459

Roy, B., Kala, C. P., Nehal, A. F., & Majila, B. S. (2004). Indigenous fermented food and beverages: A potential for economic development of the high altitude societies in Uttaranchal. *Journal of Human Ecology (Delhi, India)*, *15*(1), 45–49. doi:10.1080/09709274.2004.11905665

Rozin, P. (2005). The meaning of food in our lives: A cross-culural perspective on eating and well-being. *Journal of Nutrition Education and Behavior*, *37*, S107–S112. doi:10.1016/S1499-4046(06)60209-1

Ruanglek, V., Maneewatthana, D., & Tripetchkul, S. (2006). Evaluation of thai agroindustrial wastes for bio-ethanol production by *Zymomonas mobilis*. *Process Biochemistry*, *41*(6), 1432–1437. doi:10.1016/j.procbio.2006.01.010

Rusendi, D., & Sheppard, J. D. (1995). Hydrolysis of potato processing waste for the production of poly-ß hydroxybutyrate. *Bioresource Technology*, *54*(2), 191–196. doi:10.1016/0960-8524(95)00124-7

Russ, W., & Meyer-Pittroff, R. (2004). Utilizing waste products from the food production and processing industries. *Critical Reviews in Food Science and Nutrition*, *44*(1), 57–62. doi:10.1080/10408690490263783 PMID:15077881

Saavedra, J. M., & Perman, J. A. (1989). Current concepts in lactose malabsorption and intolerance. *Annual Review of Nutrition*, *9*(1), 475–502. doi:10.1146/annurev.nu.09.070189.002355 PMID:2669882

Sabree, Z. L., Rondon, M. R., & Handelsman, J. (2009). *Metagenomics. Genetics, Genomics*. Elsevier Inc.

Sadh, P. K., Surekha, D., & Joginder, S. D. (2018). Agro - Industrial Wastes and Their Utilization Using Solid State Fermentation : A Review. *Bioresources and Bioprocessing*, *5*(1), 1–15. doi:10.118640643-017-0187-z

Saez, L., Perez, J., & Martinez, J. (1992). Low molecular weight phenolics attenuation during simulated treatment of wastewaters from olive oil mill in evaporation ponds. *Water Research*, *26*(9), 1261–1266. doi:10.1016/0043-1354(92)90187-9

Saithi, S., & Anan, T. (2016). Phytase Production of *Aspergillus niger* on Soybean Meal by Solid - State Fermentation Using a Rotating Drum Bioreactor. *Italian Oral Surgery*, *11*, 25–30.

Sakai, S. J., Yoshida, H., Hirai, Y., & Ansari, M. (2011). International comparative study of '3R's and waste management policy developments. *Master Cycles Waste*, *13*(2), 86–102. doi:10.100710163-011-0009-x

Samant, M., Pandey, S. C., & Pandey, A. (2018). Impact of Hazardous Waste Material on Environment and Their Management Strategies. In Microbial Biotechnology in Environmental Monitoring and Cleanup (pp. 175-192). Academic Press. doi:10.4018/978-1-5225-3126-5.ch011

Sampaio, A. R. M. (2010). *Desenvolvimento, de tecnologias para produc¸ ão de etanol a partir do hidrolisado da borra de café* (MSc thesis). Braga, Portugal: Department of Biological Engineering, University of Minho.

Sandra, C. (2006). Occupational and environmental health issues of solid waste management. The World Bank Group.

Sanjeev, K. S., & Sandhu, K. D. (1990). Indian fermented foods; microbiological and biochemical aspects. *Indian Journal of Microbiology*, *30*, 135–157.

Sankaran, R. (1998). Fermented food of the Indian subcontinent. In Microbiology of fermented foods. Blackie Academic and Professional. doi:10.1007/978-1-4613-0309-1_24

Sarkar, P. K., Cook, P. E., & Owens, J. D. (1993). *Bacillus* fermentation of soybeans. *World J. Microb.Biot.*, *9*(3), 295–299. doi:10.1007/BF00383066 PMID:24420029

Sarkar, P. K., Jones, I. J., Craven, G. S., Somerset, S. M., & Palmer, C. (1997). Amino acid profiles of Kinema, a soyabean-fermented food. *Food Chemistry*, *24*, 337–339.

Sarkar, P. K., Kumar, L. D. H., Dhumal, C., Panigrahi, S. S., & Choudhary, R. (2015). Traditional and ayurvedic foods of Indian origin. *J Ethn Foods*, *2*(3), 97–109. doi:10.1016/j.jef.2015.08.003

Sarkar, P. K., Tamang, J. P., Cook, P. E., & Owens, J. D. (1994). Kinema – a traditional soybean fermented food: Proximate composition and microflora. *Food Microbiology*, *11*(1), 47–55. doi:10.1006/fmic.1994.1007

Sarkar, S., & Misra, A. (2001). Bio-preservation of milk and milk products. *Indian Food Indian*, *20*, 74–77.

Sarojnalini, C., & Singh, W. V. (1988). Composition and digestibility of fermented fish foods of Manipur. *Journal of Food Science and Technology*, *25*, 349–351.

Sato, K., Saito, H., Tomioka, H., & Yokokura, T. (1988). Enhancement of host resistance against Listeria infection by Lactobacillus casei. *Microbiology and Immunology*, *32*(12), 1189–1200. doi:10.1111/j.1348-0421.1988.tb01483.x PMID:2853287

Sauer, U. G., Deferme, L., Gribaldo, L., Hackermuller, J., Tralau, T., van Ravenzwaay, B., ... Gant, T. W. (2017). The challenge of the application of 'omics technologies in chemical rick assessment: Background and outlook. *Regulatory Toxicology and Pharmacology*, *91*(supplement 1), S14–S26. doi:10.1016/j.yrtph.2017.09.020

Savitri, B. T. C. (2007). Traditional foods and beverages of Himachal Pradesh. *Indian Journal of Traditional Knowledge*, *6*(1), 17–24.

Saxena, R., & Singh, R. (2011). Amylase production by solid-state fermentation of agro-industrial wastes using *Bacillus* sp. *Brazilian Journal of Microbiology*, *42*(4), 1334–1342. doi:10.1590/S1517-83822011000400014 PMID:24031761

Scelfo, C., Galeone, C., Bertolini, F., Caminati, M., Ruggiero, P., Facciolongo, N., & Menzella, F. (2018). Towards precision medicine: The application of omics technologies in asthma management. *F1000 Research*, *7*, 423. doi:10.12688/f1000research.14309.1

Schanes, K., Dobernig, K., & Gozet, B. (2018). Food Waste Matters—A Systematic Review of Household Food Waste Practices and Their Policy Implications. *Journal of Cleaner Production*, *182*, 978–991. doi:10.1016/j.jclepro.2018.02.030

Schatzmayr, G., Zehner, F., Taubel, M., Schatzmayr, D., Klimitsch, A., Loibner, A. P., & Binder, E. M. (2006). Microbiological for deactivating mycotoxins. *Molecular Nutrition & Food Research*, *50*(6), 543–551. doi:10.1002/mnfr.200500181 PMID:16715543

Schaub, S. M., & Leonard, J. J. (1996). Composting: An alternative waste management option for food processing industries. *Trends in Food Science & Technology*, *7*(8), 263–268. doi:10.1016/0924-2244(96)10029-7

Schink, B. (1997). Energetics of syntrophic cooperation in methanogenic degradation. *Microbiology and Molecular Biology Reviews*, *61*(2), 262–280. PMID:9184013

Schoder, D., Melzner, D., Schmalwieser, A., Zangana, A., Winter, P., & Wagner, M. (2011). Important vectors for Listeria monocytogenes transmission at farm dairies manufacturing fresh sheep and goat cheese from raw milk. *Journal of Food Protection*, *74*(6), 919–924. doi:10.4315/0362-028X.JFP-10-534 PMID:21669068

Schoen, C., Ernwiss, D., Schulz, A., Chema, D., Schweikart, J., Ernwiss, D., ... Baehr, V. (2009). Regulatory effects of a fermented food concentrate on immune functioning parameters in healthy volunteers. *Nutrition (Burbank, Los Angeles County, Calif.)*, *25*(5), 499–505. doi:10.1016/j.nut.2008.10.022 PMID:19121921

Schroeter, H., Christian, H., Balzer, J., Kleinbongard, P., Keen, C. L., Hollenberg, N. K., ... Schmitz, H. H. (2006). Epicatechin mediates beneficial effects of flavanol-rich cocoa on vascular function in humans. *Proceedings of the National Academy of Sciences of the United States of America*, *103*(4), 1024-1029.

Schüler, C., Pikny, J., Nasir, M., & Vogtmann, H. (1993). Effects of composted organic kitchen and garden waste on *Mycosphaerella pinodes* (Berk, et Blox) Vestergr., causal organism of foot rot on peas (Pisum sativum L.). *Biological Agriculture and Horticulture*, *9*(4), 353–360. doi:10.1080/01448765.1993.11978505

Scrimshaw, N. S., & Murray, E. B. (1988). The acceptability of milk and milk products in populations with a high prevalence of lactose intolerance. *The American Journal of Clinical Nutrition*, *48*(4), 1142–1159. doi:10.1093/ajcn/48.4.1142 PMID:3140651

Segrè, A., & Gaiani, S. (2011). *Transforming Food Waste into Resource*. Cambridge, UK: Royal Society of Chemistry.

Sekar, S., & Mariappan, S. (2007). Usage of traditional fermented products by Indian rural folks and IPR. *Indian Journal of Traditional Knowledge*, *6*, 111–120.

Selvakumar, P., & Pandey, A. (1998). Biosynthesis Of Glucoamylase From *Aspergillus niger* By Solid-State Fermentation Using Tea Waste As The Basis Of A Solid Substrate. *Bioresource Technology*, *65*(1-2), 83–85. doi:10.1016/S0960-8524(98)00012-1

Sethi, V., & Maini, S. B. (1999). Production of organic acids. In V. K. Joshi & A. Pandey (Eds.), *Biotechnology: Food Fermentation, Microbiology, Biochemistry and Technology* (pp. 1259–1290). New Delhi: Educational Publishers and Distributors.

Shahid, M. G., Muhammad, N., Muhammad, I., & Quratulain, S. (2016). Partial Characterization of α-amylase Produced from *Aspergillus niger* using Potato Peel as Substrate, Punjab Univ. *Journal of Zoology (London, England)*, *33*(1), 22–27.

Shankaranad, V. S., & Lonsane, B. K. (1999). Coffee husk: An inexpensive substrate for production of citric acid by *Aspergillus niger* in a solid-state fermentation system. *World Journal of Microbiology & Biotechnology*, *10*(2), 165–168.

Sharanappa, A., Wani, K. S., & Pallavi, P. (2011). Bioprocessing of food industrial waste for α-amylase production by solid state fermentation. *Int. J. Adv. Biotechnol. Res*, *2*(4), 473–480.

Sharholy, M., Kafeel, A., Gauhar, M., & Trivedi, R. C. (2008). Municipal Solid Waste Management in Indian Cities - A Review. *Waste Management (New York, N.Y.)*, *28*(2), 459–467. doi:10.1016/j.wasman.2007.02.008 PMID:17433664

Sharma, K. H., Burnwal, P. K., Dubey, L., & Rao, R. J. (2013). Optimization and Production of Cellulase from Agricultural Waste. *Research Journal of Agriculture and Forestry Sciences*, *1*(1), 18–20.

Sharma, N., & Singh, A. (2012). An insight into traditional foods of Northwestern area of Himachal Pradesh. *Indian Journal of Traditional Knowledge*, *11*, 58–65.

Sharma, R., Chisti, Y., & Banerjee, U. C. (2001). Production, purification, characterization and applications of lipases. *Biotechnology Advances, 19*(8), 627–662. doi:10.1016/S0734-9750(01)00086-6 PMID:14550014

Sharma, R., & Lal, D. (1997). Effect of dahi preparation on some water-soluble vitamins. *Indian Journal of Dairy Science, 50*, 318–320.

Shea, M. (2018). *Food waste & hunger: Solutions to the paradox*. Retrieved from www.rockefellerfoundation.org

Shi, X. (2014). *Analysis of challenges to food security and its recommended strategies*. Retrieved from: https://www.eng.mcmaster.ca/sites/default/files/uploads/analysis_of_challenges_to_food_security.pdf

Shibasaki, K., & Hessbltine, C. W. (1962). Miso fermentation. *Economic Botany, 16*(3), 180–195. doi:10.1007/BF02860037

Shyamala, B. N., & Jamuna, P. (2010). Nutritional content and antinutritional properties of pulp waste from *Daucus carota* and *Beta vulgaris. Mal, 16*(3), 397-408.

Silva, A. G., Wanderley, R. C., Pedroso, A. F., & Ashbell, G. (1997). Ruminal digestion kinetics of citrus peel. *Animal Feed Science and Technology, 68*(3-4), 247–257. doi:10.1016/S0377-8401(97)00056-4

Silva, D., Silva, M., Roberto, S., & Eleni, G. (2002). Pectinase Production By *Penicillium viridicatum* Rfc3 By Solid State Fermentation Using Agricultural Wastes And Agro-Industrial By-Products. *Brazilian Journal of Microbiology, 33*(4), 318–324. doi:10.1590/S1517-83822002000400008

Silva, G. G. D., Couturier, M., Berrin, J.-G., Buléon, A., & Rouau, X. (2012). Effects of grinding processes on enzymatic degradation of wheat straw. *Bioresource Technology, 103*(1), 192–200. doi:10.1016/j.biortech.2011.09.073

Silva, G. G. D., & Xavier, R. S. G. (2011). Successive centrifugal grinding and sieving of wheat straw. *Powder Technology, 208*(2), 266–270. doi:10.1016/j.powtec.2010.08.015

Simango, C. (1997). Potential use of traditional fermented foods for weaning in Zimbabwe. *Journal of Social Science and Medicine, 44*(7), 1065–1068. doi:10.1016/S0277-9536(96)00261-4 PMID:9089926

Singh, Chonhenchob, & Singh. (2006). Life cycle inventory and analysis of reusable plastic containers and display-ready corrugated containers used for packaging fresh fruits and vegetables. *Packaging Technology and Science, 19*, 279-293.

Singh, A., Singh, R. K., & Sureja, A. K. (2007). Cultural significance and diversities of ethnic foods of Northeast India. *Indian Journal of Traditional Knowledge, 6*(1), 79–94.

Singh, A., Singh, R. K., & Sureja, A. K. (2007b). Cultural significance and diversities of ethnic foods of Northeast India. *Indian Journal of Traditional Knowledge, 6*(1), 79–94.

Singh, J., Suhag, M., & Dhaka, A. (2015). Augmented digestion of lignocellulose by steam explosion, acid and alkaline pretreatment methods: A review. *Carbohydrate Polymers*, *117*(0), 624–631. doi:10.1016/j.carbpol.2014.10.012

Singh, T. A., Devi, K. R., Ahmed, G., & Jeyaram, K. (2014). Microbial and endogenous origin of fibrinolytic activity in traditional fermented foods of Northeast India. *Food Research International*, *55*, 356–362. doi:10.1016/j.foodres.2013.11.028

Sinha, P. R., & Sinha, R. N. (2000). Importance of good quality dahi in food. *Indian Dairyman*, *52*, 45–47.

Siso, M. G. (1996). The biotechnological utilization of cheese whey: A review. *Bioresource Technology*, *57*(1), 1–11. doi:10.1016/0960-8524(96)00036-3

Smil, V. (2010). Improving efficiency and reducing waste in our food system. *Environmental Sciences*, *1*(1), 17–26. doi:10.1076/evms.1.1.17.23766

Sobamiwa, O. (1996). Cocoa pod husk utilization in animal feeds: summaries and strategies. *Proc. 12th International Cocoa Research Conference.*

Soccol, C. R., Vandenberghe, L. S., Rodrigues, C., & Pandey, A. (2006). New perspectives for citric acid production and application. *Food Technology and Biotechnology*, *44*, 141–149.

Sohliya, I., Joshi, S.R., Bhagobaty, R.K., & Kumar, R. (2009). Tungrymbai- A traditional fermented soybean food of the ethnic tribes of Meghalaya. *Indian J. Trad. Knowl.*, *8*, 559-561.

Soni, S. K., & Sandhu, D. K. (1990). Indian fermented foods: Microbiological and biochemical aspects. *Indian Journal of Microbiology*, *30*, 135–157.

Soni, S. K., Sandhu, D., & Vilkhu, K. (1985). Studies on dosa and indigenous Indian fermented food: Some biochemical changes accompanying fermentation. *Food Microbiology*, *2*(3), 175–181. doi:10.1016/0740-0020(85)90032-2

Souza, P. M. D., Bittencourt, M. L. D. A., Caprara, C. C., Freitas, M. D., Almeida, R. P. C. D., Silveira, D., ... Magalhães, P. O. (2015). A biotechnology perspective of fungal proteases. *Brazilian Journal of Microbiology*, *46*(2), 337–346. doi:10.1590/S1517-838246220140359 PMID:26273247

Sreenivasan, E. (2013). Evaluation of Effective Microorganisms Technology in Industrial Wood Waste Management. *International Journal of Advances in Engineering and Technology*, *4*(3), 21–22.

Stajcic, N. (2013). Understanding culture: Food as a means of communication. *Hemisphere*, *28*, 5–14.

Stanbury, P. F., Whitaker, A., & Hall, S. J. (2013). *Principles of fermentation technology*. Elsevier.

Steinkraus, K. H. (1983). Lactic acid fermentation in the production of foods from vegetables, cereals and legumes. *Antonie van Leeuwenhoek*, *49*(3), 337–348. doi:10.1007/BF00399508 PMID:6354083

Steinkraus, K. H. (1995). *Handbook of Indigenous Fermented Foods* (2nd ed.). CRC Press.

Steinkraus, K. H. (1997). Classification of fermented foods: Worldwide review of household fermentation techniques. *Food Control*, *8*(5-6), 311–317. doi:10.1016/S0956-7135(97)00050-9

Stoeberl, M., Werkmeistera, R., Faulstichb, M., & Russa, W. (2011). Biobutanol from food wastes – fermentative production, use as biofuel and the influence on the emissions. *Procedia Food Waste*, *1*, 1868–1974.

Strandberg, T. E., Strandberg, A. Y., Pitkala, K., Salomaa, V. V., Tilvis, R. S., & Miettinen, T. A. (2007). Chocolate, well-being and health among elderly men. *European Journal of Clinical Nutrition*, *62*(2), 247–253.

Strotmann, C., Göbel, C., Friedrich, S., Kreyenschmidt, J., Ritter, G., & Teitcheid, P. (2017). A participatory approach to minimized food waste in the food industry-a manual for managers. *Sustainability*, *9*(1), 66. doi:10.3390u9010066

Stuart, T. (2009). *Waste: Uncovering the global food scandal*. New York: Penguin.

Subramaniyam, R., & Vimala, R. (2012). Solid state and submerged fermentation for the production of bioactive substances: A comparative study. *Int J Sci Nat*, *3*(3), 480–486.

Sudha, M. L., Baskaran, V., & Leelavathi, K. (2007). Apple pomace as source of dietary fibre and Polyphenols and its effect on the rheological characteristics and cake m making. *Food Chemistry*, *104*(2), 689–692. doi:10.1016/j.foodchem.2006.12.016

Sudharhsan, S., Sivaprakasam, S., & Karunasena, R. (2007). Physical and Nutritional Factors Affecting the Production of Amylase from Species of *Bacillus* Isolated from Spoiled Food Waste. *African Journal of Biotechnology*, *6*(4), 430–435.

Sukumar, G., & Ghosh, A. R. (2010). *Pediococcus* spp. – A potential probiotic isolated from Khadi (an Indian fermented food) and identified by 16S rDNA sequence analysis. *African Journal of Food Science*, *4*(9), 597–602.

Sumi, H., Hamada, H., Tsushima, H., Mihara, H., & Muraki, H. (1987). A novel fibrinolytic enzyme (nattokinase) in the vegetable cheese Natto; a typical and popular soybean food in the Japanese diet. *Experientia*, *43*(10), 1110–1111. doi:10.1007/BF01956052 PMID:3478223

Sun, Tian, Diamond, & Glass. (2012). Deciphering transcriptional regulatory mechanisms associated with hemicellulose degradation in *Neurospora crassa*. *Eukaryot Cell.*, *11*(4), 482–493.

Sun, S.-N., Cao, X.-F., Xu, F., Sun, R.-C., Jones, G. L., & Baird, M. (2014). Structure and thermal property of alkaline hemicelluloses from steam exploded *Phyllostachys pubescens*. *Carbohydrate Polymers*, *101*(0), 1191–1197. doi:10.1016/j.carbpol.2013.09.109

Sun, Y., & Cheng, J. (2002). Hydrolysis of lignocellulosic materials for ethanol production: A review. *Bioresource Technology*, *83*(1), 1–11. doi:10.1016/S0960-8524(01)00212-7

Sura, K., Garg, S., & Garg, F. C. (2001). Microbiological and biochemical changes during fermentation of Kanji. *Journal of Food Science and Technology*, *38*, 165–167.

Sustainable Restaurant Association. (2010). Too good to waste; Restaurant food waste survey Report 2010, UK. Author.

Swennen, K., Courtin, C. M., & Delcour, J. A. (2006). Non-digestible oligosaccharides with prebiotic properties. *Critical Reviews in Food Science and Nutrition*, *46*(6), 459–471. doi:10.1080/10408390500215746 PMID:16864139

Tamang, J.P. (2010). Benefits of traditional fermented foods. *Fermented-Foods.*

Tamang, B., & Tamang, J. P. (2009a). Lactic acid bacteria isolated from indigenous fermented bamboo products of Arunachal Pradesh in India and their functionality. *Food Biotechnology*, *23*(2), 133–147. doi:10.1080/08905430902875945

Tamang, B., & Tamang, J. P. (2009b). Traditional knowledge of biopreservation of perishable vegetables and bamboo shoots in Northeast India as food resources. *Indian Journal of Traditional Knowledge*, *8*, 81–95.

Tamang, J. P. (2001). *Kinema, Feature: Fermented soybean foods in daily life*. Food Culture.

Tamang, J. P., Chettri, R., & Sharma, R. M. (2009). Indigenous knowledge of Northeast women on production of ethnic fermented soybean foods. *Indian Journal of Traditional Knowledge*, *8*(1), 122–126.

Tamang, J. P., Dewan, S., Olasupo, N. A., Schillinger, V., & Holzapful, H. (2000). Identification and enzymatic profiles of predominant lactic acid bacteria isolated from soft variety chhurpi, a traditional cheese typical of the Sikkim Himalayas. *Food Biotechnology*, *14*(1&2), 99–112. doi:10.1080/08905430009549982

Tamang, J. P., & Nikkuni, S. (1998). Effect of temperatures during pure culture fermentation of kinema. *World Journal of Microbiology & Biotechnology*, *14*(6), 847–850. doi:10.1023/A:1008867511369

Tamang, J. P., & Sarkar, P. K. (1993). Sinki: A traditional lactic acid fermented radish tap root product. *The Journal of General and Applied Microbiology*, *39*(4), 395–408. doi:10.2323/jgam.39.395

Tamang, J. P., & Sarkar, P. K. (1996). Microbiology of mesu, a traditionally fermented bamboo shoot product. *International Journal of Food Microbiology*, *29*(1), 49–58. doi:10.1016/0168-1605(95)00021-6 PMID:8722186

Tamang, J. P., Tamang, B., Schillinger, U., Franz, C. M., Gores, M., & Holzapfel, W. H. (2005). Identification of predominant lactic acid bacteria isolated from traditionally fermented vegetable products of the Eastern Himalayas. *International Journal of Food Microbiology*, *105*(3), 347–356. doi:10.1016/j.ijfoodmicro.2005.04.024 PMID:16055218

Tamang, J. P., Tamang, B., Schillinger, U., Guigas, C., & Holzapfel, W. H. (2009b). Functional properties of lactic acid bacteria isolated from the ethnic fermented vegetables of the Himalayas. *International Journal of Food Microbiology*, *135*(1), 28–33. doi:10.1016/j.ijfoodmicro.2009.07.016 PMID:19666197

Tamang, J. P., Tamang, N., Thapa, S., Dewan, S., Tamang, B., Yonzan, H., ... Kharel, N. (2012). Microorganisms and nutritional value of ethnic fermented foods and alcoholic beverages of North East India. Indian J. Trad. Knowl., 11:7-Thakur N., Savitri and Bhalla T. C. (2004). Characterization of some traditional fermented food and beverages in Himachal Pradesh. *Indian Journal of Traditional Knowledge*, *3*, 325–335.

Tamime, A. Y., & Robinson, R. K. (1999). *Yoghurt: science and technology*. Woodhead Publishing.

Tandy, S., Healey, J. R., Nason, M. A., Williamson, J. C., & Jones, D. L. (2009). Remediation of metal polluted mine soil with compost: Co-composting versus incorporation. *Environmental Pollution*, *157*(2), 690–697. doi:10.1016/j.envpol.2008.08.006 PMID:18819736

Tang, Y. (2015). Non-genomic omic techniques. In Y. Tang, M. Sussman, D. Liu, I. Poxton, & J. Schwartzman (Eds.), *Molecular Medical Microbiology* (pp. 399–406). London: Academic Press.

Taranto, M. P., Sesma, F., de Ruiz Holgado, A. P., & de Valdez, G. F. (1997). Bile salts hydrolase plays a key role on cholesterol removal by Lactobacillus reuteri. *Biotechnology Letters*, *19*(9), 845–847. doi:10.1023/A:1018373217429

Tarazona-Dıaz, M. P., & Aguayo, E. (2013). Assessment of by-products from fresh-cut products for reuse as bioactive compounds. *Food Science & Technology International*, *19*(5), 439–446. doi:10.1177/1082013212455346

Tarazona, S., García-Canas, F., Dopazo, J., Ferrer, A., & Conesa, A. (2011). Differential expression in RNA-seq: A matter of depth. *Genome Research*, *21*(12), 2213–2223. doi:10.1101/gr.124321.111

Taurisano, Gianluca, Nicolas, & Di Donato. (2014). Reuse of April waste, recovery of valuable compounds by Eco-friendly techniques. *International J. of performation to Engg.*, *10*(4), 419-425.

Teoh, A. L., Heard, G., & Cox, J. (2004). Yeast ecology of Kombucha fermentation. *International Journal of Food Microbiology*, *95*(2), 119–126. doi:10.1016/j.ijfoodmicro.2003.12.020 PMID:15282124

Thapa, T.B. (2002). Diversification in processing and marketing of yak milk based products, *TAAAS/IYIC/Yak foundation/FAOROAP/ICIMOD/ILR/Workshop*, 484–489.

Thapa, N., Pal, J., & Tamang, J. P. (2004). Microbial diversity in ngari, hentak and tungtap, fermented fish products of North East India. *World Journal of Microbiology & Biotechnology*, *20*, 599–607.

Thassitou, P. K., & Arvanitoyannis, I. S. (2002). Bioremediation : A Novel Approach to Food Waste Management. *Trends in Food Science & Technology*, *12*(5-6), 185–196. doi:10.1016/S0924-2244(01)00081-4

Thi, N. B. D., Lin, C. Y., & Kumar, G. (2016). Waste-to-wealth for valorization of food waste to hydrogen and methane towards creating a sustainable ideal source of bioenergy. *Journal of Cleaner Production*, *122*, 29–41. doi:10.1016/j.jclepro.2016.02.034

Thingom, P., & Chhetry, G. (2011). Nutritional analysis of fermented soybean (Hawaijar). *Assam Univ J Sci Technol*, *7*(1), 96–100.

Thomas, T., Gilbert, J., & Meyer, F. (2012). Metagenomics-a guide from sampling to data analysis. *Microbial Informatics and Experimentation*, *2*(1), 3. doi:10.1186/2042-5783-2-3

Tian, M., & Qiuyan, Y. (2016). Optimization of Phytase Production from Potato Waste Using *Aspergillus ficuum*. *3 Biotech, 6*(2), 1–8.

Tian, M., Alvan, W., Tuhin, K.G., Georg, H., & Qiuyan, Y. (2017). Production of Cellulase and Xylanase Using Food Waste by Solid-State Fermentation. *Waste and Biomass Valorization*, 1-8.

Tielens, J., & Candel, J. J. L. (2014). Reducing food wastage, improving food security? *Food & Business Knowledge Platform*. Retrieved from www.knowledge4food.net

Tokatli, M., Gulgor, G., Elmaci, S. B., & Isleyen, N. A. (2015). *In-vitro* properties of potential probiotic indigenous lactic acid bacteria originating from Traditional Pickles. *BioMed Research International*, *2015*, 1–8. doi:10.1155/2015/315819

Tomkins, A. M., Alnwick, D., & Haggerty, P. (1988). Fermented foods for improving child feeding in eastern and southern Africa: a review. In Improving young child feeding in eastern and southern Africa: household level food technology; proceedings of a workshop held in Nairobi, Kenya, 11-16 Oct. 1987. IDRC.

Tonini, D., Albizzati, P. F., & Astrup, T. F. (2018). Environmental impacts of food waste: Learning and challenges from a case study on UK. *Waste Management (New York, N.Y.)*, *76*, 744–766. doi:10.1016/j.wasman.2018.03.032

Tsai, S., Ching-piao, L., & Shang-shyng, Y. (2007). Microbial Conversion of Food Wastes for Biofertilizer Production with Thermophilic Lipolytic Microbes. *Renewable Energy*, *32*(6), 904–915. doi:10.1016/j.renene.2006.04.019

Tseng, A., & Zhao, Y. (2003). wine grape pomace as antioxidant dietary fibre for enhancing nutritional value and improving stability of yoghurt and salad dressing. *Food Chemistry*, *138*(1), 356–365. doi:10.1016/j.foodchem.2012.09.148 PMID:23265499

Uc, E., & Antoine, P. T. (2014). Enzyme Production from Food Wastes Using a Biorefinery Concept. *Waste and Biomass Valorization*, *5*(6), 903–917. doi:10.100712649-014-9311-x

Uçkun, E. (2014). Glucoamylase production from food waste by solid state fermentation and its evaluation in the hydrolysis of domestic food waste. *Biofuel Research Journal*, *3*, 98–105.

Ulloa-Rojas, J. B., Verreth, J. A. J., Amato, S., & Huisman, E. A. (2003). Biological treatments affect the chemical composition of coffee pulp. *Bioresource Technology*, *89*, 267–274.

Umsakul, K., Dissara, Y., & Srimuang, N. (2010). Chemical, physical and microbiological changes during composting of the water hyacinth. *Pakistan Journal of Biological Sciences*, *13*(20), 985–992. doi:10.3923/pjbs.2010.985.992 PMID:21319457

UN General Assembly. (2015). *Resolution adopted by the General Assembly on 25 September 2015: Transforming our world: the 2030 Agenda for Sustainable Development A/RES/70/1*. Retrieved from www.un.org/ga/search/view_doc.asp? symbol=A/RES/70/1&Lang=E

Unakal. (2012). Production of α-amylase using banana waste by *Bacillus subtilis* under solid state fermentation. *European Journal of Experimental Biology*, *2*(4), 1044–1052.

UNEP. (2014). *Prevention and reduction of food and drink waste in businesses and households - Guidance for governments, local authorities, businesses and other organisations, Version 1.0*. UNEP.

Upadhyay, A., Chompoo, J., Araki, N., & Tawata, S. (2012). Antioxidant, antimicrobial, 15-LOX, and AGEs inhibitions by pineapple stem waste. *Journal of Food Science*, *77*(1), H9–H15. doi:10.1111/j.1750-3841.2011.02437.x PMID:22260109

Uton & Ostegard. (n.d.). *If we don't get help we will all die*. Retrieved from www.danchurchaid.org

Vailati-Riboni, M., Palombo, V., & Loor, J. J. (2017). What are omics sciences? In B. Ametaj (Ed.), *Periparturient Diseases of Dairy Cows*. Cham: Springer. doi:10.1007/978-3-319-43033-1_1

Van Peij, N., Gielkens, M. M. C., de Vries, R. P., Visser, J., & deGraaff, L. H. (1998). The transcriptional activator XlnR regulates both xylanolytic and endoglucanase gene expression in *Aspergillus niger. Applied and Environmental Microbiology*, *64*, 3615–3619.

Van Poppel, G., & Schaafsma, G. (1996, November). Cholesterol lowering by a functional yoghurt. In *Fd. Ing. Europe Conf. Proc., Maarssen* (pp. 12-14). Academic Press.

Van Wegen, R. J., Ling, Y., & Middelberg, A. P. J. (1998). Industrial production of polyhydroxyalkanoates using escherichia coli: An economic analysis. *Trans. IChem, E76*(3), 417–426. doi:10.1205/026387698524848

Vandermeerch, T., Alvarenga, R.A.F., Ragart, P., & Dewalf, J. (2014). Environmental sustainability assessment of food waste valorization, options, resource conserve. *Recy, 87*, 57.

Varghese, K., Lewis, H., Lockroy, S., & Williams, H. (2013). *The role of packaging in minimizing food waste in the supply chain of the future, Final report, Centre for design RMIT University*. CHEP.

Varzakas, T., Zakynthinos, G., & Verpoort, F. (2016). Plant food residues as a source of nutraceuticals and functional foods. *Foods*, *5*(4), 88–120. doi:10.3390/foods5040088

Veen, P. W. D., Ruijter, G. J. G., & Visser, J. (1995). An extreme cre A mutation in *Aspergillus nidulans* has severe eVects on D-glucose utilization. *Microbiol*, *141*(9), 2301–2306. doi:10.1099/13500872-141-9-2301

Ventour, L. (2008). Food waste report v2, in the food we waste wrap and exodus market research. Weston – Super-Mare.

Vicuña, R. (2000). Ligninolysis. *Molecular Biotechnology*, *14*(2), 173–176. doi:10.1385/MB:14:2:173

Viniegra-gonzález, G., Ernesto, F., & Cristobal, N. (2003). Advantages of Fungal Enzyme Production in Solid State over Liquid Fermentation Systems. *Biochemical Engineering Journal*, *13*(2-3), 157–167. doi:10.1016/S1369-703X(02)00128-6

VitaSoil. (2003). *Cocoa Sell Mulch* Retrieved from http://www.vitasoil.com.cocoa%20main.html

Vlaanderen, J., Moore, L. E., Smith, M. T., Lan, Q., Zhang, L., Skibola, C. F., ... Vermeulen, R. (2010). Application of Omics Technologies in Occupational and Environmental Health Research; Current Status and Projections. *Occupational and Environmental Medicine*, *67*(2), 136–143. doi:10.1136/oem.2008.042788

von Schneidemesser, E., Monks, P. S., Allan, J. D., Bruhwiler, L., Forster, P., Fowler, D., ... Sutton, M. A. (2015). Chemistry and the Linkages between Air Quality and Climate Change. *Chemical Reviews*, *115*(10), 3856–3897. doi:10.1021/acs.chemrev.5b00089

Walsh, A. M., Crispie, F., Claesson, M. J., & Cotter, P. D. (2017). Translating Omics to Food Microbiology. *Annual Review of Food Science and Technology*, *8*(1), 113–134. doi:10.1146/annurev-food-030216-025729

Wang, Q. H., Liu, Y. Y., & Ma, H. Z. (2010). On-site production of crude glucoamylase for kitchen waste hydrolysis. *Waste Management & Research*, *28*(6), 539–544. doi:10.1177/0734242X09354353 PMID:20015936

Warner, J. (2002). *Chocolate: The next health drinks?* WebMD Health Inc. Retrieved from http://www.content.health.msn.com/article/1671.52817

Watanabe, T., Hamada, K., Tategaki, A., Kishida, H., Tanaka, H., & Kitano, M. (2009). *Oral administration of lactic acid bacteria isolated from traditional south Asian fermented milk*. Academic Press.

Werbach, K., & Hunter, D. (2012). *For the Win: How game thinking can revolutionize your business*. Philadelphia: Wharton Digital Press.

Whitehair, K., Shanklin, C., & Brannon, L. (2013). Written messages improve edible food waste behaviors in a university dining facility. *Journal of the Academy of Nutrition and Dietetics*, *113*(1), 63–69. doi:10.1016/j.jand.2012.09.015 PMID:23260724

Whitworth, J. J. (2013). *Is Omics technology the future of food risk assessment?* Retrieved from: https://www.foodnavigator.com/Article/2013/10/14/Omics-technologies-and-applications-in-food-safety

Williams, H., Wikstrom, F., Otterbring, T., Lofgren, M., & Gustafsson, A. (2012). Reasons for household food waste with special attention to packaging, *J. of Cleaner Production, 24*, 148.

Wilson-Powell, G. (2017, August 18). *UK to get world's first community fridge network*. Retrieved fromwww.pebblemag.com

Wiselogel, A., Tyson, S., & Johnson, D. (1996). *Biomass feedstock resources and composition*. Washington, DC: Taylor & Francis.

Wohlgemuth, Sigma-Aldrich, Buchs, & Switzerland. (2011). Product Recovery. Comprehensive Biotechnology, 2, 591-601.

World Food Programme (WFP). (2018). *Global report on food crisis*. Retrieved from https://www.wfp.org/content/global-report-food-crises-2018

WRAP. (2007). *Understanding Food Waste*. Research Summary. WRAP, Banbury.

WRAP. (2009a). *Household Food and Drink Waste in the UK*. WRAP, Banbury.

WRAP. (2011). Fruit and Vegetable resource maps, Final Report 201., Waste and Resources Action Program (WRAP).

WRAP. (2013). *Household food and drink waste in the UK 2012*. Retrieved from: http://www.wrap.org.uk/content/household-food-and-drink-waste-uk-2012

Wyman, C. E. (1999). Production of low cost sugars from biomass: progress, opportunities, and challenges. In R. P. Overend & E. Chornet (Eds.), *Biomass: a growth opportunity in green energy and value added products* (Vol. 1, pp. 867–872). Oxford, UK: Pergamon Press.

Xie, G., & Peng, L. (2011). Genetic engineering of energy crops: A strategy for biofuel production in China. *Journal of Integrative Environmental Sciences*, *53*, 143–150.

Xu, H. L. (2001). Effects of a microbial inoculant and organic fertilizers on the growth, photosynthesis and yield of sweet corn. *Journal of Crop Production*, *3*(1), 183–214. doi:10.1300/J144v03n01_16

Xu, H. L., Wang, R., & Mridha, M. A. U. (2000). Effect of organic fertilizers and a microbial inoculant on leaf photosynthesis and fruit yield and quality of tomato plants. *Journal of Crop Production*, *3*(1), 173–182. doi:10.1300/J144v03n01_15

Yaakob, Z., Mohammad, M., Alherbawi, M., Alam, Z., & Sopian, K. (2013). Overview of the production of biodiesel from waste cooking oil. *Renewable & Sustainable Energy Reviews*, *18*, 184–193. doi:10.1016/j.rser.2012.10.016

Yadav, H., Jain, S., & Sinha, P. R. (2007a). Antidiabetic effect of probiotic dahi containing *Lactobacillus acidophilus* and *Lactobacillus casei* in high fructose fed rats. *Nutrition (Burbank, Los Angeles County, Calif.)*, *23*(1), 62–68. doi:10.1016/j.nut.2006.09.002 PMID:17084593

Yadav, H., Jain, S., & Sinha, P. R. (2007b). Evaluation of changes during storage of probiotic dahi at 78C. *International Journal of Dairy Technology*, *60*(3), 205–210. doi:10.1111/j.1471-0307.2007.00325.x

Yang, L., Huang, Y., Zhao, M., Huang, Z., Miao, H., Xu, Z., & Ruan, W. (2015). Enhancing biogas generation performance from food wastes by high-solids thermophilic anaerobic digestion: Effect of pH adjustment. *International Biodeterioration & Biodegradation*, *105*, 153–159. doi:10.1016/j.ibiod.2015.09.005

Yokotsuka, K. (1982). *Chemical Compositions of Brandies*. Institute Winery of Yamanashi University.

Yonzan, H., & Tamang, J. P. (1998). Consumption pattern of traditional fermented foods in the Sikkim Himalaya. *Journal of Hill Research*, *11*, 112–115.

Yonzan, H., & Tamang, J. P. (2009). Traditional processing of Selroti-A cereal based ethnic fermented food of the Nepalis. *Indian Journal of Traditional Knowledge*, *8*, 110–114.

Yonzan, H., & Tamang, J. P. (2010). Microbiology and nutritional value of Selroti, an ethnic fermented cereal food of the himalayas. *J. Food Biotechnology*, *24*(3), 227–247. doi:10.1080/08905436.2010.507133

Yoo, J., Alavi, S., Vadlani, P., & Amanor-Boadu, V. (2011). Thermo-mechanical extrusion pretreatment for conversion of soybean hulls to fermentable sugars. *Bioresource Technology*, *102*(16), 7583–7590. doi:10.1016/j.biortech.2011.04.092

Zhang, C., Haijia, S., Jan, B., & Tianwei, T. (2014). Reviewing the Anaerobic Digestion of Food Waste for Biogas Production. *Renewable & Sustainable Energy Reviews*, *38*, 383–392. doi:10.1016/j.rser.2014.05.038

Zhang, Q., He, J., Tian, M., Mao, Z., Tang, L., Zhang, J., & Zhang, H. (2011). Enhancement of methane production from cassava residues by biological pretreatment using a constructed microbial consortium. *Bioresource Technology*, *102*(19), 8899–8906. doi:10.1016/j.biortech.2011.06.061

Zhang, R., Hamed, M. E., Karl, H., Fengyu, W., Guangqing, L., Chris, C., & Paul, G. (2007). Characterization of Food Waste as Feedstock for Anaerobic Digestion. *Bioresource Technology*, *98*(4), 929–935. doi:10.1016/j.biortech.2006.02.039 PMID:16635571

Zhang, X., & Richard, T. (2011). Dual enzymatic saccharification of food waste for ethanol fermentation. *Proceedings of international conference on electrical and control engineering*. 10.1109/ICECENG.2011.6058308

Ziemi'nski, K., & Frąc, M. (2012). Methane fermentation process as anaerobic digestion of biomass: Transformations, stages and microorganisms. *African Journal of Biotechnology*, *11*(18), 4127–4139.

Zuluaga, V. J. (1989). Utilizacion integral de los sub products del cafe. In *Proceedings of I Seminario Internacional Sobre Biotechnologia en la Agro industria Cafetalera (SIBAC)*. Xalapa, Mexico: ORSTOM.

About the Contributors

Aparna Baban Gunjal has completed her B.Sc. and M.Sc in Microbiology and Ph.D in Environmental Science from Savitribai Phule Pune University, Pune, India. She has 04 years teaching experience. She has presented many research papers in National and International Conferences. She has published 30 research papers, 05 book chapters and 01 review article. She is Member of Asian PGPR Society of Sustainable Agriculture, USA; Biotech Research Society India; Indian Women Scientist's Association; Member of Women in Science And Engineering, India and Organization for Women in Science for the Developing World. Her research areas of interest are bioremediation; e-waste and solid waste management; solid state fermentation; plant growth promoting rhizobacteria. She has worked on Central Pollution Control Board, New Delhi, India; University Grants Commission, New Delhi, India and UPE projects. She is a reviewer for the journals viz., International Journal of Research in Environmental Science and Technology, International Journal of Environmental Sciences and Journal of Solid Waste Technology and Management and Guest Editor for Biotech Express India Magazine. She has received H. Khorana award for the best paper presentation in 2010; Pune Municipal Corporation Award in 2015; Summer Research Fellowship for Teachers by the Indian Academy of Sciences in 2017 and The Elsevier Foundation -TWAS Sustainability Visiting Expert Programme award in 2018. She has also received Travel Grants to attend the National and International Conferences.

Meghmala Sheshrao Waghmode is currently working as Lecturer in Annasaheb Magar Mahavidyalaya, Pune, India. She has 11 years of teaching experience and 10 years of research experience. Her research areas of expertise are waste management; bioremediation; nanotechnology and fungal pigments.

Neha Nitin Patil is Associate Professor and Head of Dept of Microbiology at Waghire College, Saswad, Pune, India. She has 30 years of teaching experience and 10 years of research experience. She has 10 publications to her credit and is also a life member of Biotech Research Society, India and many others. She has many awards and honors to her credit.

Pankaj Bhatt did his Bachelor in Biotechnology from Kumaun University Naintal in 2009. He completed Master in Microbiology in 2012 from Department of Biotechnology Kumaun University Nainital Uttarakhand. He have completed Ph.D in Microbiology from G.B Pant University of Agriculture and Technology Pantnagar, U.S Nagar, India. His research areas of expertise are Microbial and molecular aspect of biodegradation, pesticide biodegradation, food microbiology, plant microbe interaction. Currently he is working as Assistant Professor in Dolphin (P.G) College of Biomedical and Natural Sciences Dehradun, India.

* * *

Kiran Ahuja has done outstanding research in the fields of optical and wireless communication. She received her B.Tech & M.Tech degree in Electronics and Communication Engineering from Punjab Technical University, Jalandhar, India and Ph. D in Electronics & Communication Engineering in wireless communication and networks field from Thapar University, Patiala, Punjab, India. She worked as faculty at DAVIET, Jalandhar since 2006. Presently, pursuing post-doctoral fellowship at NIT, Jalandhar under N-PDF scheme of SERB, India.

Siridevi B. was born in Maharashtra, India, in 1992. She has received the M. Tech degree in Biotechnology from the RV College of Engineering, Visvesvaraya Technological University Belgaum in 2016. Since then, she has been with Spice and Flavor science department at CSIR-CFTRI, Mysuru, Karnataka, where she is currently the Junior Researcher.

Dixit V. Bhalani has completed his B.Sc and M.Sc. from Sardar Patel University Vallabh Vidhya Nagar Gujarat. He has got Gold medal in M.Sc and Awarded by DST-INSPIRE fellowship by Department of Science and Technology (DST) India, He has joined his Ph.D. at AcSIR-CSIR CSMCRI Bhavnagar Gujarat. He has published more than six international research articles in peer-reviewed journals.

Prabhakumari C., Deputy Principal Scientist of CEPCI Laboratory and Research Institute has been an active researcher in the area of Biotechnology in the cashew waste management and enzymology.

Kulbhushan Chand received the B.Tech. degree in Electronics and Communication Engineering from the Guru Nanak Dev University, India, in 2010, and the M.Tech. degree in Electronics and Communication Engineering from the Lovely Professional University, India, in 2013. He is currently working towards the Ph.D. degree in Electronics and Communication Engineering with Dr. Arun Khosla at the Dr. B R Ambedkar National Institute of Technology Jalandhar, India. His research interests include biofeedback, embedded systems, wearable physiological sensors and computing, healthcare and gamification.

Arvind Kumar Singh Chandel completed his B.Pharm from Barkatullah University Bhopal M.P. after that he has qualified Graduate Pharmacy Aptitude Test (GPAT) and NIPER JEE and joined M.S. Pharma at NIPER HAjipur after that he awarded by CSIR -JRF-GATE and SRF from CSIR HRDG India for pursuing his PhD degree, He joined his PhD. in ACSIR CSMCRI Bhavnagar Gujarat India, He has Published more than sixteen international articles in peer-reviewed journals.

Nivas Desai was born in Maharashtra, India, in 1982. He received the M.Sc. degree in Botany in 2004 and PhD degrees in Botany from Shivaji University, Kolhapur, India in 2011. In 2011, he joined the Biological Oceanography Department of CSIR-National Institute of Oceanography, Dona Paula, Goa of as a researcher. In 2014 he received DST-Fasttrack Young Scientist Project for 'optimization of growth and lipid production from marine cyanobacteria under varied environmental and nutritional conditions' from DST. Since August 2017, he has been with the Department of Spice and Flavor science at CSIR-CFTRI, Mysuru, where he is working as Research Associate. His current research interests include extraction, characterization and encapsulation of bioactive compounds. He is a Life Member of the Mangrove Society of India (MSI). He has more than 25 national and international publications and he has participated in 25 national and international conferences.

Baban B. Gunjal is B. Tech (Gold Medalist) and M.Tech from IIT Kharagpur, West Bengal, India. He has teaching and research experience of 35 yeas and 60 publications to his credit. He has guided many P.G. students for their research work. He is member of many professional societies.

Bhavika Jain is a Final Year Student pursuing Bachelors of Technology in Instrumentation and Control Engineering at Dr. B.R. Ambedkar National Institute of Technology.

Tushar Joshi has completed B.Sc. (Biotechnology) from Kumaun University, Nainital in 2011. He completed M.Sc. (Biotechnology) from Bangalore University in 2013. Currently, he is enrolled as a research scholar at Department of Biotechnology, Bhimtal Campus, Nainital. His research area is drug discovery against plant bacteria.

Arun Khosla is working as an Associate Professor in the Department of Electronics and Communication Engineering and has research expertise in Soft Computing, Artificial Intelligence, and Machine Learning, Assisted Technologies.

Vismaya N. Kumar is a BCIL trainee working at CEPCI Laboratory and Research Institute in the field of food waste management. She is pursuing her PhD in the area of Biotechnology.

M. Baby Mariyatra studied in School of Chemistry, Bharathidasan University Tiruchirappalli for her M.Sc. and PhD degrees during 1998-2006. She passed CSIR-UGC NET for lecturership in 2000. She carried out her postdoctoral research in IIT-Bombay and University of Dublin, Trinity College, Ireland. At present, she is working as Assistant professor in the department of Chemistry, St. Xavier's college, Palayamkottai.

Aneesh Mathachan is the Director of CDRL, St. Joseph's College and has research areas of expertise in disease prevention and control using novel therapeutic strategies and effective waste remediation techniques. He is awarded with the Young Scientist Award, UGC Research Grant in 2016, Best Teacher award in 2015, etc.

Pushpa Murthy has 20 years of contributions to coffee and Coffee biotechnology. She is engaged with classic research in last two decades in the field of Coffee biotechnology, processing and production have dominated the development of coffee technology ever since. She established collaborative work with Coffee Board of India as well as NFRI Institute, Japan which has helped to discover future forecast. She has done a generation of ECF to the host institute along with main lab project, CSIR Five year plan project on Spices which has provided new technologies on plantation plants such as Chilli, Turmeric, Ginger etc. Her contribution in the technology development for turmeric is noteworthy. As a result, it became possible to maintain food quality and safety for the product. Dr. Murthy is now certainly working on the development of improved process for preparation of quality cocoa beans (Theobroma cocoa) through biotechnological approaches. Recently, she has worked on flavour profiling of coffee and cocoa through microbial technology. She has shown, future prospects of work shall be devoted more towards social causes of eco-friendly sustainable, techniques to preserve the natural resources with bio-technological/ green approaches.

Matthew Chidozie Ogwu is a researcher in the School of Biological Sciences, Seoul National University, Korea Republic. Matthew holds a master's degree in Plant Diversity and Conservation with distinction as well as a bachelor's degree in Plant Biology and Biotechnology (First Class Honours). He is a Lecturer at the University of Benin, Nigeria. His research focusses on sustainable development issues arising from various scales of biogeographical interactions. Matthew has published over 50 peer-reviewed scientific works and presented his results at different international conferences. He has received many awards and scholarships.

Ashutosh Paliwal is Ph.D student currently working at Department of Biotechnology, Kumaun University, Nainital.

Veena Pande is M.Sc. (Biochemistry) from G.B. Pant University of Agriculture and Technology, Pantnagar in 1990. She has got her Ph.D. in Botany (Specialization: Plant Biotechnology) Kumaun University, Nainital in 2004. She is Head, Department of Biotechnology, and has experience of over 15 years in teaching and research. She has published/presented over 85 papers in many high quality international journals, books and international conferences.

Veni Pande has completed her B.Sc (ZBC) in 2012 and M.Sc. (Microbiology) from Kumaun University Nainital, India, Currently she is a research scholar at Department of Biotechnology, Bhimtal Campus, Nainital. Her major research areas are bioremediation and biodegradation.

Anupam Pandey is a Research Scholar currently working at ICAR-Directorate of Coldwater Fisheries Research (DCFR), Bhimtal, Uttarakhand. He has a research experience of four years in the field of biotechnology.

Ashok Pandey is currently Distinguished Scientist at the Centre for Innovation and Translational Research, CSIR-Indian Institute of Toxicology Research, Lucknow, India and Executive Director (Honorary) at the Centre for Energy and Environmental Sustainability – India. Formerly, he was Eminent Scientist at the Center of Innovative and Applied Bioprocessing, Mohali and Chief Scientist & Head of Biotechnology Division and Centre for Biofuels at CSIR's National Institute for Interdisciplinary Science and Technology, Trivandrum. His major research and technological development interests are industrial & environmental biotechnology and energy biosciences, focusing on biomass to biofuels & chemicals, waste to wealth & energy, industrial enzymes, etc.

Satish Chandra Pandey is a Ph.D. scholar in Department of Biotechnology, K.U. Nainital, India and also working as a lecturer in Department of Zoology, SSJ Campus K.U. Nainital, India. He holds a Master's degree in Microbiology from CBSH,G.B. Pant University of Agriculture and Technology, Pantnagar, India. His major research is on human parasites, molecular parasitology, microbiology and drug discovery against the human parasites.

Sindhu Raveendran is currently working as DST WOS-B Scientist at MPTD, CSIR-NIIST, Trivandrum, India. She has about 153 publications/communications with an h- index of 26 (Google Scholar Citations). She is a recipient of several National and International awards and fellowships.

Mukesh Samant has completed Ph.D from CDRI, Lucknow, He has 6 years of teaching experience and more than16 years of research experience. His major research areas are molecular parasitology, immunology, microbiology and drug discovery. He has a number of research publications to his credit and also many awards. Presently he is working as an Assistant Professor at Kumaun University, Department of Zoology, SSJ Campus, Almora.

Diksha Sati has completed her B.Sc. (ZBC) in 2013 and M.Sc. (Zoology) in 2015 from Kumaun University, SSJ Campus, Almora. Currently she is a Research Scholar in Department of Zoology, Kumaun University, SSJ Campus, Almora. Her major research area is stress tolerance in PGPR.

Celin Tennis Raju is doing her B.Sc. Chemistry in St. Xavier's College, Palayamkottai, Tamil Nadu, India. She is very passionate in research indulging social problems and identifying the remedial measures for the problems which brings a change in the society.

Suresh Thorat is the former Head Dept of Food Science and Technology, Mahatma Phule Krishi Vidyapeeth Rahuri. He has guided 11 students for Ph.D and has 33 years of teaching experience. His research area of expertise is food safety.

Ankita Tripathi is a PhD scholar at Department of Biotechnology, Bhimtal Campus, Kumaun University, Nainital, Uttarakhand.

Priyanka Tripathi is a PhD student currently working at National Institute of Pathology (ICMR), New Delhi. She has a research experience of around four years in the field of biotechnology.

Index

www.ingramcontent.com/pod-product-compliance
Lightning Source LLC
LaVergne TN
LVHW081341300926
849161LV00008B/2040